Y0-BFE-678

ELEMENTARY PRACTICAL STATISTICS

A series of undergraduate mathematics texts under the editorship of Carl B. Allendoerfer

MODERN MATHEMATICS: AN INTRODUCTION *by Samuel I. Altwerger*

BASIC MATHEMATICS REVIEW: TEXT AND WORKBOOK *by James A. Cooley*

GEOMETRY, ALGEBRA, AND TRIGONOMETRY BY VECTOR METHODS *by Arthur H. Copeland, Sr.*

FUNDAMENTALS OF COLLEGE ALGEBRA *by William H. Durfee*

TRIGONOMETRY AND THE ELEMENTARY TRANSCENDENTAL FUNCTIONS *by Tomlinson Fort*

RETRACING ELEMENTARY MATHEMATICS *by Leon Henkin, W. Norman Smith, Verne J. Varineau, and Michael J. Walsh*

APPLIED BOOLEAN ALGEBRA: AN ELEMENTARY INTRODUCTION *by Franz E. Hohn*

ELEMENTARY CONCEPTS OF MATHEMATICS, 2d ed., *by Burton W. Jones*

ARITHMETIC: AN INTRODUCTION TO MATHEMATICS *by L. Clark Lay*

INTRODUCTION TO PROBABILITY THEORY *by James R. McCord, III and Richard M. Moroney, Jr.*

MODERN ALGEBRA WITH TRIGONOMETRY *by John T. Moore*

ELEMENTARY PRACTICAL STATISTICS *by A. L. O'Toole*

ELEMENTARY PRACTICAL STATISTICS

A. L. O'Toole
Drake University

THE MACMILLAN COMPANY, NEW YORK
COLLIER-MACMILLAN LIMITED, LONDON

© Copyright, A. Lawrence O'Toole, 1960 and 1964

All rights reserved. No part of this book may be reproduced in any form without permission in writing from the publisher, except by a reviewer who wishes to quote brief passages in connection with a review written for inclusion in a magazine or newspaper.

First Printing

Library of Congress catalog card number: 64–11038

The Macmillan Company, New York
Collier-Macmillan Canada, Ltd., Toronto, Ontario

Printed in the United States of America

DESIGNED BY R. A. KASELER

HA
29
O85

Foreword

This general course in elementary statistics aims to do for the social world and human relations something similar to what the elementary laboratory courses in the natural sciences do for the physical world and man's relations to it.

Elementary instruction in the natural sciences has not been intended for only the scientifically minded students, but for all students. The idea has not been to make scientists of everyone; no one expects the average citizen to apply a laboratory test to each aspirin tablet or vitamin capsule before he or she swallows it. Nevertheless, everyone needs to have some familiarity with the laws of nature and with the experimental method of improving human knowledge of the natural world and man's relations to that world.

Similarly, the purpose in teaching elementary statistics is not to make statisticians of everyone, but to help give everyone some familiarity with an increasingly important method for improving human knowledge of the world in which we live and of our relationships in and to that world. If there is anyone who believes it is not necessary to be able to think in terms of statistical data and elementary statistical concepts, perhaps it will be sufficient to cite for him just one example of the importance of statistics: Many millions of citizens in this country today have their wages, the basis for the livelihood of themselves and their families, tied to the government cost-of-living indexes. Because of the significance of statistics for all, this course in the elements of the subject is not designed for only mathematically gifted students, but for all young people.

The main objective of this elementary course should not be encyclopedic information, but rather the development of a way of thinking about human situations and problems. It is not proposed here that the student should swallow in one gargantuan gulp all that is known about statistical methods, only to find that he has swallowed a mass that he can not assimilate and has given himself mental indigestion. Nor will it be possible to treat with great rigor all the aspects of statistical methods that are needed in the course; it is similarly impossible to be rigorous in other elementary courses in mathematics and science. A course based on this book should aim only to give the student a meaningful and useful familiarity with, and appreciation of, some of the most important elementary concepts of the statistical method as applied to practical human problems. More than many other fields of study, elementary statistics can have the motivational and rewarding characteristics that make education useful, interesting, and memorable for many people.

The only mathematical prerequisite needed for success in this course is some acquaintance with—not mastery of—algebraic symbols, formulas, equations, and graphs, such as might be obtained in high school or college from either a course in algebra or a course in general mathematics. Many students with little experience in algebra and little fondness for the traditional type of mathematics course can enjoy this course and profit from it. Perhaps they should be afforded a chance to have this kind of mathematical experience. After all, the purpose in teaching elementary mathematics is not solely to make mathematicians, but also to make effective social participants.

Out of consideration for the wide range of students who, I think, might find the course interesting and useful, I have not attempted to write the material in the extremely brief, cryptic style characteristic of many mathematics textbooks. Quite the contrary; I deliberately have employed an expansive style which is filled with illustrative and explanatory details—even a bit of repetition here and there to help the reader follow the developments easily.

In general, lessons learned through dynamic experiences are by far the most impressive and intelligible to us, and they have the most lasting effect on us. Rich and dynamic experiences in all the steps of investigation of realistic problems seem to be necessary if mere verbalism is to be avoided and good understanding of the statistical method is to be attained.

It is hoped that, in this course, students will spend part of their time studying realistic problem situations, not merely words. Thus they will learn through their own experiences and investigations. Every word of their statistical vocabulary should be intimately related to their actual experiences, so that it will be clear to them that statistics is a method for dealing with relations and situations of their daily lives, things with which they are most familiar.

Alfred North Whitehead defined education as "the acquisition of the art of the utilization of knowledge." Perhaps there are few more widely desirable aspects of this art nowadays than an understanding of the basic principles of the science of the collection and interpretation of statistical data.

No claim is made here that this course by itself will work miracles for either school or students. But it is believed that the course will prove to be a profitable, interesting, and enjoyable adventure in education for many students.

Preface

This book is intended for a first course in statistics to which one half-year (or two quarters) will be devoted. It is designed to serve as an elementary introduction to all fields of application of the statistical method. For many students, this course probably will be all the formal study of statistics they need in life. Others will desire to pursue more advanced work either in the mathematical theory of statistics or in one of the many fields of specialized applications of statistical methods, such as, for example, agriculture, biology, business, chemistry, education, engineering, medicine, psychology, and sociology; they would be well prepared to do that after this general introduction to statistics.

I hope that students using this text will be permitted to choose their own problems for investigation. Because there has been no way for me to know in advance exactly what areas of statistical methodology will be of most interest to a given class, more material has been included in this book than any specific group of students is likely to need in a first course. On the other hand, I have not attempted to include all the simple and useful statistical ideas that are available, and another author might have chosen a somewhat different set of topics.

Each group of students and their teacher should work those problems in the exercises that will be most interesting and useful in learning how to deal with data obtained in their own investigations and in understanding the methods used by other groups in their investigations.

In some of the exercises, problems of a strictly algebraic nature have been included. These problems are intended for students who wish to develop their ability in algebra and for students who are interested in the mathematical theory of statistics; statistics offers practically unlimited opportunities for the study of the best parts of algebra in useful and meaningful ways. Other students may omit the problems that involve algebraic proofs and developments. But all students ought to try to use enough simple algebraic symbolism to help them generalize and fix the big ideas of elementary statistics in their memories and thus avoid the intolerable amount of arithmetic computation so often characteristic of unorganized work in this area of thinking.

At least one numerical illustration of every statistical process discussed in this book is given in complete detail. These numerical illustrations may serve as models for problem solving and for developing an understanding of the statistical ideas. From these models, students can draw ideas that will help them invent ways to find good answers to their own questions about the problems and the situations they are investigating.

Acknowledgments

I wish to express sincere thanks to the following companies for their generosity in providing hitherto unpublished information as a basis for interesting practical examples: The Florsheim Shoe Company, Chicago, Illinois, for Tables 6.1, 6.2, 6.22, 6.23, and 6.25; to Cluett, Peabody & Co., Inc., Troy, New York, for Tables 6.14, 6.15, 6.20, 6.21, and 6.24; and to Alfred Decker & Cohn, Inc., Chicago, Illinois, for Table 6.26.

I am indebted to the late Professor Sir Ronald A. Fisher of Cambridge University and to Messrs. Oliver & Boyd Ltd., Edinburgh, for permission to reprint Table IV and Table V.A from their book *Statistical Methods for Research Workers* as my Tables 7.3 and 8.17.

I am similarly indebted for permission to reproduce probability tables to Dr. K. R. Nair for Table 5.4, Professor Leslie H. Miller for Table 6.11, Mrs. Frieda Swed Cohn and Dr. Churchill Eisenhart for Table 6.17, Professor Donavon Auble for Tables 7.5 to 7.8, and the late Professor Edwin G. Olds for Table 8.10.

My thanks are expressed here also to H. Burke Horton for Table A.15; to the editors of the *Journal of the American Statistical Association*, the *Annals of Mathematical Statistics*, and *Sankhya* for permission to reprint material from those journals; and to all the writers whose data I have used for interesting illustrations, including Sir B. Seebohm Rowntree for Tables 1.1 and 3.17, Professor S. A. Courtis for Table 6.3, Mr. D. J. Davis for Table 6.12, Professor D. C. Rife for Tables 8.1, 8.2, and 8.3, Emma J. McDonald for Table 9.11, and Wallace M. Hazel and Warren K. Eglof for Table 9.12.

Des Moines, Iowa   A. L. O'TOOLE

Contents

ELEMENTARY PRACTICAL STATISTICS

Chapter 1

STATISTICAL SAMPLING

1.1 INTRODUCTION

There are several ways in which students may attempt to gain understanding and appreciation of the statistical method of problem-solving. By far the best way for most people to learn the statistical method of thinking is to choose their own problems for statistical investigation and collect the data themselves. Then they will not just read about statistics; they will study statistical methods in action.

It is likely that in nearly every problem you choose for investigation you will find it necessary to use samples. You can save yourself a lot of grief and disappointment in your investigations if you know some of the problems and techniques involved in good sampling. Such knowledge can help to protect you against collecting worthless data and against making deceptive estimates and wrong decisions from worthless data.

No matter which one of the possible approaches you take to the study of statistical methods, you will benefit by giving considerable attention at the beginning to the concept of the random sample. First, here are some illustrations of practical applications of sampling. There are many thousands of applications, but space can be taken here for only a few examples.[1]

1.2 PRACTICAL APPLICATIONS OF SAMPLING

Your happiness and your understanding of affairs that influence your life many times every day depend in one way or another upon statistical sampling. The cornflakes that you ate for breakfast this morning tasted the same as those that you ate last week and last month. Why? Because the maker has a quality-control sampling plan that tells him immediately

[1]*Suggestion:* Read Chapter 1 rapidly. Do not expect to understand it all clearly at the beginning of the course. It is a sort of preview to introduce you to some of the most important ideas in statistics.

if any of the things involved in the manufacturing process get out of adjustment. This sampling plan protects him from the large losses that he might suffer if he manufactured a lot of off-standard cornflakes that people would not like. And it protects you by guaranteeing that the product you buy will be what past experience leads you to expect it to be.

The fact that you can drive your automobile into a repair shop whenever the car needs service and know that the standard parts will fit your car is so because the maker of the car and of the spare parts maintains a quality-control sampling system. The same is true of your radio and television set and of most of the thousands of other kinds of equipment that people use today.

If you look in the "help wanted" section of your daily or Sunday newspaper you may see some manufacturer's advertisement offering jobs to statistical quality-control workers. Quality control makes a lot of good jobs for young men and women in industry at present. The statistical methods that you are going to learn in this course are among the most important things that quality-control men and women need to know.

Manufacturers use statistical sampling to find out the likes and the dislikes of the people who buy products. This enables them to improve their products so that more people will like them and buy them. The men and women who do this kind of statistical work are called market researchers. There are many thousands of young men and women employed in market research today.

In the early part of this century, it was generally accepted that if a medicine did not have a horrible taste it probably was no good. Currently, the antibiotics and the other new "miracle" drugs, and other medicines, are manufactured in forms that have the tastes people actually like best, such as strawberry, raspberry, and cherry.

Eli Lilly and Company is a leading drug manufacturer. This company has what is called the Lilly Junior Taste Panel.[2] The panel is composed of boys and girls from five to thirteen years of age. It is used to determine preferred color, odor, and taste for medicinal products intended for boys and girls.

For example, in a test to decide which of three available liquids would be preferred in a new children's product called Suspension Co-Pyronil, the Lilly Junior Taste Panel voted as follows:

Preparation	*Number Who Prefer It*
E	43
L	20
T	37
Total	100

[2]*Physician's Bulletin* XIX, No. 7 (July 1954), pp. 195–197, issued by Eli Lilly and Company, Indianapolis 6, Indiana.

As a result of this test, preparation L was eliminated, and E and T were then tested against each other with fifty boys and girls of the panel. The result was:

Preparation	Number Who Prefer It
E	32
T	18
Total	50

The company says that after such a test is completed and the children's responses are tabulated statistical analysis is used to determine the significance of the data obtained. A 5 per cent level of significance is required. In other words, the company wants the odds to be at least 95 to 5 that its decisions in these matters are correct. Preparation E was preferred clearly (significant at the 5 per cent probability level) and it is incorporated in the product now.

Gilbert Youth Research is a market research company with headquarters in New York City that conducts national sample surveys among young people between the ages of five and twenty-five. It determines their attitudes, opinions, preferences, and likes and dislikes respecting consumer problems and other matters of interest to young people and to business people whose success or failure is dependent in whole or in part on young people as consumers.

Gilbert Youth Research was started several years ago by a couple of young people in the Middle West while they were still in school. They began by making sample surveys among their fellow students. They gradually expanded the scope of their activities, and soon were earning considerable money from their research. When they left school they made a full-time business for themselves out of their youth surveys.

A manufacturer of coconut for cakes and pies wanted to sell the coconut in vacuum-packed tin cans. Now, as you may have noticed, the tin cans in which food products are sold usually have a thin coating of plastic on the inside to protect the food from contamination caused by chemical reaction with the metal. It is possible to apply the lining to the can in almost any desired color.

One of the market researchers assigned to work on this product prepared two cans. One can had a blue-tinted plastic lining and the other had a golden, or yellowish, lining. He took a scoopful of coconut out of a bin at the plant and put part of the coconut into each of the two cans. Then he showed the two open cans of coconut to a sample of housewives, asking each to indicate the can that she would choose. She was asked also to give the reason for her preference.

About two-thirds of the women preferred the coconut in the can with the blue lining. They said that the coconut in that can was whiter than the

coconut in the other can. In other words, the blue-tinted lining empha-
sized the whiteness, purity, and freshness of the coconut. The light reflected
from the yellow lining in the other can gave the coconut in it a slightly
dull, yellow or brownish appearance.

This was sufficient proof to the manufacturer that his product would
sell better in the can with the blue lining—provided he could be sure that
the evidence was obtained from a truly representative sample of housewives.

1.3 DISTORTED SAMPLING

Following is a famous example of what may be called distorted sampling.
In 1936 a magazine called the *Literary Digest* took a poll before the national
presidential election in November of that year. The magazine editors sent
out by mail over ten million ballots (questionnaires). They obtained the
names and addresses from telephone directories, automobile registration
records, and the like. More than two million ballots were filled out and
returned to the magazine. The returned ballots were distributed as follows:

Ballots favoring Landon (the Republican)	1,293,669	57.1%
Ballots favoring Roosevelt (the Democrat)	972,897	42.9%
Total	2,266,566	100.0%

The magazine editors predicted from these data that Landon would
be elected President of the United States. In the actual election that followed a few days later, Roosevelt defeated Landon by an overwhelming
margin. The actual votes cast for the two candidates were:

Votes for Roosevelt	27,476,673	62.2%
Votes for Landon	16,679,583	37.8%
Total	44,156,256	100.0%

The electoral college voting was even more in Roosevelt's favor. Out
of 531 electoral votes, Landon received only 8 (5 in Maine and 3 in Ver-
mont).

Why did the *Literary Digest* make such a serious error in estimating how
the election would turn out? Certainly their sample was large enough—
over two million ballots. This example illustrates one of the most important lessons that the statistician needs to learn, namely, that just because
a sample is large it is not necessarily a good sample. Some of the best samples
contain relatively small numbers of cases. A relatively small sample often
may be quite satisfactory for one's purposes if it is a representative sample.
But a seriously distorted sample usually is worse than worthless no matter
how large it is; for it can deceive you into costly mistakes.

The *Literary Digest* had been a very popular magazine for more than
twenty years. It was a predecessor of *Life* magazine. But the error that it

made in estimating the results of the presidential election in 1936 ruined the *Literary Digest* and it went out of business a short time after that election. The magazine had spent about half a million dollars on the public opinion poll. In spite of the large size of the sample and in spite of the large amount of money spent on the survey, the sample was nonrepresentative and the estimates made from it were significantly wrong.

There probably were several kinds of distortion in the *Literary Digest* sample. In the first place, by drawing the sample from lists of telephone subscribers and automobile owners, the *Literary Digest* excluded from the sample many of the poorer people in the country. These poorer people tended to favor Roosevelt in their voting. In the second place, the sample was distorted further still by the fact that about four-fifths of the people who received ballots did not return them to the magazine. The people who do not return mailed questionnaires often tend to have attitudes different from those who mark and return the questionnaires.

Whenever the sampling plan prevents some members of the population from being drawn into the sample, seriously erroneous estimates are likely to result. Likewise, whenever some of those requested to express their opinion do not comply, seriously erroneous estimates are likely to result. Substituting cooperative people for those who refuse to comply is not a safe way to correct the distortion, although many investigators who use sample surveys make such substitutions and gamble on being able to derive good estimates from the tainted data.

Consequently, thoroughly scientific sampling requires (1) that the sampling be done in such a way that each member of the population has the same probability as every other member of the population of being drawn into the sample,[3] and (2) that true data be obtained for every member drawn into the sample.

1.4 UNDISTORTED SAMPLING

Here is an example of fairly good sampling. In April 1950, a large advertising agency had several clients who were spending millions of dollars a year on advertising by means of television. Some of these clients were wondering whether they were getting their money's worth out of the programs. Other clients were debating among themselves as to whether or not they ought to spend the large sums of money that would be needed

[3]There are more complex probability samples in which the probabilities are not the same for all individuals. But the probability of each individual being drawn into the sample must be known and must be different from zero. In some probability-sampling plans, it is not necessary that the exact probabilities be known before the data are collected; sometimes information needed for determining the exact probabilities can be collected along with the sample data. Nevertheless, the exact probabilities must be known before the reliability of the sample results can be determined. You are not likely to need such complex sampling plans in this course.

if they went into television advertising. They had to consider also the serious consequences that might follow if they did not get into television early; all the most desirable hours for broadcasting would be bought up by other companies.

One of the questions to which the agency and all the clients wanted the answer was: "How many families in the United States have television sets in their homes?" So many different guesses were being made by the people most concerned that the situation was one of great confusion and frustration.

Out of the 46 million homes in the United States, a probability sample of about 1500 homes was selected for the advertising agency. Representatives of the advertising agency visited each of the 1500 designated homes in May 1950, and determined whether or not the occupants had a television set. Only a few of these families could not be reached or refused to give the information.

It required only a couple of weeks for the market research men and women in the agency to analyze the data and prepare a report. They informed the management of the agency that approximately 12.5 per cent of the occupied dwelling units in the United States were equipped with television sets in May 1950. This was believed to be a reasonably reliable estimate, although it was obtained by means of a relatively small sample of the homes in the country. It served as a basis on which the management of the agency and the clients of the agency made important business decisions involving many millions of dollars.

1.5 CENSUS AND SAMPLE

Now, you remember that in April 1950, the United States Government took a complete census of the country. One of the things recorded by the census takers was the presence or absence of a television set in the home. The census indicated that 12.3 per cent of the occupied dwelling units in the United States had television sets in April 1950.

But here is a very important fact to remember. It took so long to process the census data that the result on television ownership was not ready to be announced to the public until June 1951—that is, more than a year after the advertising agency and its clients had their own reliable and accurate estimate of television ownership. Business people and others often cannot afford to wait a year for information. Frequently, data as much as a year old are no longer useful. Statistical sampling often enables organizations to obtain the needed information quickly.

Table 1.1 is another illustration of the fact that usually it is not necessary to take a complete census in order to obtain sufficiently reliable and accurate estimates on important social and economic affairs. Rowntree took a complete census of the 12,155 working-class families in the city of York, England. After he had computed the actual percentage of income spent for rent by the families, Rowntree performed an experiment to find out how much error he would have made if he had visited only 10 per cent of the homes instead of all.

He pulled out every tenth census schedule from the stack of 12,155 schedules. Then he computed the estimates that he would have obtained if he had visited only this 10 per cent of the families. You can find the sizes of the sampling errors that would have occurred if you compare the figures in the second and third columns of Table 1.1. The largest sampling error would have been 1.7 per cent, the error for income class C.

Rowntree also drew samples of one in twenty, one in thirty, one in forty, and one in fifty families. The corresponding estimates of the percentage of income spent on rent are shown in columns 4, 5, 6, and 7 of Table 1.1. The sampling errors are quite small in most instances, even for samples as small as one family out of fifty. This illustrates the savings in time and money that sampling makes possible in providing reliable and accurate estimates of things that people want to know.

TABLE 1.1

Per Cent of Income Spent on Rent*

Income Class†	Complete Census	Samples				
		1 in 10	1 in 20	1 in 30	1 in 40	1 in 50
A (1748)	26.5	26.6	25.9	27.0	28.3	27.1
B (2477)	22.7	22.9	23.5	23.3	22.3	22.6
C (2514)	19.8	18.1	17.2	18.3	17.2	18.0
D (1676)	15.8	16.0	14.4	15.8	17.1	16.9
E (3740)	11.3	11.0	10.1	10.7	11.2	11.5

**Source:* B. Seebohm Rowntree, *Poverty and Progress.* London: Longmans, Green & Co., 1941, Table 10, p. 489. Reproduced here with permission of the author.

†The numbers in parentheses are the numbers of families in the income classes.

The United States Bureau of the Census takes a sample of the working people of the country once every month. Look at the most recent issue of the *Monthly Report of the Labor Force.* You will see that by means of this statistical sampling plan the United States Government is able to estimate from month to month how, many people are employed in each major type of industry and in each major occupation, and how many unemployed there are.

The Department of Agriculture draws samples of farms frequently. By using sample surveys, the Department obtains estimates of how many pigs and cattle, how much wheat and corn, and the like, are being produced on the farms of the nation. In this way the statisticians are able to forecast the food supply for the coming year and to make other estimates needed by the Government and by private industry.

The Bureau of Labor Statistics uses samples to collect data and make estimates of the cost of living every month. The wages of millions of workers in America are raised or lowered on the basis of Bureau of Labor Statistics estimates of the rise or fall of the cost of living. The Bureau calls this the Consumer Price Index.

Many thousands of statistical workers are employed by the city, state, and Federal governments. In fact, almost all government employees are more successful in their jobs if they have some knowledge of statistical methods of thinking.

1.6 PUBLIC OPINION POLLS

No doubt you are familiar with some of the public opinion polls that are reported regularly in newspapers and magazines. These polls are based on samplings of parts of the population. Unfortunately, however, many of the opinion polls use types of sampling that are defective. Many do not use probability samples. That is, they do not take positive action to guarantee that each member of the population they are studying has the same probability as every other member of that population of being in the sample.

The argument that the poll takers usually give when they ask people to accept estimates based on their polls is that they have made many good estimates in the past in connection with elections and the like. That is a weak argument. The *Literary Digest* built up its tremendous reputation by that means during the twenty years from 1916 to 1936. But look at what happened to the magazine's poll in 1936! Other poll takers claimed great success from 1936 to 1948; then nearly all the public opinion polls went wrong during the presidential election in November 1948.

1.7 CONSUMER RESEARCH

Likewise, few of the consumer surveys that are made by commercial concerns use thoroughly scientific sampling systems. That is, few use probability samples. The reason that probably would be given you if you asked these concerns why they do not use better sampling systems is that the cost would be too high. That is not necessarily a good reason in every case.

Sometimes companies that make sample surveys want information only for general purposes and do not intend to take action of a serious nature on the basis of the survey. In such cases, a crude sampling system may be satisfactory, and it might be unwise to spend a large amount of money on a scientific sampling plan. In other cases, where important action is likely to be taken on the basis of estimates obtained from the survey, it might be advisable to spend a few thousand dollars more to obtain a good sample and protect the organization from the possible loss of millions of dollars as a result of action taken on a distorted estimate.

Be sure that you learn during this course how to recognize poor sampling systems and to be somewhat skeptical of the results obtained by unsound sampling plans. You will be a much wiser and better citizen if you develop the habit of examining the sampling plans before you accept the results of all the public opinion polls and consumer preference surveys that are published nowadays or are used for political or propaganda or advertising purposes.

1.8 PROBABILITY SAMPLES

A sound sampling system does not need to use the weak argument that it provided good estimates most of the time in the past. Probability samples have a strong argument, namely, that the reliability of an estimate provided by a probability sample can be demonstrated by the mathematical theory of probability. In other words, the reliability of a probability sample is based on mathematical theory and, for that reason, every probability sample has a "built-in" degree of reliability. Consequently, you can have as much confidence in the first probability sample drawn as you can in the tenth or the hundredth sample.

1.9 HOW TO DRAW GOOD SAMPLES

How can you be sure that the samples you draw in the investigations that you will choose for yourselves in Chapter 2 will be good ones? The best way to learn good sampling procedures is to go ahead and draw some good samples. That is what we shall do now. We shall begin with some simple samplings in order to demonstrate the principles of sampling theory.

There are two kinds of difficulties in sampling, namely, theoretical difficulties and practical difficulties. First, let us study the theoretical part of sampling. For this purpose, it will be advisable not to draw samples of people. People are the source of many of the practical difficulties of sampling. For example, people can be away from home when you call to interview them, or they may refuse to answer your questions, or they may say "I don't know." They may misunderstand your question and therefore give you the wrong answers, or they may lie deliberately. None of these practical difficulties can interfere with our sampling experiment if we use a group of red beads and white beads in a bag.

Take a bag and put into it 2000 red beads and 3000 white beads. If you draw a large number of random samples of beads from this bag and then study the results, you can discover most of the important principles of the theory of statistical sampling. If you learn these principles now and if you apply them later in your own practical sampling situations, it is likely that you will collect good data and that you will be able to draw reliable conclusions from your samples.

1.10 UNIVERSE, ELEMENTARY UNIT, AND PARAMETER

Before we proceed to discuss sampling further, we need to learn three new statistical terms. First, we need a word to represent the total group of things from which the sample is to be drawn. Sometimes the total group will be composed of individual human beings. At other times, the total group may be composed of families rather than individuals. Sometimes, the total group may be composed of all the parts turned out by a machine in a day or in a part of a day. At other times, the total group may be composed of all the assembled machines, such as airplanes or automobiles, produced by a particular manufacturer in a specified period of time. The total group might be composed of all the cards in an office filing cabinet. Or it might be composed of all the kernels of wheat in a carload. Or it might be the 5000 red beads and white beads in a bag.

The word that many statisticians use to designate any possible total group in which they are interested is *universe*. Some statisticians use the term *population* for this purpose. *Universe* is preferred here, because *population* is commonly associated with people, and, as we have seen already, the total group in a practical statistical investigation often is not composed of people.

The individuals or the items or the things whose characteristics are to be studied in the statistical investigation are called the *elementary units* of the universe. For example, if you decide to investigate the attitude of the men in a college toward compulsory military training, each man is an elementary unit of the universe of men in the college. In a consumer survey of family expenditures for a certain type of food, each family is an elementary unit in the universe composed of families. In a quality-control study of television picture tubes, each picture tube is an elementary unit in the universe of picture tubes made by the company.

The purpose in drawing a sample in a practical situation often is to estimate some important characteristic of the universe from which the sample was drawn. Statisticians use the word *parameter* to mean an important characteristic of the universe. For example, in a television audience survey, the universe may consist of all the potential viewers in the area to which the program is broadcast. Each individual in the area who is a potential viewer of the program is an elementary unit of the universe. The parameter that is to be estimated might be either the number or the proportion of the potential audience that actually watched the program.

1.11 PROBABILITY OF AN ELEMENTARY UNIT'S BEING IN A SAMPLE

One of the most important questions that can be asked about a sampling situation is this: What is the probability that any particular elementary unit in the universe will be included in the sample drawn from the universe?

In order to understand this important principle of probability, let us imagine that the 5000 red and white beads in the bag (universe) are numbered from 1 to 5000. Mix the beads thoroughly in the bag. Then blindfold someone and ask him to prepare to draw a bead from the bag. What is the probability that the bead numbered, say, 653 will be the bead that he draws? You probably will agree that the correct answer is 1/5000, that is, 0.0002. In other words, there is one chance in five thousand that any designated bead will be the one drawn in this situation. Furthermore, you probably will agree that each bead in the bag has the same probability of being drawn, namely, 1/5000 or 0.0002.

Of course, if the beads varied in size and weight, some of them would have a better chance, that is, greater probability, of being drawn. In our sampling experiment, however, we are assuming that the only difference between any of the beads in the bag is that some are red and the others are white.

Let us agree then that for our purposes in this book we may think of the probability of the occurrence of an event as being the proportion of the times that the event is expected to occur, as indicated by all the available relevant information.

The idea of probability is the most fundamental in statistics, because probability provides the key link between a sample and the universe from which the sample was drawn. Without the idea of probability there would be no logical or scientific way in which to draw conclusions about a universe after studying a sample drawn from that universe.

1.12 A SAMPLING EXPERIMENT

It is suggested that the members of the class organize in pairs. Then, let each pair of students find a convenient time in which to draw 20 samples of 100 beads from a universe of 5000 beads of which 2000 are red and 3000 are white.

Be sure that the beads in each sample are returned to the universe and thoroughly mixed into the universe before your next sample is drawn.

Make a record of the number of red beads in each of the 20 samples that you draw. Record the proportion of red beads in each of your 20 samples. Compute the total number of red beads in the 20 samples and the average number of red beads per sample. Also, find the total of the 20 proportions of red beads in the samples and the average of these 20 proportions. Your record should be similar to that shown in Table 1.2.

TABLE 1.2
Record of 25 Samples of 50 Beads Drawn from a Universe Containing 2000 Red Beads and 3000 White Beads

Sample Number	Number of Red Beads in the Sample	Proportion of Red Beads in the Sample
1	18	.36
2	22	.44
3	25	.50
4	21	.42
5	10	.20
6	19	.38
7	23	.46
8	18	.36
9	13	.26
10	21	.42
11	16	.32
12	22	.44
13	15	.30
14	20	.40
15	26	.52
16	19	.38
17	24	.48
18	20	.40
19	27	.54
20	21	.42
21	14	.28
22	20	.40
23	19	.38
24	24	.48
25	17	.34
Total	494	9.88
Average	19.76	0.3952

Students' Names   John Henderson and Helen Argyle

The records for the whole class should be tabulated in a form similar to that shown in Table 1.3. It shows the way a similar experiment turned out when 8 pairs of students drew a total of 200 samples with 250 beads in each sample. Their universe consisted of 1800 red beads and 3200 white beads. Each pair of students drew 25 samples.

TABLE 1.3

The Number of Red Beads in Each of 200 Samples of 250 Beads from a Universe Containing 1800 Red Beads and 3200 White Beads

Sample No.	Number of Red Beads in Sample							
	Pair 1	*Pair 2*	*Pair 3*	*Pair 4*	*Pair 5*	*Pair 6*	*Pair 7*	*Pair 8*
1	. 95	99	85	86	88	90	91	81
2	83	90	95	97	98	104	103	107
3	93	80	83	93	96	76	101	81
4	84	97	89	85	99	95	97	85
5	103	85	96	91	100	92	84	88
6	92	95	86	91	90	94	81	94
7	88	82	81	95	90	95	104	85
8	83	92	93	87	94	98	82	102
9	93	98	83	87	81	88	78	87
10	98	84	83	83	96	89	82	93
11	92	76	96	78	74	84	89	88
12	87	92	93	111	77	90	88	87
13	99	102	103	92	90	109	98	86
14	80	87	88	91	106	101	97	89
15	86	84	92	89	93	94	89	87
16	90	93	85	100	94	90	85	89
17	99	80	86	90	72	97	82	94
18	88	97	80	96	82	91	78	73
19	91	101	84	94	87	100	86	91
20	86	92	90	79	95	75	106	91
21	90	95	92	78	97	92	87	95
22	96	77	96	94	88	102	85	99
23	94	101	81	84	100	91	69	88
24	89	86	93	99	86	86	83	89
25	93	81	91	79	96	79	89	84
Total	2272	2246	2224	2249	2269	2302	2214	2233
*Average**	90.88	89.84	88.96	89.96	90.76	92.08	88.56	89.32

*The average number of red beads in the 200 samples is 90.045.

1.12.1 Sampling errors. The proportion of red beads in the universe from which the 200 samples recorded in Table 1.3 were drawn was $\pi = 1800/5000 = 0.360$. If that proportion of the beads in a sample containing 250 beads were red, the number of red beads in the sample would be $(0.360)(250) = 90$. The difference between this 90 and the number of red beads drawn in any one sample is called a *sampling error*. (Notice that only 11 of the 200 samples recorded in Table 1.3 contained exactly 90 red beads.)

The students computed the proportion p of red beads in each of the 200 samples. The 200 values of p are given in Table 1.4. Only 11 of the 200 samples produced a proportion p of red beads exactly the same as the proportion π of red beads in the universe from which the samples were drawn, namely, $\pi = 0.360$. In other words, there were sampling errors in most of the proportions of red beads.

TABLE 1.4

The Proportion of Red Beads in Each of 200 Samples of 250 Beads from a Universe Containing 1800 Red Beads and 3200 White Beads

Sample No.	*Proportion p of Red Beads in a Sample*							
	Pair 1	*Pair 2*	*Pair 3*	*Pair 4*	*Pair 5*	*Pair 6*	*Pair 7*	*Pair 8*
1	.380	.396	.340	.344	.352	.360	.364	.324
2	.332	.360	.380	.388	.392	.416	.412	.428
3	.372	.320	.332	.372	.384	.304	.404	.324
4	.336	.388	.356	.340	.396	.380	.388	.340
5	.412	.340	.384	.364	.400	.368	.336	.352
6	.368	.380	.344	.364	.360	.376	.324	.376
7	.352	.328	.324	.380	.360	.380	.416	.340
8	.332	.368	.372	.348	.376	.392	.328	.408
9	.372	.392	.332	.348	.324	.352	.312	.348
10	.392	.336	.332	.332	.384	.356	.328	.372
11	.368	.304	.384	.312	.296	.336	.356	.352
12	.348	.368	.372	.444	.308	.360	.352	.348
13	.396	.408	.412	.368	.360	.436	.392	.344
14	.320	.348	.352	.364	.424	.404	.388	.356
15	.344	.336	.368	.356	.372	.376	.356	.348
16	.360	.372	.340	.400	.376	.360	.340	.356
17	.396	.320	.344	.360	.288	.388	.328	.376
18	.352	.388	.320	.384	.328	.364	.312	.292
19	.364	.404	.336	.376	.348	.400	.344	.364
20	.344	.368	.360	.316	.380	.300	.424	.364
21	.360	.380	.368	.312	.388	.368	.348	.380
22	.384	.308	.384	.376	.352	.408	.340	.396
23	.376	.404	.324	.336	.400	.364	.276	.352
24	.356	.344	.372	.396	.344	.344	.332	.356
25	.372	.324	.364	.316	.384	.316	.356	.336
Total	9.088	8.984	8.896	8.996	9.076	9.208	8.856	8.932
*Average**	.364	.359	.356	.360	.363	.368	.354	.357

*Rounded to three decimal places. The exact average value of p for the 200 samples is 0.36018.

The 200 sampling errors in the proportion p for these 200 samples are shown in Table 1.5. For example, for the first sample, $0.380 - 0.360 =$

+0.020. For the second sample, $0.332 - 0.360 = -0.028$, and so on. Errors in proportions that are greater than 0.360 are considered to be positive errors; their plus signs are omitted in Table 1.5. Errors in proportions that are less than 0.360 are considered to be negative errors, and minus signs are attached to them in that table.

TABLE 1.5

Sampling Errors in Proportion of Red Beads in 200 Samples of 250 Beads from Universe of 1800 Red Beads and 3200 White Beads

Sample No.	Sampling Error in the Proportion p							
	Pair 1	*Pair 2*	*Pair 3*	*Pair 4*	*Pair 5*	*Pair 6*	*Pair 7*	*Pair 8*
1	.020	.036	−.020	−.016	−.008	.000	.004	−.036
2	−.028	.000	.020	.028	.032	.056	.052	.068
3	.012	−.040	−.028	.012	.024	−.056	.044	−.036
4	−.024	.028	−.004	−.020	.036	.020	.028	−.020
5	.052	−.020	.024	.004	.040	.008	−.024	−.008
6	.008	.020	−.016	.004	.000	.016	−.036	.016
7	−.008	−.032	−.036	.020	.000	.020	.056	−.020
8	−.028	.008	.012	−.012	.016	.032	−.032	.048
9	.012	.032	−.028	−.012	−.036	−.008	−.048	−.012
10	.032	−.024	−.028	−.028	.024	−.004	−.032	.012
11	.008	−.056	.024	−.048	−.064	−.024	−.004	−.008
12	−.012	.008	.012	.084	−.052	.000	−.008	−.012
13	.036	.048	.052	.008	.000	.076	.032	−.016
14	−.040	−.012	−.008	.004	.064	.044	.028	−.004
15	−.016	−.024	.008	−.004	.012	.016	−.004	−.012
16	.000	.012	−.020	.040	.016	.000	−.020	−.004
17	.036	−.040	−.016	.000	−.072	.028	−.032	.016
18	−.008	.028	−.040	.024	−.032	.004	−.048	−.068
19	.004	.044	−.024	.016	−.012	.040	−.016	.004
20	−.016	.008	.000	−.044	.020	−.060	.064	.004
21	.000	.020	.008	−.048	.028	.008	−.012	.020
22	.024	−.052	.024	.016	−.008	.048	−.020	.036
23	.016	.044	−.036	−.024	.040	.004	−.084	−.008
24	−.004	−.016	.012	.036	−.016	−.016	−.028	−.004
25	.012	−.036	.004	−.044	.024	−.044	−.004	−.024
Total	.088	−.016	−.104	−.004	.076	.208	−.144	−.068
*Average**	.004	−.001	−.004	−.000	.003	.008	−.006	−.003

*Rounded to three decimal places. The exact average sampling error for the 200 samples is 0.00018.

1.12.2 Theory of sampling. Why is so much attention given here to the sampling experiments? It is because the most important part of all

statistical theory is the theory of sampling errors. Statisticians have discovered that errors in random sampling follow certain laws or formulas.
By using these mathematical laws or formulas they can predict—before
any samples are drawn from a universe—just what sizes of sampling errors
may be expected to occur. Some of the most useful of these mathematical
laws or formulas are described in Chapters 4 to 9. Here we can examine
briefly the ways in which some of them work.

1.12.3 How we use the laws of sampling errors. It was suggested that
the class draw about 200 samples with 100 beads in each, from a universe
consisting of 2000 red and 3000 white beads. Now, it is possible to tell in
advance almost exactly what kind of set of sampling errors you are going
to obtain, provided you do a careful job of drawing the samples—mix the
beads in the bag thoroughly before you draw each sample, count accurately
and record the number of beads of each color in each sample, and do not
lose beads or in any other way change the probabilities of red beads and
white beads being drawn from the bag.

About 68 per cent of the samples will have sampling errors between
-0.049 and $+0.049$ in the proportion p of red beads. When you construct
a table similar to Table 1.5, count the number of sampling errors in p that
are in the interval from -0.049 to $+0.049$ and see if this prediction is not
very nearly correct. Only about 5 per cent of your samples will produce
sampling errors in p outside the interval from -0.096 to $+0.096$. Check
this prediction when you have tabulated all your sampling errors. Only
about 1 per cent of your sampling errors in p will fall outside the interval
from -0.126 to $+0.126$. Check this prediction too. All of these predictions
can be made from the theory and the formulas to be explained in detail
in Chapters 3 to 7. And remember that these predictions were made long
before you drew the samples.

Let us suppose that the students who drew the 200 samples from which
Table 1.4 was constructed had drawn only one sample, namely, the first
recorded in that table. This sample produced as estimate $p = 0.380$. In
Chapter 4 you will learn how to compute *confidence intervals*. A 95 per cent
confidence interval to be associated with the estimate $p = 0.380$ is the interval from 0.321 to 0.439. This interval includes the true value, $\pi = 0.360$.

If you examine the 200 values of p in Table 1.4 and count the number
of samples that produced values of p such that the true value of π, 0.360, does
not lie in the 95 per cent confidence intervals, you will find 10 samples,
that is, 5 per cent of the 200 samples. The 10 samples that produced such
values of p are those that produced the proportions 0.276, 0.288, 0.292,
0.296, 0.300 and 0.424, 0.424, 0.428, 0.432, 0.444.

Figure 1.1 shows this graphically. The 200 values of p in Table 1.4 are
indicated by the heavy black dots. The 200 confidence intervals are indi-

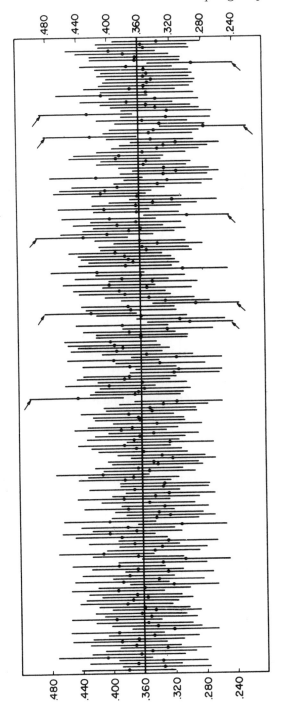

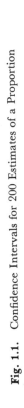

Fig. 1.1.   Confidence Intervals for 200 Estimates of a Proportion

cated by the 200 vertical line segments drawn through the black dots. The true value, 0.360, is indicated by the horizontal line drawn across the graph from the point 0.360 on the vertical scale. You will notice that all except 10 of the vertical line segments overlap the horizontal line representing 0.360. Five of the vertical line segments that do not overlap the horizontal line for 0.360 are entirely above that line; the other five vertical line segments that do not overlap the horizontal line for 0.360 are entirely below the line.

These results are not just accidents that occurred in this example and may never occur again. A basic principle in sampling theory tells us to expect these results whenever very large numbers of large random samples are drawn from a universe.

In an example of market research mentioned earlier, 65 per cent of the housewives in a sample of 1000 said they would prefer to buy coconut in the can with the blue lining; 35 per cent preferred the other can or said they could see no difference between the two cans of coconut. But how much confidence can the producer of this product have in the result of this sampling? How likely is it that sampling error as a result of chance in the selection of these women resulted in the 65 per cent preference for the blue lining? Perhaps the true proportion of women who would prefer coconut in a blue-lined can is 50 per cent. It can be shown that 99 per cent confidence limits corresponding to $p = 0.65$ are approximately 0.61 and 0.69, when the size of the sample is 1000. The lower limit is far above 0.50. Consequently, in this situation the coconut producer can feel very safe in believing that the true value of the proportion in the universe is considerably greater than 0.50.

Here is another example. Chemists in the laboratory of a leading coffee manufacturer thought that they had developed a new formula for roasting and blending coffee that would be a big improvement over the coffee the company was making at that time. But the management of the company knew it was possible that the people who drink coffee might not agree with the chemists. A mistake in this matter could cost the company millions of dollars. New machinery would need to be installed in order to make the coffee according to the new formula; consequently, the company could not continue to make the old type of coffee as well as the new.

A sample survey of 1000 coffee drinkers showed that 54 per cent preferred the coffee made by the new formula and 46 per cent preferred that made by the old. If you had been the statistician for this company, what advice would you have given the management on the basis of this evidence alone?

Ninety-nine per cent confidence limits in this example are approximately 0.50 and 0.58. There is about a 1 per cent risk involved in the statement that the confidence interval contains the true value of π for the universe of coffee drinkers. Consequently, there is only about 1 chance in 200, or a risk of one-half of 1 per cent, that the true value of π for the universe of coffee drinkers is less than 0.50. That is a small risk and, ordinarily, the evidence would be considered strong enough to justify the decision that the value of π for this universe of coffee drinkers is greater than 0.50.

Therefore, you might tell your employer that there is strong evidence that slightly more than half of the universe of coffee drinkers prefer the coffee made by the new formula. No doubt, the management considered many other factors before making its decision; the company did not, in fact, change over to the new formula at that time.

The confidence limits that were used here and the decision that was stated with reference to the value of π involved the assumption, of course, that the sample of coffee drinkers was a random sample drawn from the universe of coffee drinkers. If the group of 1000 was not a probability sample but was merely the most conveniently available group of coffee drinkers— for example, a group of employees of the company or the customers of one particular department store—then there would be no known method for determining the reliability of the estimate obtained from the sample.

Here is a slightly different way in which to determine the degree of confidence to be attached to an estimate obtained from a random sample. In the coffee example, what management wanted was definite assurance that considerably more than 50 per cent of coffee drinkers would prefer the new formula. Otherwise, it would have been a costly mistake to make the changeover. Therefore, let us make the hypothesis that the true proportion π of the universe of coffee drinkers who prefer the new formula is exactly 0.50. In other words, we state that there are as many coffee drinkers who prefer the coffee made by the old formula as there are coffee drinkers who prefer the coffee made by the new formula. Now we ask ourselves this crucial question: What is the probability of drawing a random sample of 1000 coffee drinkers in which the sampling error due to random chance in selection would produce a value of p as great as 0.54 for the proportion preferring the coffee made by the new formula?

It can be shown that only about one-half of 1 per cent of all the possible random samples containing 1000 coffee drinkers from a universe of coffee drinkers in which $\pi = 0.50$ would produce values of p as great as or greater than 0.54. In other words, the probability is only about 1/2000 or 0.005 that a value of p as great as or greater than 0.54 is due to sampling error in our example. That is a small probability. Consequently, the difference between $p = 0.54$ and $\pi = 0.50$ is considered to be statistically significant, and the hypothesis that $\pi = 0.50$ is rejected.

However, even though the difference between p and π in this example is considered to be *statistically* significant, that difference may not be *practically* significant. The accounting department may be able to advise the management that unless at least 55 per cent of the universe of coffee drinkers definitely prefer the new formula, the increase in sales would not pay for the cost of the changeover within a reasonable period. Therefore, we must distinguish between statistical significance and practical significance in dealing with samples. Statistical significance corresponds to small probability that the result observed is due to sampling error; a decision about the practical significance of a statistical result frequently may involve other factors as well.

A statistician does not become annoyed by the fact that he cannot tell anyone the exact value of the parameter he is trying to estimate. An estimate with a pair of confidence limits not too far apart is sufficient for most statistical purposes. Good approximations represented by estimates in which he can have a known degree of confidence are all that the statistician needs, and that businessmen, government officials, and others need.

Remember that the universe comes first, the sample second, and the estimate third. Some people who call themselves statisticians make the serious mistake of obtaining the "sample" first, by using the most available and cooperative people. Then they claim that an estimate computed from such a sample is a reliable estimate for the universe of all the people in the United States or all the people in some other large group.

For example, teachers sometimes require the students in their classes to fill out questionnaires or to participate in experiments. Then the teachers make estimates from these data and publish conclusions, as if the estimates and the conclusions were based on probability samples representative of all the young people in the nation. Nothing could be sillier. A group of 25 or 30 students in a particular class in a particular college is unlikely to be a random sample of all the young people in the United States.

Consequently, whenever you are preparing to undertake a statistical investigation from which it is necessary to obtain estimates whose reliability is known, be sure that you have defined the universe carefully before you try to draw the sample. If you have not defined the universe carefully, how will you know which elementary units are eligible to be in the sample and which are not? In practical sampling situations, some unexpected difficulties usually arise, and if you have defined the universe clearly, they are often more easily surmounted.

1.13 WHAT IS A LARGE SAMPLE?

The term "large sample" has been used several times in this chapter and it will be necessary to distinguish between "large samples" and "small samples" at many places in the remainder of this book. The principal reason for making this distinction is that some of the mathematical theory of statistical methods is simpler for large samples than it is for small samples. Sometimes the reverse is true. Consequently, the distinction between large samples and small samples in this book is based on mathematical theory rather than on practical considerations such as the time or the cost required to obtain a sample of elementary units from a universe.

Fortunately, in types of situations in which the mathematical theory is simpler for large samples than it is for small, the size of the sample need not be very large in order that the simpler mathematical theory may be applicable. And in types of situations in which the mathematical theory is simpler for small samples than it is for large samples, excellent approximations may be easily obtained for large samples.

For the purposes of the mathematical theory in this book, any sample that contains at least thirty-two elementary units will be considered a "large sample." If a sample does not contain at least thirty-two elementary units, it will be considered a "small" one. Occasionally, it may be desirable to specify that the sample must be a "very large sample," and by this it shall be meant that it must contain at least a hundred elementary units.

Because so much mathematical theory that is available to us today is applicable to small samples, some people have gained the erroneous impression that it is no longer necessary to obtain large samples in any practical statistical investigations. The truth is that although we now have much mathematical theory that is just as valid for small samples as the mathematical theory for large samples, the allowable amount of sampling error, that is, the confidence interval, for an estimate based on a small sample may be so large as to render the estimate practically useless. Or, to express this idea another way, differences occurring in small samples frequently must be exceedingly large in order to be considered statistically significant by the commonly used tests of significance. Consequently, in most practical situations, for high reliability of estimates and for tests of significance that are sensitive to small differences we usually need to obtain data for large or even very large samples.

1.14 THE NATURE OF STATISTICAL METHODS

In most situations, we may think of complete information about a universe as being the data obtained by a complete and perfect census or examination of all the elementary units in the universe. On the other hand, our incomplete information about a universe may take the form of data obtained from a relatively small sample of the elementary units in the universe.

Statistical methods are practical methods for drawing conclusions or in some way improving our knowledge by using incomplete information about a universe—not any or every kind of incomplete information, but preferably only those kinds to which statistical methods of estimation and inference are known to be applicable. It is possible to justify mathematically and logically the making of estimates and the drawing of inferences from incomplete information provided the samples are probability samples. Usually, it is impossible to know whether statistical methods of estimation and inference are genuinely useful if the sample is not a probability sample.

A conclusion drawn or knowledge gained about a universe from a sample, that is, from incomplete information, never is certain to be perfectly accurate. It must be regarded as probable rather than certain and approximate rather than perfectly accurate.

The three principal problems that arise when we have only incomplete information about a universe are (1) how to estimate the unknown characteristics or parameters of the whole universe; (2) how to determine the reliability of the estimates based on the incomplete information, that is, how to indicate limits of the probable errors in the estimates; and (3) how to test specific hypotheses about the unknown characteristics or parameters of the universe.

Through understanding of the pattern of results likely to occur in the process of repeated drawings of random samples from a universe we are able to decide what information should be extracted from one random sample drawn from that universe. In other words, the basis for interpretation of one such sample is the understanding of how it is likely to fit into the pattern formed by all the possible random samples that might have been drawn from the given universe. For this reason experience gained by drawing and analyzing a large number of random samples from a universe is valuable in the study of statistics.

For example, Table 3.10 is a frequency distribution for the sample results given in Table 1.3, and Figure 3.3 is a graph for the same data. If, now, you pick out one result from the list of sample results given in Table 1.3, you can see where that particular result fits into the pattern indicated by the frequency distribution in Table 3.10 and the graph in Figure 3.3.

The study of statistical methods of estimation and inference is much more important today than it was before the twentieth century. Much of the available sampling theory has been worked out since 1900. This development allows use of the incomplete information contained in relatively small probability samples as a basis for estimates and inferences of known reliability about the characteristics of universes.

An essential characteristic that distinguishes the statistical method of thinking from the traditional mathematical method of thinking is the complete dependence of the statistical method on the modern concept of probability and the complete absence of this idea in traditional deductive mathematics.

In the pure mathematics of arithmetic, geometry, and algebra—with which you may be familiar—the method of thinking, that is, the means of acquiring new knowledge, is that of deductive logic. There, no conclusion is considered as true until a deductive proof of it has been constructed.

In statistics, the method of acquiring new knowledge is that of inductive inference. The inductive logical procedure is a method of reasoning from observed facts to the inferences that those facts warrant. Here, a statement is considered as acceptable or unacceptable insofar as it meets or fails to meet the tests of inductive logic; the test of truth or acceptability is not deductive proof.

The key element in the process of inductive reasoning in statistics is the idea of probability; a statement is acceptable if it is based on observations obtained in such a way that the probability is very great that the statement is true and very small that the statement is false.

Deductive reasoning is based on postulates or assumptions of knowledge or fact; in pure mathematics, for example, the postulates are assumed to be true and new knowledge is deduced from them. Inductive reasoning takes into account postulates of ignorance—an unfamiliar procedure for pure mathematicians.

In statistics, a postulate of ignorance asserts that certain things are not known. Furthermore, the validity of the inductive argument requires that those things must be unknown; this is essential in any correct decision about a state of uncertainty. Probability describes accurately—with mathematical precision—a state of uncertainty. It is not a defect in a probability statement that the statement would be different if the observed data were different.

States of uncertainty are not common in the processes of reasoning on which the traditional pure mathematics is based. In fact, the sophisticated modern aspects of the concept of probability were unknown to the ancient Greek and Islamic mathematicians. Consequently, their method of proof was the method of deductive logic. Of course, no one objects to deductive proof in situations where that method is applicable. But in the physical, biological, and social sciences, and in the daily lives of everyone, most of the knowledge needed for guidance in decision-making must be gained through the inductive method of statistics.

All knowledge gained by that method must be phrased in terms of probability. And it is a serious error for anyone to suggest that probability statements *must* be replaced by deductive proof. In such attempts, one may be doomed in advance to failure, because, by its very nature, a legitimate inference reached by the inductive procedure of statistics may not be susceptible to deductive proof; the way in which the principle of probability enters into the inductive process makes it unnecessary, if not erroneous, to seek a deductive proof of a conclusion reached via statistical methods.

Perhaps the remarks made above will help you understand what is meant when it is said that statistics is a method for making decisions under conditions of uncertainty and a method for drawing conclusions from incomplete information.

1.15 REMARKS

Do not be surprised or worried if you do not understand everything that has been mentioned in this first chapter. Each of the following chapters will help to make these things clear to you. Consequently, let us hurry on to Chapter 2 and begin to think as statisticians and research workers. You will learn more from one real research job that you plan and carry out yourselves than you could learn by reading about a hundred problems that other people solved but in which you are not interested. Be sure that you enjoy yourselves while you are carrying out all the steps in the investigations that you begin in Chapter 2.

EXERCISES

1. Some newspaper editors and radio station operators took a poll before a recent presidential election by asking in the newspapers and on the radio that people mail in signed post cards stating for whom they intended to vote. (a) Do you think that a representative sample of all the voters in the United States is likely to be obtained in that way? Do you think that a distorted estimate is likely to be the result of such a method of sampling? (b) Is there any mathematical method inherent in such a sampling procedure for determining the degree of confidence that can be attached to the estimate obtained from the sample?

2. Not long ago a farm newspaper made a survey of the buying intentions of Wisconsin farm families. The editor mailed a questionnaire to each of 1500 farm families who subscribed to the newspaper, asking which, if any, of a long list of items they intended to buy in the near future. Only 269 completed questionnaires were sent back to the newspaper. The newspaper distributed to its advertisers and others a report showing the results of the survey for each product. For example, the report stated that 27 per cent of the farm families in Wisconsin intended to buy a tractor soon and that 17 per cent intended to buy an electric clock. What do you think about the reliability of the estimates obtained in that survey? Are the mathematical formulas for computing confidence limits applicable to the estimates obtained?

REFERENCES

Frederick Mosteller, *et al.* "The Pre-election Polls of 1948," *Bulletin* 60. N.Y.: Social Science Research Council, 1949.

Richard Ruggles and Henry Brodie, "An Empirical Approach to Economic Intelligence in World War II," *Journal of the American Statistical Association* 42 (1947), pp. 72–91.

Frederick F. Stephan, "History of the Uses of Modern Sampling Procedures," *Journal of the American Statistical Association* 43 (1948), pp. 12–39.

Chapter 2

PLANNING A STATISTICAL INVESTIGATION

2.1 INTRODUCTION

Some students claim that they can study better with the radio tuned to a musical program than they can in a perfectly quiet room. Other students say that any radio program or other noise distracts them and that they prefer to study in a quiet atmosphere. Among those who like to have music while they are studying, some think that classical music is best for this purpose. Others prefer popular semiclassical music. Still others say that they can learn most quickly and easily when they have as background a program of fast, loud jazz music.

It would not be difficult to carry out a statistical investigation to test these claims about the effectiveness of music as an aid to study. You might select four random samples of students. Then, ask one of the four groups to go to a very quiet room. Ask another group to go to a room in which you have arranged for classical music to be played. Place a third group in a room in which popular semiclassical music will be played. Place the fourth group in a room where fast, loud jazz music will be played.

Give all four groups of students the same assignment—say, in history—and permit them to study it for one hour. If you want to obtain data on the speed of reading as well as on the quantity of information learned, ask each student to record the time at which he or she finished reading the assigned material.

At the end of the hour, give all four groups of students a test containing, say, fifty questions about the material in the assignment. Then tabulate and analyze the test scores to see if there are any significant differences among the four groups insofar as knowledge of the assigned material is concerned. Similarly, analyze the data on speed of reading the assignment to see if there are any significant differences among the groups of students in this respect.

What are the hypotheses that this experiment is designed to test? One way in which to state a hypothesis for this problem is: Music is not effective as a learning aid. Or, you might state the hypothesis as follows: There is no significant difference between the results of studying in a quiet room and studying in a room where music is being played. Another hypothesis would be this: There is no significant difference between the effectiveness of one kind of music and the effectiveness of any other kind of music as an aid to studying an assignment. Similar hypotheses can be stated about the effect of music on the speed of reading a new assignment in history. You will learn how to test hypotheses such as those stated here in sections 5.1 to 5.2 and 7.6 to 7.6.2.

There is a report of one experiment on this subject. The conclusions drawn from the data were (1) there are no significant differences in the amounts of information learned by groups studying in a quiet room, a room where classical music is being played, a room where popular semi-classical music is being played, and a room where fast, loud jazz music is being played; and (2) students studying in a room where fast, loud jazz music is being played read the material significantly faster than students studying in any of the other three situations.

2.2 THE STEPS IN PROBLEM SOLVING

A complete experience in problem solving involves some or all of the following steps: (1) becoming aware of a problem situation that needs to be investigated; (2) formulating the problem; (3) planning the investigation; (4) collecting and recording the data; (5) organizing and presenting the data in tables and graphs; (6) analyzing the data to discover important facts and relationships; (7) interpreting the results of the analysis, for example, making estimates of the parameters in the universe and stating the reliability of the estimates; (8) making a report on the project.

Each of these eight steps is important and should be treated in any course that aims to give the student experience and awareness of the usefulness of statistics in our lives. Of course, the steps listed above are not always carried out one at a time and in the order given. For instance, some facts and relationships may become so obvious while the data are being collected that no formal mathematical analysis is necessary.

2.3 BECOMING AWARE OF A PROBLEM

There are many problems that a person might wish to solve, some trivial, others so complex and fundamental in character that they cannot be solved until more powerful scientific methods are discovered. Here are a few suggestions to guide you and your classmates in selecting problems for study in this course.

First, the problem should be interesting and important to many members of the class. Second, the problem should be large enough to provide interesting work for several students. On the other hand, the problem should not be so big or so complicated that the details cannot be carried out thoroughly and completely in the limited time available for this course.

Here is a word of caution. In general, the beginner's tendency is to tackle problems too big rather than too small when he is allowed to choose them. One reason for this is that in the beginning he usually has a rather vague understanding of the problem and is not aware of all the complications and relationships that are involved. Consequently, you should attempt what you will be able to do thoroughly—what *can* be done, not merely what you would like to do.

2.3.1 Examples of suitable problems. The number of problem situations suitable for statistical investigation by the class is practically unlimited. Here are a few illustrations that may help to start your thinking.

1. The age at which young people marry (see Table 3.8).
2. An opinion or attitude poll on some topic of current interest.
3. The relative popularity of various athletic activities.
4. Speed in arithmetic processes or in mechanical manipulations (see Table 6.3).
5. The proportion of left-handed students in the school.
6. Is there any correlation between hand preference in one-hand operations and eye preference in one-eye operations such as aiming a rifle or looking through a telescope, a microscope, or a hole in a wall?
7. Weights or heights of students.
8. Student accidents.
9. The ways in which students spend their money.
10. A "before and after" test of the attitude, preference, or some other characteristic(s) of students in connection with some event or activity scheduled for the school or the community (see sections 5.1 and 5.1.1.)
11. A test of the significance of the effect of some "experimental" condition or event, using two matched samples (see sections 5.1 and 5.1.1).
12. A test of the significance of the difference between two large groups of people with respect to some topic of interest to you, using two independent samples (see sections 5.2 and 7.6 to 7.6.2).
13. A test of a hypothesis that the distribution of some characteristic among the student body is a distribution of some specified form such as, for example, a normal distribution or a rectangular distribution (see section 6.6).
14. A test of the randomness of some series of events (see section 6.7).
15. The determination of a linear prediction equation for height and age or for weight and age or for some other pair of characteristics of students (see Chapter 9).

16. The accuracy and/or the precision of some type of measurement made frequently in physics, chemistry, biology, psychology, sociology, or economics (see section 3.7).

Some of the tables and some of the exercises in this book may suggest interesting situations for you to investigate. The best problems for you are those about which you would like to know something and which it would be fun to investigate.

2.4 THREE TYPES OF CLASSIFICATION FOR ELEMENTARY UNITS

We obtain our data by recording certain properties or characteristics of the elementary units that are drawn into a sample from a universe. It is necessary to distinguish between three different types of classification that may be made. First, elementary units may have truly quantitative characteristics. Second, elementary units may have truly nonquantitative characteristics. Third, elementary units may be classified in ways that are neither truly quantitative nor truly nonquantitative. Let us call this third type of classification "partly quantitative classification."

The reason why we need to understand and recognize the differences between these three types of classification of elementary units is that a statistical method of analysis and interpretation appropriate for data based on one type of classification may not be appropriate for data based on another type of classification. In a practical situation, we may be taking a big risk of deceiving ourselves if we use an inappropriate method of analysis and interpretation for our data or if we do not use the best method of analysis in situations in which two or more methods are applicable.

2.4.1 Truly nonquantitative classification. We might classify each individual in a sample of students as male or female. Such a distribution might be listed in a series as follows:

Student (Elementary Unit)	Sex (Nonquantitative Classification)
A	Male
B	Female
C	Female
D	Male
E	Male

The distribution might be summarized in a frequency table as follows:

Sex	Number of Students
Male	3
Female	2
Total	5

Other examples of truly nonquantitative classification of elementary units in a sample are: political affiliation (Republican, Democrat); marital status (married, single, widowed, divorced, separated); condition of health (good, fair, poor); handedness (right-handed, left-handed, ambidextrous); defective, nondefective; vaccinated, not vaccinated; accepted, rejected; citizen, alien; soldier, sailor, airman, marine; blonde, brunette, redhead; silk, wool, cotton, synthetic; geographical regions, for example, east of the Mississippi, west of the Mississippi, the North, the South. You can add other truly nonquantitative classifications to this list for yourself.

Characteristics that are classified in this way usually are called *attributes* of the elementary units. When elementary units are classified in this way, the difference between two elementary units is described verbally, that is, by a word or words, rather than numerically.

Another feature that distinguishes the truly nonquantitative type of classification from the truly quantitative type of classification is that classifying elementary units in a truly nonquantitative way involves merely *counting* the elementary units of each kind, whereas classifying elementary units in a truly quantitative way involves *measuring* each elementary unit with some measuring device.

Furthermore, in a truly nonquantitative classification of elementary units, there is usually no natural or logically necessary order in which the different categories should be listed before the data are collected. After the data have been collected, we may, if we choose to do so for ease of interpretation or for esthetic reasons, list the categories in the order of the frequency with which they belonged to the elementary units in the sample. For example, there is no natural or logically necessary order in which the types of movie in Table 11.1 should be listed.

The following table illustrates one way to organize the data in a sample for the purpose of investigating the probability of relationship between two truly nonquantitative classifications, namely, the handedness and sex of students. A method for analyzing such a tabulation is given in section 8.2.

Handedness and Sex of 390 Students

	Right-Handed	*Left-Handed*
Male	174	26
Female	177	13

2.4.2 Truly quantitative classification. Consider each of the following tabulations:

Weights of Students

Student *(Elementary Unit)*	*Weight* *(Quantitative Classification)*
A	121 lbs.
B	112 lbs.
C	92 lbs.
D	103 lbs.

Temperatures of Patients in a Hospital

Patient *(Elementary Unit)*	Temperature *(Quantitative Classification)*
A	99.3°F
B	98.6°F
C	101.4°F
D	98.2°F
E	99.0°F

In the first table, each of the students is described by a number to which a constant unit of measurement (the pound, avoirdupois) is attached, and the measurement scale has a zero point. Similarly, in the second table, each patient is described by a number to which a constant unit of measurement (the degree, Fahrenheit) is attached, and the measurement scale has a zero point.

For every truly quantitative classification, there is a one-dimensional measurement scale that has a constant unit of measurement and a zero point. The unit of measurement is constant for all parts of the scale and at all times. The difference of one pound between a weight of, say, 117 pounds and a weight of 116 pounds is exactly the same as the difference of one pound between a weight of 67 pounds and a weight of 66 pounds. And the pound avoirdupois remains constant as time passes.

The unit of measurement that is used may be arbitrary. For example, in the United States, we usually measure a person's weight in pounds. In France, a person's weight usually is expressed in grams or kilograms. Similarly, in our daily affairs, we usually measure temperature in degrees Fahrenheit. In physics and chemistry laboratories, temperature often is measured in degrees centigrade.

The relationship between the Fahrenheit and centigrade scales for measuring temperature is $F = 9C/5 + 32$, where $F =$ the temperature in degrees Fahrenheit and $C =$ the temperature in degrees centigrade. Some corresponding temperature readings are

centigrade	$-17\ 7/9$	0	20	40	100
Fahrenheit	0	32	68	104	212

Notice that the ratios of corresponding intervals on the two temperature scales are equal. For example,

$$\text{Centigrade} \qquad\qquad \text{Fahrenheit}$$

$$\frac{40 - 20}{20 - 0} = 1 = \frac{104 - 68}{68 - 32}$$

and

$$\frac{100 - 20}{40 - 20} = 4 = \frac{212 - 68}{104 - 68}$$

This property of equal ratios in two truly quantitative scales of measurement can also be demonstrated for the scales that measure weights in pounds avoirdupois and in grams.

Observe that the zero point of the Fahrenheit scale of temperature has a meaning entirely different in terms of temperature from that of the zero point on the centigrade scale. In other words, for measuring temperature the scales that are commonly used (Fahrenheit and centigrade thermometers) have *arbitrary zero points*. On the other hand, zero pounds has exactly the same meaning in terms of weight as zero grams. We say that the two commonly used scales for measuring weight (pounds and grams) have *true zero points*.

A measurement scale that has (1) a constant unit of measurement and (2) a true zero point is called a *ratio scale*. A measurement scale that has (1) a constant unit of measurement and (2) an arbitrary zero point is called an *interval scale*.

If the measurement scale on which a classification of the elementary units in a sample is based has a constant unit of measurement, we may regard the classification as a truly quantitative one, no matter whether the measurement scale has a true zero point or an arbitrary zero point. The nature of the zero point (arbitrary or true) of the measurement scale does not affect the statistical method of analysis of the data.

Examples of characteristics for which truly quantitative classifications of elementary units ordinarily are used are: a person's height, or age, or wages; the number of beads in a bag or the number of kernels on an ear of corn; the price of a dress; the attendance at a football game; the speed of an automobile. You can think of other examples of truly quantitative characteristics or properties of elementary units.

In a truly quantitative classification of elementary units, the difference between any two elementary units in a sample is described by a number with a constant unit of measurement attached to it. One-dimensional characteristics or properties of this kind will be called *truly quantitative characteristics* or *variables*. Data that are organized by using a truly quantitative classification system for the elementary units will be called a *distribution of a variable*.

2.4.3 Partly quantitative classification. In some practical statistical situations, it may not be feasible to obtain truly quantitative information about each elementary unit in a sample, or we may know in advance that truly quantitative information about each elementary unit in the sample is not necessary for the type of statistical analysis that we intend to carry out after the data have been collected.

Suppose we know merely that with respect to some characteristic in which we are interested, elementary unit A is better than or is greater than or in some specific way precedes elementary unit B, and that elementary unit B is better than or is greater than or in the same specific way precedes elementary unit C, and so on for all the other elementary units in the sample. For example, if we are investigating the annual incomes of a group of people, we might know only that A's income is greater than B's, that B's income is greater than C's, and so on.

Whenever we have this kind of information about the elementary units in a sample, we can list the elementary units in the logical, natural order of their relative importance or precedence. Furthermore, if we choose to do so, we can assign a rank to each elementary unit in the sample, as in the following example.

Annual-Income Classification of Five People

Person (*Elementary Unit*) in Order of Precedence or Importance	Rank of the Elementary Unit (*Person*) on the Basis of Income
John Smith	1
James Brown	2
Henry Long	3
Arthur Jones	4
William Anderson	5

Several features of this classification should be considered. First, a person's income is a kind of characteristic that could be measured in a truly quantitative way, namely, in dollars and cents. That is, even if we use a partly quantitative classification for the elementary units in a sample with respect to their incomes, we know that there is an underlying truly quantitative measurement scale or classification system for the incomes of elementary units.

Although the table tells us that John Smith's income is greater than James Brown's income, it does not tell us anything about how much greater the income of John Smith is than the income of James Brown. Also, the table gives us no information about the relative size of the difference between the incomes of Smith and Brown as compared with the difference between the incomes of, say, Long and Jones. In other words, the table provides no justification for belief that the income differences between the five men are equal. That is why the classification of elementary units in this example is said to be only partly quantitative.

There is a natural, logical order of importance or precedence in a classification of elementary units such as that given in the table. This follows from the fact that the partly quantitative classification used in the table has at least a vague or crude relationship to the underlying truly quantitative measurement scale that could have been used to classify the elementary units according to their incomes. On the other hand, it is important to remember that the mere assignment of numerical ranks to the elementary units in a sample does not convert a partly quantitative classification into a truly quantitative one, because the differences between ranks are not expressed in terms of a constant unit of measurement.

Here are additional examples of summaries of partly quantitative classification of elementary units.

Income Category (*Partly Quantitative Classification*)	*Number of Families* (*Elementary Units*)
High	12
Above average	35
Average	57
Below average	32
Low	8
	Total 144

Military Precedence (*Partly Quantitative Classification*)	*Number of Soldiers* (*Elementary Units*)
Sergeant	5
Corporal	10
Private	42
	Total 57

Results of an Examination (*Partly Quantitative Classification*)	*Number of Students* (*Elementary Units*)
Passed	19
Failed	6
	Total 25

Suppose that your class has taken an examination consisting of 100 true-or-false questions in statistics. Let A's score on the examination be 85, let B's score be 82, and let C's score be 79. Should we consider the scores on this examination to be a truly quantitative classification of the students in the class?

Let us agree that there exists, theoretically at least, an underlying variable that represents knowledge or ability in statistics. However, in order to justify considering the actual scores on the test to be a truly quantitative classification of the students, we would need to be able to prove that the scores were determined by a one-dimensional measurement scale that has a constant unit of measurement. We would have to prove that giving the correct answer to any one question on the examination was exactly equivalent in amount of statistical knowledge to giving the correct answer to any other question on the examination. If we could prove that about the scores on the examination, we would be able to say that A is exactly as much superior to B in statistical knowledge as B is superior to C.

We probably should not consider the scores obtained by people on most scales of ability, achievement, aptitude, or personality or other educational, psychological, and sociological phenomena as truly quantitative classifications of the people. Actually, in many instances, such scores probably are little better than simple ordering or ranking types of classification and should be treated as partly quantitative data in statistical analysis. In most of the above-mentioned areas of investigation, we either do not know how at present or do not feel that it is necessary to use truly quantitative classifications for the elementary units in our samples. Furthermore, some tests in these areas are mixtures of basically different characteristics and, therefore, such scales probably are not one-dimensional. We ought to remember these things when planning an investigation and beginning to apply statistical methods of analysis to data.

The ordinary operations of arithmetic, namely, addition, subtraction, multiplication, division, and the extraction of roots, are legitimate and produce truly quantitative results when they are applied to the data in a truly quantitative classification. On the other hand, if the ordinary operations of arithmetic are applied to the data in partly quantitative classifications, the results of the computations are not always and necessarily truly quantitative or as meaningful as they would be if they had been obtained by application of the same arithmetical manipulations to truly quantitative data. Nevertheless, there are many situations in which some arithmetical operations are legitimate with partly quantitative data; we shall see one good example when we study rank correlation in Chapter 8, and there will be many other examples in this book. It probably is correct to say, however, that, in general, statistical methods of analysis that are appropriate for partly quantitative data involve less arithmetical manipulation of the data than do the statistical methods of analysis that are best for truly quantitative data.

In each of your investigations during this course, you ought to plan to collect data about at least one variable and data about at least one attribute. Furthermore, when you are planning to collect data about a variable characteristic of elementary units, consider in each instance the relative advantages and disadvantages of using truly quantitative classification and partly quantitative classification. In this way, you will learn how to work with the three types of classification systems.

Another suggestion: when you are choosing characteristics to be recorded for each elementary unit in the sample, consider the possibility of choosing two characteristics that might be related. You will then be able to test a hypothesis about the existence of a relationship between the two characteristics for the universe as well as making estimates for the two characteristics separately. Some methods for studying relationships are given in Chapters 8 and 9.

2.4.4 Discrete variables and continuous variables. A variable for which some values within the limits of the range are both theoretically and practically impossible is called a *discrete variable* or a *discontinuous variable*. In Table 1.3, the number of red beads in a sample must be a whole number, that is, an integer; fractional values are impossible. In Tables 1.4 and 1.5, certain fractional values are possible and certain others are impossible. Other examples of discrete variables are the number of people in a family, the number of students in a class, and the number of rows of kernels on an ear of corn.

A discrete variable changes from one value to another by a step, that is, by skipping certain numbers. Frequently each step is unity, as in the case of the number of beads in a sample or the number of people killed in automobile accidents in a year. But sometimes each step between different values of the variable is a fraction. For example, in Tables 1.4 and 1.5, each step is $1/250 = 0.004$.

A variable for which all values within the limits of the range are at least theoretically possible is called a *continuous variable*. Examples are a person's height, weight, age, or temperature. Time is a continuous variable. The speed of an automobile is a continuous variable.

Although theoretically a continuous variable can have any value whatsoever in the range, observed values of the variable are more or less accurate approximations of the true values. For example, a person's height can be measured only approximately. The amount of error in a recorded measurement of a person's height depends upon the precision of the measuring instrument and upon the carefulness of the person who makes the measurement. Nevertheless, a person's height is a continuous variable. If you were 5 feet 5 inches tall a year ago and if you are 5 feet 6 inches tall today, you passed through every fractional part of the inch that you gained. You did not skip any fractions of the inch between 5 feet 5 inches and 5 feet 6 inches.

There is another important distinction to be made. Sometimes the underlying variable in a situation is a continuous variable, but for practical reasons a discrete variable is used in its place. For example, the circumference of a man's neck, the length of his arm, the width of a woman's foot, and the length of a woman's foot are continuous variables. But shirt manufacturers and shoe manufacturers cannot supply an infinite number of sizes of shirts and of shoes. Practical considerations force them to make only a few sizes of shirts and of shoes. See Tables 6.20 to 6.23.

A 33-inch shirt sleeve must be accepted by all men whose arm lengths theoretically could require sleeves anywhere from $32\frac{1}{2}$ inches to $33\frac{1}{2}$ inches in length. The smallest step between two different available sleeve lengths is 1 inch. Similarly, when a woman buys a pair of shoes, she has a choice only in half-size intervals. If a size $5\frac{1}{2}$ shoe is too short for her, then she must try a size 6.

Consequently, although the underlying variables are continuous, the actually observed and recorded variables in Tables 6.20 to 6.23 are discrete.

Ability usually is considered to be a continuous variable from a theoretical point of view. In practice, when a person's ability is measured by a test the number of test items answered correctly usually is considered to be the measure of a person's ability. The number of test items answered correctly must be a whole number. The score on such a test is a discrete variable.

Income and wealth may be thought of theoretically as continuous variables. However, when we measure income or wealth in a practical situation, we are limited by the kind of money that is being measured. The smallest denomination of money in the United States is the cent. Consequently, when we measure and record incomes or wages or prices or bank deposits, we must use a discrete variable for which the smallest possible step between two different values of the variable is one cent.

2.5 FORMULATING THE PROBLEM

Usually, we do not see the details of a problem at first. We need to explore the situation and to ask ourselves questions about it. As remarked earlier, two of the principal purposes of statistical investigations usually are the estimation of parameters of universes and the testing of hypotheses. Consequently, it usually will be desirable to formulate your problem in such a way that the data to be collected will provide an estimate of a parameter, along with a suitable pair of confidence limits, or so that a procedure will be available for testing the hypothesis that you have made. Here are some suggestions that may be helpful to you in formulating your problems.

2.5.1. List the relevant characteristics. Often, it is helpful in formulating a problem to make a list of the *relevant nonquantitative characteristics* of the elementary units involved in the problem, a list of their *relevant quantitative characteristics*, and a list of the *relational aspects* of the problem.

The following quotation illustrates the way in which efficient thinking focuses attention on quantitative, nonquantitative, and relational aspects of a problem. In the quotation you should be able to count about eighteen items for which either partly quantitative data or truly quantitative data can be obtained.

> The principal factors determining revenue, income, and rates in the business of broadcasting are the population and wealth of the community in which the station is located (including the number of receiving sets per capita), the area and population to which the station delivers a satisfactory signal (determined, in turn, by its frequency, power, hours of operation, limitations due to interference either from other stations or from static, and conductivity of the soil), the number of other stations serving in whole or in part the same area, and network affiliations.[1]

[1]Docket No. 6051, Federal Communications Commission, Washington, D. C., September 1941.

2.5.2 Define the terms. Try to define terms carefully so that they are not ambiguous. For example, look up the Census Bureau definitions of "occupied dwelling unit," "family," and "household." Different definitions of the term "unemployed" may account for some of the large differences in estimates of unemployment that are supplied by different organizations.

Suppose that you decided to investigate the extent of blindness in your community. How would you define a blind person? Some investigators of blindness define a "blind" person as anyone who has less than 10 per cent vision as measured by a particular standard test of vision.

The kind of definition that is usually needed in statistical investigations is called an *operational definition*. In other words, a good definition for statistical purposes usually contains an operational rule that leaves no room for doubt or question as to the way in which any elementary unit in the universe should be classified. The definition that was given for blindness is an example of an operational definition. One investigator studying the handedness of very young children said he found that whichever hand was used in wiping a table with a dust cloth, rolling the cloth into a ball, and throwing it away was a reliable indication of hand preference.

Interviewers often run into practical difficulties when they are in the field conducting interviews or measurements in connection with statistical surveys. Frequently, these difficulties are caused by failure to give the interviewers good operational definitions. For example, an inadequate definition of the universe for an investigation may cause the interviewers to include in the sample elementary units that should have been considered ineligible or to exclude from the sample elementary units that actually were eligible.

Here are two good definitions that some students made in planning an investigation of the movie preferences of boys and girls. They defined the universe as consisting of "all the boys and girls registered in Roosevelt senior high school for the current term." And for their survey they defined the term "motion picture" as including "only feature-length sound-motion pictures, produced by commercial motion-picture companies, and licensed and distributed to motion-picture theatres for regular exhibition."

2.5.3 Choose the categories. Frequently, the formulation of a problem for statistical investigation involves the selection of suitable categories, that is, the construction of a classification system. Classification is not always easy. However, if the categories are not skillfully worked out, the whole investigation may be ruined. Do not just choose the first categories that come to mind. Look through this book and other texts on statistics to find hints and suggestions for setting up a good system of categories. Better still, go to your library and see if you can find any references to research that has already been done on the problem that you have chosen. The Census Bureau and other Government organizations, as well as many private institutions, collect and publish data on a wide variety of matters. You may find that some of these organizations have for many years been using a classification system that will suit your purposes for planning your investigation and that will give you an excellent basis for interpreting data after they are collected. For example, you may be able to make tests to indicate whether or not the group that you study is similar in its characteristics to the groups that others have studied.

The student committee, mentioned above, that was planning the study of the movie preferences of boys and girls found in their library a report of a study that Howard M. Bell made of the motion picture preferences of boys and girls in Maryland. Bell used the following nine types of movies as the basis for his classification system: musical comedy, historical, action and Western, love story, mystery, gangster and G-men, comedy of manners, news and education, other types (see Table 11.1). These categories seemed to serve satisfactorily in Bell's sample survey. Furthermore, the committee decided to use Bell's categories so that they would be able to compare the boys and girls in their school with the boys and girls in Maryland with respect to movie preferences.

2.6 STATISTICAL HYPOTHESES

A committee of students planning an investigation to estimate the proportion of students in the school who regularly watched a particular weekly television show thought that they probably should classify the students in the sample by sex as well as by whether or not they watched the show regularly. They made this decision because they thought that there might be a significant difference between the proportion of women in the school who watched the show regularly and the proportion of men in the school who watched the show regularly.

Notice that a hypothesis entered into their formulation of the problem. The proper formulation of hypotheses is one of the most important steps in the planning of a statistical investigation. The good statistician formulates his hypothesis in a way such that there will be a test of significance available for application after the data have been collected and analyzed.

2.6.1 Null hypotheses. There are two possible ways to state a hypothesis. The first is to state the hypothesis in a form that is believed to be true and then try to prove it. The other is to state the hypothesis in a form that is believed to be false and then try to show that the data in a sample force us to reject or nullify this false hypothesis. In all the sciences, it frequently is easier to obtain sufficient evidence to convince almost anyone that he ought to reject a false hypothesis than it is to obtain sufficient evidence to convince everyone of the validity of a true hypothesis. You may recall that in demonstrative geometry the indirect method of proof often was much easier than the direct method of proof for propositions.

When a statistical hypothesis is stated in a form that is believed to be false, statisticians usually call it a *null hypothesis*. Null hypotheses are hypotheses formulated for the express purpose of being rejected. For example, if we desire to prove that a claim is false, we formulate the null hypothesis that the claim is true. If we wish to prove that two universes are not equal in some characteristic, we formulate the null hypothesis that the two universes are equal with respect to that characteristic.

If the evidence provided by the data that we collect in a statistical investigation is not sufficiently strong to convince us that we should reject the stated null hypothesis, then all that we can do at that time is state that we do not reject the null hypothesis. In at least some situations, we may feel that the stated null hypothesis is false, but that at the moment we do not have sufficient evidence to reject it. We may reserve judgment.

Even if we fail to reject a stated null hypothesis, we may have added some worth-while information to our fund of knowledge about the situation. For example, we might decide that further investigation is advisable and we may have gained some new ideas about how to improve our methods of experimentation or investigation.

Let us suppose that there are three and only three possible hypotheses to explain a particular event or situation. If we select one of the hypotheses and then produce evidence strong enough to compel us to reject that hypothesis, we still may not know which of the other two hypotheses is the true one. But if we can select one of the two remaining alternatives and produce evidence that compels us to reject *that* hypothesis, then the one hypothesis remaining will be accepted as the true explanation of the observed event or situation.

Consequently, the method of setting up hypotheses and then collecting data to disprove them one by one, if possible, is a process of elimination. In case there are only two possible hypotheses to account for an event or a situation, the rejection of one of them leads to the acceptance of the other. Therefore, we should try to state the null hypothesis for an investigation in a way such that there is only one possible alternative, the hypothesis that we really desire to accept. In the planning stage of the investigation we should choose both a null hypothesis and an alternative hypothesis.

For example, in the problem of the colored lining for the coconut cans, the manufacturer knew because of his long experience in selling products to consumers that one color probably would be a better aid than the other in selling his coconut. Therefore, it was important to him to discover the consumer's preference. But it was immaterial to him which color was the preferred one, because he could produce cans with either color equally well at no difference in cost. In other words, he felt that the proportion π preferring the coconut in the blue-lined can probably was not the same as the proportion $1 - \pi$ preferring the coconut in the yellow-lined can, that is to say, π was not exactly equal to 0.50, and it was all the same to him whether

π turned out to be less than 0.50 or greater than 0.50. Therefore, for that investigation we would choose $\pi = 0.50$ as the null hypothesis and $\pi \neq 0.50$ as the alternative hypothesis.

In the problem of the two coffee formulas, the engineers for the manufacturer felt strongly that considerably more than 50 per cent of the coffee drinkers would prefer coffee made by the new formula. Their attention and interest and desire were focused on the hypothesis that the proportion π preferring the new formula was greater than 0.50. In other words, $\pi > 0.50$ was the hypothesis that they desired to accept. Consequently, the appropriate null hypothesis for that investigation was $\pi \leq 0.50$.

This null hypothesis $\pi \leq 0.50$ may at first glance seem to be very complicated in that it requires us to reject every possible value of π from 0 to 0.50 before we can accept the alternative hypothesis that $\pi > 0.50$. But the fact is that if we reject the hypothesis that $\pi = 0.50$ because the value of p in the sample is sufficiently greater than 0.50, then we automatically must reject all the other possible values of π from 0.50 down to 0. The explanation of this is simple. If the value of p in the sample is so much greater than 0.50 that our sample is quite improbable under the assumption that $\pi = 0.50$, then, as we will see later, it is even more improbable that our sample came from a universe in which π has a value such as 0.49 or 0.40 or 0.23 or 0.07.

In brief symbolic form, these different types of null hypotheses and alternative hypotheses may be stated in pairs as follows:

(1) Null hypothesis: $\pi = 0.50$ Alternative: $\pi \neq 0.50$
(2) Null hypothesis: $\pi \leq 0.50$ Alternative: $\pi > 0.50$
(3) Null hypothesis: $\pi \geq 0.50$ Alternative: $\pi < 0.50$

You can write similar pairs of hypotheses for situations in which you would be interested in an average for a universe rather than a proportion.

Similarly, if we wish to show merely that the proportion π_1 of men who favor a proposal in a universe of men is not the same as the proportion π_2 of women who favor the proposal in a universe of women or, in other words, that $\pi_1 \neq \pi_2$, then we should choose as the null hypothesis for this investigation the form $\pi_1 = \pi_2$. For, if we reject the null hypothesis that $\pi_1 = \pi_2$, then we will be able to accept the alternative hypothesis that $\pi_1 \neq \pi_2$.

On the other hand, if we desire to show that the value of π_1 for the universe of men is greater than the value of π_2 for the universe of women, that is, that $\pi_1 > \pi_2$, then we should choose as the null hypothesis the form $\pi_1 \leq \pi_2$ and hope that the data in the sample will enable us to reject this null hypothesis. If we desire to show that the value of π_1 for the universe of men is less than the value of π_2 for the universe of women, that is, that $\pi_1 < \pi_2$, then we should choose as the null hypothesis for the investigation the form $\pi_1 \geq \pi_2$ and hope that the data in the sample will enable us to reject this null hypothesis.

These different types of null hypotheses and alternative hypotheses may be stated in pairs as follows:

(4) Null hypothesis: $\pi_1 = \pi_2$ Alternative: $\pi_1 \neq \pi_2$
(5) Null hypothesis: $\pi_1 \leq \pi_2$ Alternative: $\pi_1 > \pi_2$
(6) Null hypothesis: $\pi_1 \geq \pi_2$ Alternative: $\pi_1 < \pi_2$

You can write similar pairs of hypotheses for investigations in which you would be interested in comparing the averages for two universes rather than the proportions for two universes.

2.6.2 Procedure for testing a null hypothesis. In addition to choosing a null hypothesis and an alternative hypothesis for an investigation, the investigator must also choose a procedure for testing the null hypothesis by using the data to be collected during the investigation. The test procedure that is appropriate for a particular investigation usually depends in part upon the type of null hypothesis and the type of alternative hypothesis that have been chosen.

For example, if a pair of hypotheses of the form (1) or of the form (4) given in the preceding section has been chosen for the investigation, then it is necessary to use in the test the probabilities at both ends or tails of the probability distribution or curve (for example, the normal probability curve). Consequently, such a test procedure may be described as a *two-tails test* of the null hypothesis. (See Figures 2.1 and 2.4.)

If a pair of hypotheses of the form (2) or of the form (5) has been chosen for the investigation, then it is appropriate to use in the test only the probability at the right-hand end or tail of the probability distribution or curve.

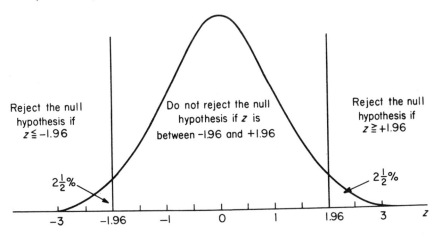

Fig. 2.1. Two-Tails Test at the 5 Per Cent Probability Level of a Null Hypothesis Such as (1) $\pi = 0.50$, (2) $\mu = 100$, (3) $\pi_1 = \pi_2$, or (4) $\mu_1 = \mu_2$

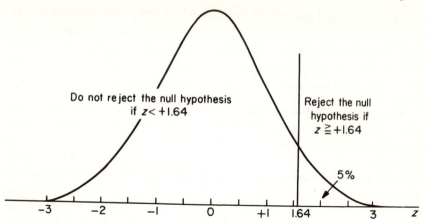

Fig. 2.2. One-Tail Test at the 5 Per Cent Probability Level of a Null Hypothesis
Such as (1) $\pi \leq 0.50$, (2) $\mu \leq 100$, (3) $\pi_1 \leq \pi_2$, or (4) $\mu_1 \leq \mu_2$

Such a test procedure may be described as a *one-tail test* of the null hypothesis. (See Figure 2.2.)

If a pair of hypotheses of the form (3) or of the form (6) has been chosen for the investigation, then it is appropriate to use in the test only the probability at the left-hand end or tail of the probability distribution or curve. Such a test procedure may be described as a *one-tail test* of the null hypothesis. (See Figure 2.3.)

The selection of a procedure for testing a null hypothesis includes the specifying of the probability level at which you will feel safe in rejecting

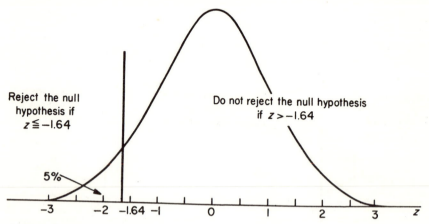

Fig. 2.3. One-Tail Test at the 5 Per Cent Probability Level of a Null Hypothesis
Such as (1) $\pi \geq 0.50$, (2) $\mu \geq 100$, (3) $\pi_1 \geq \pi_2$, or (4) $\mu_1 \geq \mu_2$

the null hypothesis, such as, for example, the 0.05 probability level or the 0.01 probability level. This is the same as specifying the proportion of the area that is to be in the one tail or in the two tails of the probability curve for the test of the null hypothesis. It is customary to use the small Greek letter α to stand for the probability level at which the null hypothesis is to be tested.

Figures 2.1 to 2.4 show that when a one-tail test of a null hypothesis is used, the value of z computed from the observed data need not be as large in absolute value to permit rejection of the null hypothesis as it would need to be if a two-tails test of a null hypothesis were made at the same probability level. For example, the computed value of z for a two-tails test of a null hypothesis at the 0.05 normal probability level must be at least as large as $+1.96$ or at least as small as -1.96 to permit rejection of the null hypothesis. But, for a one-tail test of a null hypothesis at the 0.05 normal probability level the computed value of z need be only as large as or larger than $+1.64$ if the one-tail test is made in the right-hand tail of the probability curve, or only as small as or smaller than -1.64 if the one-tail test is made in the left-hand tail of the probability curve. To express this idea in another way, the value 1.96 for z is at the 0.05 normal probability level in a two-tails test and the value 1.96 for z is at the 0.025 normal probability level in a one-tail test.

You will find further discussion of hypotheses and many illustrations of procedures for testing hypotheses as you proceed through this book. It is a good idea to construct a graph similar to one of the Figures 2.1 to 2.4 whenever you need to apply a test of a null hypothesis in a practical situation. Only after you have chosen the test procedure and the probability level

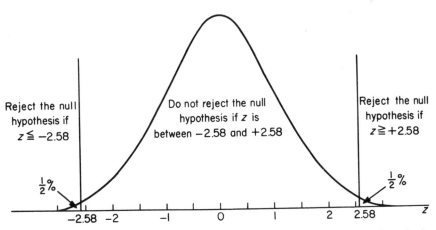

Fig. 2.4. Two-Tails Test at the 1 Per Cent Probability Level of a Null Hypothesis Such as (1) $\pi = 0.50$, (2) $\mu = 100$, (3) $\pi_1 = \pi_2$, or (4) $\mu_1 = \mu_2$

for the test of the null hypothesis is it possible to determine an appropriate size of sample to be used in the investigation.

2.7 APPROPRIATE SAMPLE SIZE

From the purely mathematical point of view the proper size of sample depends upon (1) the degree of reliability needed in the estimates that are to be derived from the sample data, (2) the probability level α at which any tests of hypotheses or tests of significance are to be made, (3) the design of the experiment or investigation, and (4) the method used to draw the sample. From a practical point of view, the proper size of the sample may depend also upon the amounts of time and money available for the project.

Some fairly simple methods that often can be used for determining an appropriate sample size are discussed and illustrated in Chapters 4 and 7. Other material that may be useful as a basis for determining the sample size in some situations may be found in Chapters 5, 6, 8, and 9. You will need some familiarity with the standard deviation, the binomial probability law, the normal probability law, and the nature of the distributions of sampling errors in order to understand how to determine the proper sample size to use in an investigation.

One of the common mistakes that investigators make in determining an appropriate size of a sample is to think of the over-all size of the sample and forget that what are called "breakdowns" of the sample may be required for analyzing and interpreting the data. For example, it might be decided that the data obtained from a random sample containing 100 students would provide sufficient reliability in some particular estimate based on the whole sample. Later it might be desirable to break down the sample by sex. Then there would be data for only about 50 men from which to make an estimate, and the reliability of the estimate for men might be considered too low. The same would be true for the estimate for women.

In some investigations, in addition to breaking down the sample by sex, it might be desirable to break down the subsample for each sex into school class. If the total sample for the four classes consisted of 100 students, then there probably would be less than 10 women in the subsample of women in the senior class. A subsample containing less than 10 women (or men) might be too small to be worth analyzing separately because of the possibility of large sampling error in any estimate obtained from such a small sample.

A good way for determining how large a sample to use in an investigation is to decide first what breakdowns of the sample are likely to be important in interpreting the data after they have been collected. Then make sure that the number of elementary units to be drawn into the sample is large enough to provide the desired reliability in the estimate to be obtained from each breakdown and also in any tests of significance or tests of hypotheses that are to be made. Build up the total sample in that way.

It follows that questionnaires and classifications should be so planned that the number of elementary units answering the questions or falling into the classes will be large enough to justify statistical analysis or interpretation in terms of the universe.

In the investigation referred to at the beginning of section 2.6, the students who were planning the investigation decided that they would break down the sample by sex but not by class. Because the students on this committee had not had any practical experience with the mathematical formulas for computing sample size, they requested advice from their teacher about how to determine the sample size for their investigation. The committee and the teacher decided that high reliability was not necessary, because no action involving the possibility of serious consequences was to be taken on the basis of estimates obtained from the sample. They decided that 95 per cent confidence limits 0.10 above and below each estimated proportion would be satisfactory. Furthermore, the students were reminded that they would be able to use formula 4.11 or formula 4.12 to demonstrate that their estimates had the desired degree of reliability. They decided also that they would use formula 5.2 to test the significance of the difference between the two proportions at the $\alpha = 0.10$ significance level. After the teacher showed the students how to use formulas 4.14 and 4.15, it was decided to include in the sample 100 men and 100 women. There were 2057 students in the school, almost equally divided among men and women.

The committee prepared an alphabetical list of the men in each of the four classes and an alphabetical list of the women in each of the four classes. Then they used a table of random digits similar to Table A.15 to draw eight random numbers in the range from 1 to 10, that is, one random number for each of the eight lists of students. These eight random numbers were used to determine the first person in each of the eight alphabetical lists who was to be in the sample. Then they marked every tenth person in each list after the first one marked for the sample.

Statisticians call this kind of sampling systematic sampling with a random start. This sample also may be called a stratified sample; it was stratified by sex and by school class. Notice that although the committee did not intend to break down the sample data by class, they stratified the universe by class in order to insure that each class had an appropriate number of representatives in the sample.

2.8 METHODS FOR DRAWING SAMPLES

There are many methods for drawing samples, and a method that may be appropriate for one statistical investigation might not be appropriate for another. Some of the most frequently useful methods for drawing samples will be described for you now.

For the purposes of elementary practical statistics, we shall consider a sample drawn from a finite universe to be a *random sample* if care is taken to insure that every elementary unit in the universe has an equal chance of being chosen for the sample. Usually it is not sufficient to be able to say merely that no particular elementary unit in the universe was excluded deliberately from the sample. Good sampling practice usually requires that positive action be taken to make certain that every elementary unit in the universe has its proper probability of being included in the sample. For the student investigation of preferences in television shows mentioned above, positive action consisted in (1) obtaining an accurate list of the students from the office of the Dean of Students and (2) using a table of random digits to determine which students in the list would be in the sample.

2.8.1 Sampling with replacement after each draw. Consider the universe of 5000 red beads and white beads in which 1800 of the beads are red and 3200 are white. The value of π, the proportion of red beads in the universe, is 0.36.

Mix the universe of beads thoroughly. Draw one bead from the universe. The probability P that this bead will be red is $1800/5000 = 0.36$, which is the same as the value of π. Record the color of the bead that was drawn. Then put the bead back into the universe. Mix the universe thoroughly. Then draw a second bead from the universe. The probability P that the second bead will be red is still $1800/5000 = 0.36$, which is the same as the value of π. Record the color of the second bead that was drawn. Then put the bead back into the universe. Mix the universe thoroughly.

Draw a third bead from the universe. The probability P that the third bead will be red is still $1800/5000 = 0.36$, which is the same as the value of π. Record the color of the third bead that was drawn. Then put the bead back into the universe. Mix the universe thoroughly. Continue to draw one bead at a time and always replace it in the universe before drawing another bead until you have drawn a sample of 200 beads and recorded their colors.

At each successive drawing, the composition of the universe is exactly the same as it was at the beginning. And the probability P of drawing a red bead is exactly the same at each drawing, namely, $P = 0.36 = \pi$. This kind of sampling procedure is called *sampling with replacement after each draw*. In this type of sampling, the same bead could be drawn more than once.

There is another easy way to draw a sample of beads with replacement of the bead after each draw. If you have a reserve supply of red beads and white beads, you can replace the bead obtained in each draw by putting into the universe a bead of the same color from the reserve supply. In this way, you will be able to see the whole sample of beads when you have completed the sampling process.

2.8.2 Sampling without replacement after each draw. If you do not replace the first bead that is drawn from the universe before drawing the second bead, the probability of drawing a red bead in the second draw is not the same as it was in the first draw.

The probability of drawing a red bead in the first draw is $P_1 = 1800/5000 = 0.36 = \pi$. Let us suppose that a white bead was drawn in the first draw. Then if that bead is not replaced in the universe, there will be only 4999 beads in the universe and 1800 of them will be red. Consequently, the probability of drawing a red bead in the second draw will be $P_2 = 1800/4999$, which is not the same as $P_1 = 1800/5000$.

Let us suppose that the bead drawn in the second draw is red. If neither of the two beads already drawn is replaced in the universe, there will be only 4998 beads in the universe for the third draw. And 1799 of the beads in the universe will be red. Consequently, the probability of drawing a red bead in the third draw will be $P_3 = 1799/4998$, which is not the same as the probability of drawing a red bead was in either of the previous two draws.

If you continue to draw one bead at a time in this way, you can build up a sample of 200 beads. The composition of the universe is changed by each draw and the probability of drawing a red bead is changed by each draw.

This kind of sampling procedure is called *sampling without replacement after each draw*. In this kind of sampling procedure, it is obviously impossible for the same bead to be drawn more than once.

A simpler and faster way to do sampling without replacement after each draw from a universe such as the universe of red and white beads is to scoop out the sample of 200 beads all at once.

Many of the samples that people draw in practical situations are samples drawn without replacement after each elementary unit is drawn. For example, if you draw a sample of the students in your school for the purpose of conducting an opinion poll or an attitude survey, it is not likely that you will put a student's name back into the universe after his or her name has been drawn. You probably will not want the same student to be interviewed more than once for the sample.

A sample containing n elementary units selected from a universe containing N elementary units is called a *simple random sample drawn without replacement* if the sampling procedure is one in which all the possible combinations of n elementary units that may be formed from the N elementary units in the universe have the same probability of being included in the sample.

What is the probability of each elementary unit's being included in a simple random sample drawn without replacement? For purposes of illustration, let us take the simple universe consisting of 10 people whom we may designate by the letters A, B, C, D, E, F, G, H, I, and J. Suppose that we desire to estimate the average age, or average income, or average weight of this universe by selecting a small sample of the individuals in the universe.

There are ten different ways in which a sample of one individual may be drawn from this universe of ten individuals. Clearly, the probability that any particular one of the individuals, say, C, will be the one drawn is 1/10.

Suppose that we are going to use a sample of two individuals from the universe of ten individuals. There are 45 different ways in which a sample of two individuals may be drawn, without replacement, from a universe of ten individuals. These 45 combinations are:

AB	BC	CD	DE	EF	FG	GH	HI	IJ
AC	BD	CE	DF	EG	FH	GI	HJ	
AD	BE	CF	DG	EH	FI	GJ		
AE	BF	CG	DH	EI	FJ			
AF	BG	CH	DI	EJ				
AG	BH	CI	DJ					
AH	BI	CJ						
AI	BJ							
AJ								

What is the probability that any particular individual, say, A, will be included in a simple random sample of two individuals drawn, without replacement, from this universe of ten people? Examination of the 45 combinations listed above will show that there are 9 combinations that include A. Each of the 45 combinations is as likely as any other to be drawn in simple random sampling. Consequently, the probability that A will be included in a sample of two from a universe of ten elementary units is $9/45 = 1/5 = 2/10$. Each of the other individuals has the same probability as A, namely, 2/10, of being drawn in a sample of two from the universe of ten individuals. For example, there are exactly 9 combinations in the above list that include B. And because each combination in the list is as likely as any other combination to be drawn for the sample, the probability that B will be included in a sample of two from this universe of ten individuals is $9/45 = 1/5 = 2/10$. You can repeat this process for a few other individuals and thus convince yourself that the probability of any particular elementary unit's being included in a sample of two from a universe of ten elementary units is 2/10.

Similarly, if we wish to know the probability that any particular individual will be included in a simple random sample of three individuals drawn, without replacement, from this universe of ten individuals, we might first list all the different combinations of three individuals, for example, ABC, ABD, ABE, etc. We would find that 120 different combinations of three individuals are possible in a universe of ten individuals. Moreover, we would find that each individual is included in exactly 36 of the 120 combinations. Furthermore, in simple random sampling, any one of these 120 combinations has the same chance or probability as any other of being drawn in a sample of three. Therefore, the probability that any particular individual, say, F, will be included in a simple random sample of three drawn, without replacement, from a universe of ten is $36/120 = 3/10$.

In general, the probability that any specified elementary unit in a universe containing N elementary units will be included in a simple random sample of n elementary units, drawn without replacement, is n/N. An algebraic proof of this can be found in section A.4 (Appendix). This means that the probability of each elementary unit in the universe being included in a simple random sample, drawn without replacement, is known and is not zero. Consequently, a simple random sample drawn without replacement is a probability sample. Simple random sampling, without replacement of each elementary unit that is drawn before another elementary unit is drawn, is one of the most frequently used sampling procedures in practical work.

Let us suppose that we have in front of us a list giving, in alphabetical order, the names of 1673 students. Suppose, further, that we want to draw a sample of approximately 100 students from this list.

We might draw a simple random sample without replacement after each draw. We could do this by first numbering the names from 1 to 1673 and then drawing 100 random numbers in the range from 1 to 1673 by using a table of random digits such as Table A.15. The names corresponding to these 100 random numbers would indicate to us a simple random sample of the students whose names are on the list. Each student would have the probability $100/1673$ of being drawn into the sample.

In drawing a simple random sample from a list, there is danger that the elementary units drawn into the sample will not be well spread throughout the entire list. For example, there is danger of drawing too many people whose names begin with the same letter or who may be of the same national origin. And there may be large parts of the list from which no elementary units are selected.

In many practical sampling situations, it is easy to devise a way to make sure that the elementary units drawn into the sample are thoroughly spread over the whole list, so that there are no unduly large clusters of elementary units and no unduly long gaps. One of the simplest and best ways to do this is to use what is called *stratified random sampling*.

2.8.3 Stratified random sampling. First we might divide the list of names into approximately 100 parts, because we want approximately 100 students to be in the sample. Dividing 1673 by 100 gives 16.73. Because 100 does not divide into 1673 evenly, we may choose either 16 or 17 as the number of names to be in each part of the list. Suppose that we decide to use the number 17 here.

Then there will be 99 parts in the list. These 99 sections can be marked off easily by drawing a line below every 17th name on the list. For example, count off the first 17 names and then draw a line between the seventeenth and the eighteenth names. Similarly, draw a line between the thirty-fourth and thirty-fifth names on the list, and so on until you have drawn a line between the 1666th and the 1667th names on the list. There will be only seven names left for the 99th section or part of the list.

Each of these sections or parts of the list is called a *stratum*. The process of separating a list into sections or parts such as these is called *stratifying* the list. There are 99 strata in the example above.

Now, draw a random number in the range from 1 to 17 by using a table such as Table A.15. Suppose, for purposes of illustration, that the random number turns out to be 6. Then the sixth student from the beginning of the list is the only student in the first stratum who is to be in the sample.

Draw a random number in the range from 1 to 17 again. This time, let us suppose that the random number turns out to be 13. Then the thirteenth student from the beginning of the second stratum is the only student in the second stratum who is to be in the sample. Continue in this way until you have drawn a random number in the range from 1 to 17 ninety-nine times. Then you will have determined 98 or 99 students for the sample. And the 98 or 99 names will be well spread throughout the list.

The ninety-ninth stratum here contains only seven names. You may imagine that there are ten blank spaces at the end of that stratum. If the ninety-ninth random number is greater than seven, then it corresponds to one of the blank spaces, and no student from the ninety-ninth stratum is to be in the sample. In that case, the sample will contain only 98 students.

If we had chosen to put 16 students rather than 17 into each stratum, there would have been 105 strata. There would have been 16 students in each of the first 104 strata and 9 in the last.

In stratified random sampling the universe is first stratified. Then a random process is used to draw one or more elementary units from each stratum. Each student in the example had probability $P = 1/17$ of being included in the sample.

Of course, in many practical sampling situations, more than one elementary unit is drawn from each stratum for the sample and the different strata need not be the same size. For example, in a consumer research project you might stratify the families in the United States by geographic regions, for example, the Northeast, the North Central, the Southeast, the South Central, and the West. Obviously, there are not likely to be the same number of families in each of those five geographic regions. However, it is easy to draw for the sample a certain number of families from each geographic region so that each family in the United States has the same probability as any other family of being drawn into the sample. Furthermore, you might stratify the families in each geographic region into urban, rural–nonfarm, and rural–farm categories. And you might further stratify the urban families into city-size groups.

Choose, if you can, for stratification purposes the characteristics that will have the greatest influence on the data to be obtained from the elementary units that happen to be drawn into the sample. Of course, in many statistical investigations it is impossible to know in advance with certainty what characteristics are most important for stratification purposes.

For the investigator who uses a probability sample, it is not a serious matter if he happens to overlook some important stratification characteristic or if the information used as a basis for his stratification is somewhat out of date; it means merely that the confidence interval for the estimate he obtains from the sample will be slightly larger than it would be if he had used the best stratification. In other words, stratification is not in any way an essential element of probability sampling; randomness is.

Age, education, sex, nationality, occupation, income, home ownership, size of family, and political affiliation are examples of characteristics that sometimes are important for stratification purposes. Of course, you cannot use a particular stratification characteristic to stratify a universe unless you have on hand at least approximate information about the way in which the universe is distributed according to that characteristic. For example, sufficient census data are available for stratification of the population according to some characteristics but not for stratification according to other characteristics. Here again the "quota" sampler is handicapped much more seriously than is the user of a probability sample.

Stratification of a universe with respect to a relevant characteristic can be expected to be most effective in reducing the size of the confidence interval for the estimate to be obtained from the sample if the elementary units within each stratum are as nearly alike as possible with respect to the characteristic and if there are relatively large differences between the various strata with respect to the characteristic.

It is possible, in fact, though highly unlikely in practical sampling situations, to produce lower reliability in an estimate by using simple stratified sampling. However, such a loss is likely to be trivial unless the number of elementary units in each of the strata is very, very small.

Even when elaborate stratification is carried out with meticulous care, the mathematical gain in reliability of the estimates derived from the sample often is trivial as compared with the reliability that would have been obtained without stratification of the universe. Moreover, it is a rather complicated mathematical job to find out how much improvement in reliability is obtained by stratifying the universe.[2]

---

[2]*Disproportionate stratified sampling* is a form of stratified sampling in which the probability of an elementary unit's being included in the sample is not the same for all the strata. Disproportionate stratified sampling procedures sometimes yield considerably higher reliability in the estimates than do simple (proportionate) stratified sampling procedures with the same total size of sample.

Stratification often can be carried out without increasing the cost of the sampling plan, and it frequently has the effect of reducing the cost of field work involved in collection of data.

2.8.4 Systematic sampling with a random start.

Drawing random numbers is a somewhat tedious task, and it requires considerable time if there are a great many strata. Consequently, statisticians often use another method than that of stratified random sampling described above.

Consider again the problem of drawing a sample of approximately 100 students from a list containing the names of 1673 students in alphabetical order. We begin by determining how many elementary units to put into the first stratum. Because 1673/100 is approximately 17, let the first stratum consist of the first 17 names on the list.

Now draw a random number in the range from 1 to 17 by using Table A.15. Let us suppose here, for purposes of illustration, that the random number turns out to be 11. In that case, the eleventh name from the beginning of the list is the name of the first student to be included in the sample. Mark this name well. Then, by a systematic step process, mark every seventeenth name in the list, starting from the eleventh name. The names that are marked in this procedure are the names of the students who are to be included in the sample. This method of sampling is called *systematic sampling with a random start*. Notice that only one random number was required. It is a very easy and rapid method for drawing samples.

Investigators who make surveys in which they interview housewives, or others, in their homes often use systematic sampling to determine which homes in a block to use for interviews. Market researchers sometimes telephone to homes and ask what radio program or television program, if any, the members of the household are listening to or looking at. One way to determine the telephone numbers to be called is to draw a systematic sample from the names listed in the telephone directory.

There are many organizations from which it is possible to purchase lists of names and addresses for research purposes. But such lists often are not complete and accurate lists of the desired universe and, consequently, they often are not satisfactory for probability sampling. It often is extremely difficult to prepare a complete and accurate list, and it is usually difficult to keep a list up to date over a period of time.

Sometimes investigators draw a systematic sample without a random start. Frequently, those who do this use the first name or home or item on the list and then take the systematic step through the rest of the list. Such a procedure destroys the probability character of the sample. Therefore it may be bad sampling procedure. If you deliberately choose the first name or any other particular name on the list for the sample, then the other names in the first stratum have no chance whatsoever to be in the sample. But if you use a random start, then all the elementary units in the first stratum have equal probability of being drawn into the sample. And, consequently, all the elementary units in the other strata have equal probability of being drawn into the sample.

Consider a simple illustration. In large cities such as New York and Chicago, there are many tall apartment buildings. Frequently these buildings are designed so that all the apartments in a vertical line have the same number of rooms but there may be apartments of several different sizes on the same floor. Imagine a twelve-storied building with ten apartments on each floor, the four corner apartments on each floor being seven-room apartments and the remaining six apartments on each floor being four-room apartments. Obviously, the larger apartments are intended for larger families and rent at higher rates.

Now, if you are making a survey among housewives in that neighborhood and if you use a list of apartment numbers, such as is usually posted in the entrance lobby of the building, to draw a systematic sample of every tenth apartment in that building, with a random start, then all the apartments in your sample will be seven-room apartments or all will be four-room apartments. If size of family or amount spent for rent is related to the topic of your investigation, then you probably will have a seriously distorted sample insofar as that building is concerned. On the other hand, by choosing a new random number for each floor, you could reduce the risk of this source of error. Similarly, in sampling the homes in city blocks, there often is danger that systematic sampling will include either too many or too few of the more expensive corner houses.

If you can justify the assumption that the elementary units in a list (or in a building or in a city block) are arranged in random fashion, then systematic sampling is equivalent to simple random sampling. Unfortunately, the assumption of random distribution in the list or universe often is not justifiable. For example, in alphabetical lists of people, members of the same family and people of the same national origin tend to form clusters.

We may summarize this section as follows. The principal advantages of the systematic sampling procedure are its simplicity, the ease and speed with which the elementary units for the sample may be determined, and the fact that the procedure sometimes yields an estimate that has higher reliability, that is, a smaller confidence interval, than the estimate produced by a simple random sample of the same size.

The principal disadvantage of the systematic sampling procedure is the danger that in sampling from data that happen to be periodic in nature the systematic step may coincide with a period or a multiple or submultiple of a period. It is important to remember that there may be periodicity in the universe even though a systematic sample drawn from the universe does not indicate the periodicity. This is a serious shortcoming of the systematic sampling method, and it may cause serious but unsuspected error in the results obtained by people who use this sampling procedure indiscriminately. Stratified random sampling does not have this shortcoming and, therefore, it is generally a safer sampling procedure.

2.8.5 Other kinds of sampling. A kind of sampling plan frequently used in investigations of consumer attitudes and in many other kinds of investigation is called *cluster sampling* or *two-stage sampling*. For example, in sample surveys of the families in a large city, investigators often draw a probability sample of the blocks in the city and then draw a probability sample of the families within each of the sample blocks. In such investigations, the families in each block in the city constitute a cluster. In national sample surveys, investigators sometimes first draw a probability sample containing, say, 100 counties from the more than 3000 counties in the United States and then draw a probability sample of households within each of the sample counties. Here, the households in each county in the United States constitute a cluster. In sampling 1000 cards from a very large number of file cards in drawers, the cards in each drawer might be considered to be a cluster. First, a probability sample of, say, 50 drawers might be drawn, and then a systematic sample of 20 cards might be drawn from each of the 50 sample drawers. Stratification often is used in combination with a cluster-sampling plan.

In general, the confidence interval to be attached to an estimate obtained from a cluster sample of size *n* will be smallest when each cluster contains as varied a mixture as possible with respect to the characteristic that is being investigated and the different clusters are as nearly alike in composition as possible with respect to the characteristic. These requirements for the greatest effectiveness of cluster sampling are the reverse of the requirements for the greatest effectiveness of stratification.

Unfortunately, in many practical situations in which it is desirable to use cluster sampling, each useful cluster of elementary units into which it is feasible to divide the universe is likely to be relatively homogeneous with respect to the characteristic that is to be investigated and there are likely to be relatively large differences between these clusters in the universe. Consequently, in such situations, cluster sampling usually yields an estimate that has a larger confidence interval than the estimate that would be obtained from a simple random sample of the same size. It is not surprising in such situations to find that the confidence interval for an estimate obtained from a cluster sample of size *n* is one and one-half or two times as great as the confidence interval that would have been obtained from a simple random sample of size *n* drawn from the same universe.

This means that if a specified size of confidence interval is desired for the estimate to be produced by a sample, then the size of the sample may need to be considerably larger if a cluster sample is used than if a simple random sample is used. Nevertheless, in many practical situations it may be possible to obtain data from a cluster sample of, say, 2000 families in the United States at lower cost and in less time than to obtain data from a simple random sample containing 1000 families. In other words, cost and administrative factors involved in collection of data frequently make it advisable to use cluster sampling.

In some investigations, a large sample of the elementary units in the universe is investigated first in a simple or general way and later a small subsample drawn out of the large sample is studied in greater detail. This method is called *double sampling* or *two-phased sampling*. For example, an investigator who wishes to study factors related to the occupations of the individuals in a universe might, if suitable data for stratifying the universe by occupation are not available, first draw a very large random sample from the universe merely to discover the way in which the universe is distributed by occupation. He could use these occupation data to stratify the large sample and then draw from the large sample a small, stratified sample for intensive study by means of long, detailed interviews about the factors in which he is interested.

Another type of sampling plan is called *sequential sampling*. In this type of sampling, the size of the sample need not be determined in advance. As each elementary unit, or small group of elementary units, is drawn into the sample, the accumulated data are examined and a decision is made to terminate or to continue the sampling. In this way, a conclusion often can be reached with a high degree of confidence from a relatively small number of elementary units.

2.8.6 Should a probability sample be used? To determine whether a particular sample obtained in an investigation is a probability sample from a specified universe, examine the sampling plan by which the elementary units actually were procured for the sample; do not try to make this determination by examining the characteristics of the elementary units that happen to be in the sample after the data have been collected.

If (1) the sampling plan for the investigation provided that the elementary units to be in the sample would be selected from the specified universe by a mechanical, random procedure, such as tossing a coin or drawing random numbers from a table of random digits, and (2) the elementary units in the sample actually were determined by such a random procedure, then the sample is a probability sample. In other words, the probability characteristic of a sample refers to the procedure by which the elementary units in the sample actually were selected and not to the characteristics of the particular elementary units that happen to be in the sample.

Similarly, in this book, the expression "representative sample from a specified universe" means that the sample is a probability sample drawn from the specified universe. Consequently, any probability sample drawn from a universe is a representative sample from that universe. And if a sample is not a probability sample from a specified universe, there is no justification for calling it a representative sample from that universe no matter what examination may show the characteristics of the elementary units in the sample to be. Advocates of the so-called "quota samples" often describe their samples as representative samples from a universe, but their use of the term "representative" does not mean probability or random; it merely means stratified, when it has any meaning at all.

True, in many situations in which statistical information is needed it may not be feasible to use probability samples. However, the burden of proof that it was not feasible to use a probability sample must rest on the shoulders of the investigator who uses a nonprobability sample. He ought to be able and ready to prove that the cost of a probability sample would have been prohibitive under the circumstances and that the risk of serious systematic error in using the nonprobability sample was small enough to be borne. Any other explanation for the use of a nonprobability sample may be only an alibi for shoddy statistical work.[3]

2.9 THE PROPOSAL

Usually, the plan for a statistical investigation must be prepared in the form of a proposal. The proposal is submitted for approval to the head of the investigating organization and to all the other parties who have an interest in the investigation, especially those who are expected to pay the cost of the investigation and those who are going to make use of the results.

The contents of the proposal will depend to some extent upon the nature of the investigation that is planned. Ordinarily, however, the proposal should include the following things: (1) A statement of the problem as you have formulated it. (2) The purpose of the investigation, that is, why you recommend that the investigation be made, and the kind of information that you expect to obtain. (3) A draft of the questionnaire or a description of other instruments that you plan to use in collecting the data. (4) A brief explanation of the sampling plan that you intend to use. (5) A statement of the time required to make the investigation, that is, a detailed schedule for the investigation, including the date on which field work will begin and the date on which the final report will be ready. (6) A detailed breakdown of the estimated cost of the survey, including the cost of (a) planning the investigation and preparing the questionnaire, (b) collecting the data, and (c) processing the data and preparing the final report.

[3] William G. Cochran, Frederick Mosteller, and John W. Tukey, "Principles of Sampling," *Journal of the American Statistical Association*, 49 (1954), pp. 13-36. See p. 32.

For the projects that you are to carry out in this class, the proposal ought to indicate whether or not you plan to use a probability sample. If you decide to use a nonprobability sample, you should state the reason why and explain how the sample will be obtained. If you decide to use a probability sample, then the proposal ought to indicate (1) any estimates you intend to obtain from the sample, (2) any hypotheses you intend to test, (3) any tests of significance you intend to make, (4) the degree of reliability to be required in the estimates, (5) the probability level at which the tests of hypotheses and/or the tests of significance will be made, and (6) the techniques by which the sample size will be determined and by which the tests of hypotheses and/or the tests of significance will be made. Usually, also, it is desirable to indicate the ways, if any, in which the universe will be stratified.

Investigations being made by the class ought to be carried out in as businesslike a manner as possible. Consequently, you should prepare a proposal for the investigation that you have formulated and planned. Submit the proposal to your teacher for approval. Your teacher may suggest others with whom you can discuss your proposal. Discussion of the proposal with your supervisor and with the other interested parties almost always results in an improvement of the investigation. Someone may make a good suggestion to you about improving the wording of a question. Another may suggest a way to reduce the cost of the survey. Someone else may notice that if a slight change were made in the questionnaire important breakdowns and cross-tabulations could be made.

2.10 PRETEST THE INSTRUMENTS AND INSTRUCTIONS

Any questionnaire or other instrument that is to be used in a statistical investigation should be given a preliminary test. This is an important step. Frequently it saves an investigation from being a failure, because serious ambiguities, omissions, or other defects in the instrument or in the instructions are discovered while there is still time to correct them.

If a questionnaire is to be used in your investigation, prepare some tentative copies of it and of the definitions and instructions. Then, each member of the committee should pretest the questionnaire and the instructions by conducting a few interviews with people whose names are not on the list for the sample.

2.11 PREPARE FOR THE FIELD WORK

After the results of the preliminary test have been studied thoroughly, the questionnaire or other instrument and the instructions should be prepared in final form.

In large-scale surveys and especially if a long and complicated questionnaire is to be used, the last step before the questionnaire is printed is precoding of the questions. In this step a code number is assigned to each answer space. If the answers are to be transferred to punch cards, as is usually the case in large government and commercial surveys, the code numbers assigned to the answer spaces correspond to the columns and the rows on the punch cards. For example:

Which, if any, of the following magazines have you looked through during the past week?

Life	23-1	☐
Newsweek	-2	☐
Saturday Evening Post	-3	☐
Time	-4	☐
Business Week	-5	☐

For checklist questions, such as, for example, a question requiring a "Yes" or "No" answer, a specific column and a specific row can be preassigned for each possible answer in the checklist. But for questions to which there is no way to know in advance the number or the kinds of different answers that are likely to be received, only the column for coding the answers can be preassigned. The rows will be assigned after the questionnaires have been completed.

The precoding of a large questionnaire may not be a simple matter. Anyone who is planning an investigation in which the data will be punched on cards for machine tabulation ought to consult someone who is familiar with machine-tabulation methods.

It is very important that thought be given to methods for obtaining the good will and cooperation of all who are to be involved in the investigation in any way. Make arrangements well in advance with school officials or others whose cooperation is necessary. Also, use good judgment in selecting the time for gathering data.

Prepare yourself well before you attempt to interview a person. Know what you are going to say as an introduction. Usually, a brief, straightforward introductory statement is best. Have confidence in yourself. Wear a smile. Never enter an interview with a discouraged or grouchy manner.

Investigators can cause distortion in the data they collect if they permit their own pet theories to influence the way they ask the questions or if their own prejudices cause them to discriminate in any way against those they interview. You should have no axe to grind while you are collecting data. You should exercise the utmost fairness and honesty.

And you should follow the instructions scrupulously. In an investigation intended to determine the nutritional value for children of an extra 15 ounces of whole milk each day for a period of four months, 20,000 children in 67 schools participated. Teachers in the 67 schools were instructed to use a purely random method to divide their 20,000 pupils into two groups, the control group and the experimental group, a method of research to be described later in this book. The 10,000 children in the experimental group were given 15 ounces of extra milk daily. The 10,000 children in the control group were not given any extra milk. The height and the weight of each of the 20,000 pupils were measured and recorded at the beginning and at the end of the four-month test period.

This investigation was carried out in England in the depression year 1930 when there were many undernourished children from poor families in the schools. With the best of intentions many of the teachers tended to put into the experimental group the obviously poor, frail, undernourished children who might benefit most from the free milk, and to put into the control group the best-clothed, sturdiest, and healthiest pupils.

No doubt, many, if not all, of these teachers were unconscious of the fact that they were not using a strictly random method for dividing the children into the two groups. Nevertheless, they wrecked that research project. Because of the lack of randomness in selection of the children for the two groups, no valid estimates could be obtained from the data and no valid tests of significance or tests of hypotheses could be made.

People who tamper with statistical investigations often do so under the delusion that they are improving the investigation. One of the disadvantages of very large statistical investigations is that it is extremely difficult to administer the project so efficiently that no one will attempt, consciously or unconsciously, to alter the research method.

Remember that to have a perfect probability sample, data must be collected for 100 per cent of the elementary units that are drawn for the sample. However, being human we do not expect perfection in sampling or in anything else that we do. But perfection should be our aim; if we fall a bit short of 100 per cent, we may still feel that we have done a reasonably good job. You probably will be surprised when you learn how difficult it often is to collect a good sample, even in a relatively simple type of investigation. Let us agree that if you obtain data for at least 95 per cent of the elementary units in your sample you may feel that you have obtained a good or satisfactory sample, though not a perfect one. Then you may treat the sample as a probability sample but with cautions or warnings attached to the estimates and the conclusions that you draw, to indicate the lack of perfection in the sample obtained.

2.12 EDITING AND CODING THE RECORD FORMS

Before tabulation of the results of a statistical investigation that uses a form (a questionnaire, for example) to record the data for each elementary unit in the sample, each completed form should be checked thoroughly to make sure that all the information recorded on it is in good condition. This is especially necessary if the information on the record forms is to be transferred to punch cards before tabulation of the results. For example, interviewers sometimes write a respondent's answer on the wrong line or in the wrong space on a line; the person who checks this questionnaire should put the answer in its proper place. Sometimes answers that are ridiculous or that have no bearing on the investigation are written on the form; such answers should be deleted so that they will not be counted.

This procedure of checking every item of information on a completed form to see that everything on the form is in good condition for tabulation is called *editing* the form. If several people are editing the record forms for the same survey, it is very important that they all operate under the same rules. The person or the committee in charge of the research project must give the editors very clear and specific instructions about how to handle each type of situation they will meet in editing the record forms.

After the record forms have been edited, answers to questions on the form that were not completely precoded may need to have row numbers assigned to them. Row numbers are assigned to these questions after the number and kinds of different answers in a random subsample of the completed record forms have been ascertained and classified by the person or committee in charge of the research project. Then a coder, or a group of coders, is assigned to the task of coding all the answers to the questions that were not completely precoded. When this work has been completed and each coder's work has been checked, preferably by another coder, the record forms are ready for tabulation.

2.13 INFORMAL STATISTICAL THINKING

Here is a brief story quoted from *The New Yorker* for February 18, 1956.

> Overheard in the ballroom of the Colony Club. "I always put on a lot of makeup when I go shopping. I've noticed that when you have plenty of makeup on, you get served faster."[4]

This seemingly trivial little story embodies the basic principles of the statistical method of testing a hypothesis. The lady made this decision as a result of some observations which unconsciously, no doubt, she considers to be a random sample of observations. The number of observations probably was small; nevertheless, unless and until further observations lead her to change her mind, she will use this hypothesis as a practical basis for deciding how much makeup to wear when she goes shopping in the future.

The lady used no elaborate formulas, she made no computations with

[4]Reprinted by permission. Copr. © 1956 *The New Yorker Magazine, Inc.*

pencil and paper, she used no tables, and she did not refer to any books. It required practically no mental effort for her to do this bit of practical problem solving by the statistical method. It is true that she probably could not tell anyone the exact probability that her decision is correct. The fact is that she does not need to know the exact probability; all she needs to know is that the probability is considerably greater than $\frac{1}{2}$.

There probably are a thousand times as many opportunities for each of us to use this informal kind of statistical thinking as there are for us to use the formal and elaborate kind of statistical method.

EXERCISES

1. What is meant by an attribute of an elementary unit in a universe? Are you studying any attributes in your investigation? If "Yes," what kind of classification system will be applied to each elementary unit in the sample?

2. What is meant by a variable characteristic of the elementary units in a universe? Are you studying any variables in your investigation? If "Yes," what kind of classification system will be applied to each elementary unit in the sample?

3. The classification system used in baseball under which each pitch is classified as a ball or a strike is: a truly nonquantitative classification☐, a partly quantitative classification ☐, a truly quantitative classification ☐.

4. Both the Fahrenheit scale and the centigrade scale for measuring temperature are interval scales. True ☐ False ☐

5. The metric system of measurement for weight is a ratio scale. True ☐ False ☐

6. Prove that the two scales that measure weight in pounds and in grams have the property of equal ratios.

7. What is meant by saying that in a statistical investigation the definitions usually should be operational definitions? List any operational definitions that you made while planning your investigation.

8. What is meant by the term "null hypothesis"? List the null hypotheses and the alternative hypotheses, if any, that you made while planning your investigation. What is the null hypotheses for the example cited in section 2.13?

9. Is the sample that you planned for your investigation a probability sample? What is the probability of each elementary unit in the universe being included in your sample? Is it a sample drawn with replacement after each draw? Is it a stratified sample? How did you decide what the size of your sample should be?

10. The purpose in drawing a sample from a universe usually is to learn something about the universe as a whole rather than about the individual elementary units in the sample. True ☐ False ☐

11. The procedure by which the sample is obtained from the universe determines in part, at least, the method of analysis and interpretation that is appropriate. True ☐ False ☐

12. The randomness and the representativeness of a sample are inherent characteristics of the procedure by which the sample actually is obtained. True ☐ False ☐

13. Positive steps must be taken by the investigator in planning the investigation and in collecting the data if he wishes the sample to be a probability sample. True ☐ False ☐

14. Statistical methods are distinguished from other mathematical methods by the fact that statistical methods involve the notion of probability. True ☐ False ☐

15. The "fruit-stand plan" in which the best goods are placed on display as a "sample" to be easily observed is an example of: probability sample ☐, representative sample ☐, simple random sample ☐, stratified random sample ☐, systematic sample with a random start ☐, nonrepresentative or distorted sample ☐.

16. Some classifications have no essential order. True ☐ False ☐

17. In a taste test involving seven flavors of ice cream, it is likely that you would be able to indicate the order in which you prefer the seven flavors but you probably would not be able to state exactly how much "value" difference there is between each pair of flavors. Data resulting from such a taste test would be in the form of: a truly quantitative classification ☐, a partly quantitative classification ☐, a truly nonquantitative classification ☐.

REFERENCES

William G. Cochran, *Sampling Techniques.* New York: John Wiley & Sons, Inc., 1953.

Leon Festinger and Daniel Katz, *Research Methods in the Behavioral Sciences.* New York: The Dryden Press, Inc., 1953.

Cyril H. Goulden, *Methods of Statistical Analysis.* New York: John Wiley & Sons, Inc., 1952. (agriculture, biology)

Morris Hansen, William N. Hurwitz, and William G. Madow, *Sample Survey Methods and Theory.* Vol. I. New York: John Wiley & Sons, Inc., 1953.

Quinn McNemar, *Psychological Statistics.* New York: John Wiley & Sons, Inc., 1955.

Mildred B. Parten, *Surveys, Polls and Samples.* New York: Harper & Brothers, 1950.

S. L. Payne, *The Art of Asking Questions.* Princeton, N. J.: Princeton University Press, 1946.

W. J. Youden, *Statistical Methods for Chemists.* New York: John Wiley & Sons, Inc., 1951.

Chapter 3

ESTIMATES BASED ON THE AVERAGE AND THE VARIABILITY OF A SAMPLE

3.1 INTRODUCTION

In Table 3.1 the hourly earnings of twelve truck drivers are listed. What are some of the questions that might be asked about the universe from whch the sample of twelve truck drivers was drawn?

TABLE 3.1

Hourly Rates of Pay of Twelve Light-Truck Drivers in Boston

Driver's Initials	Hourly Rate of Pay	Driver's Initials	Hourly Rate of Pay
T. A.	$1.80	V. P.	$2.33
H. B.	1.50	R. Q.	1.95
M. C.	1.55	J. S.	1.80
A. J.	1.80	F. T.	2.15
E. K.	1.35	H. V.	1.55
B. M.	2.05	L. W.	1.65

Do all the truck drivers in the universe receive the same rate of pay, or do they differ greatly in pay rate? What is the range of the pay rate among the universe of truck drivers? What is the best available estimate of the average hourly earnings of the universe of truck drivers from which the sample was drawn? Can we state confidence limits for this estimate of the average hourly earnings of the truck drivers in the universe?

In order to answer these questions, we need to know how to estimate the average hourly rate of pay of the truck drivers in the universe and we need to know how to estimate the amount of variability among the pay rates of drivers in the universe.

63

3.1.1. Serial distributions. Tables 1.3, 1.4, 1.5 and 3.1, 3.2, 3.3 are illustrations of what statisticians call *serial distributions*. In serial distributions, the data are arranged simply as a series of numbers. The data in a serial distribution may be arranged in alphabetical order, order of magnitude, or in any other convenient order. Frequently, the data are shown in the order in which they happened to be collected. If the data are arranged in order of magnitude, as in the second column of Table 3.2, the distribution is said to be an *ordered array* or simply an *array*.

TABLE 3.2

Hourly Rates of Pay of Twelve Light-Truck Drivers in Boston

Driver's Initials	Hourly Rate of Pay	Driver's Initials	Hourly Rate of Pay
V. P.	$2.33	T. A.	$1.80
F. T.	2.15	L. W.	1.65
B. M.	2.05	H. V.	1.55
R. Q.	1.95	M. C.	1.55
J. S.	1.80	H. B.	1.50
A. J.	1.80	E. K.	1.35

3.2 THE RANGE OF A VARIABLE

Table 3.2 shows that the range of the values of the variable is from $1.35 to $2.33. This information gives us a basis for stating that there are large differences in the pay rates of the drivers in the universe; some truck drivers in the universe earn almost twice as much per hour as others. The difference between the pay rates of the highest paid and the lowest paid workers in the sample is $0.98.

The range of the variable in a distribution is an important bit of information. It gives us some idea of the magnitude of the differences that exist among the elementary units in the universe from which the sample was drawn.

3.3 ESTIMATES OF THE CENTRAL VALUE IN THE UNIVERSE

One of the main purposes of statistical methods is to summarize in one estimate, or in a few estimates, all the information that the data for the sample of elementary units can tell us about the universe from which the sample was drawn.

In many distributions there is a tendency for most of the values of the variable to cluster around the middle of the range of the variable. Consequently, one of the most important estimates that a statistician can make about a universe usually is the *central value* of the variable in the universe. Of course, this estimate of the central value must be determined from the data in a sample.

There are several methods, some better than others, for making an estimate of the point in the range of the variable around which most of the values of the variable for the elementary units in the universe tend to cluster. Some methods for estimating the central value of the variable in the universe are more appropriate for use with certain types of data than they are with other types of data, and some methods for making this estimate produce results that have greater reliability than the results obtained by other methods; some kinds of estimates of the central value of the variable in the universe are likely to have smaller sampling errors than other kinds.

3.3.1 The midpoint of the range. Just as the range of the values of the variable in a sample gives us at least a rough estimate of the range of the values of the variable in the universe, so also the midpoint of the range in the sample gives us at least a rough estimate of the central value of the variable in the universe. The *midpoint of the range* is simply the average of the smallest value and the largest value of the variable in the sample.

For the small sample in Table 3.1, the range is from \$1.35 to \$2.33 and, consequently, the midpoint of the range in the sample is (\$1.35 + \$2.33)/2 = \$3.68/2 = \$1.84. Therefore, \$1.84 is our first estimate of the central value of the variable (hourly rate of pay) in the universe of truck drivers.

3.3.2 The mode. Another easy way to obtain an estimate of the important central clustering point in the range of the variable in the universe is to find the value of the variable that occurs most frequently in the sample. The pay rate \$1.80 occurs three times in Table 3.1. Consequently, we might take \$1.80 as our estimate of the central clustering point in the range of the variable in the universe. The pay rate \$1.80 might be considered to be the most common or the most typical value of the variable. Statisticians call it the *mode* or the *modal value* of the variable in the sample. This is our second estimate of the central value of the variable in the universe of truck drivers.

Sometimes in small samples no value of the variable occurs more often than any other. In such cases, there is no mode in the sample and therefore some other method for estimating the central clustering point in the universe must be used. In most situations in which there is no mode in the sample, it is because there are not enough elementary units in the sample to show the real nature of the distribution in the universe. Occasionally, however, statisticians find very large samples that have two or three modes.

3.3.3 The median. Another method for obtaining an estimate of the central clustering point in the range of the values of the variable in the universe is to determine a value in the range of the variable that splits the sample distribution into two groups with the same number of elementary units and the same number of values of the variable in each group. The value of the variable that splits the distribution in this way is called the *median* of the sample.

An easy way to find the median for the twelve truck drivers is to begin with Table 3.2 in which the twelve pay rates are arranged in order of magnitude. Begin at the top of the column of pay rates and count downward until half of the number of values of the variable—six, in this case—have been counted. The sixth value of the variable, counting downward from the top of the column, is $1.80, the pay rate of A. J. Now, begin at the bottom of the column and count upward to the sixth value of the variable. It is $1.80, the pay rate of T. A.

If we draw a straight line horizontally across the table between the lines on which A. J. and T. A. occur, this line will mark the middle of the distribution in the sample. The median is the value of the variable midway between the value that is immediately above the horizontal line and the value that is immediately below the horizontal line. In this example, the two values are the same, namely, $1.80. In other examples you probably will not have exactly the same value immediately above and immediately below the line. The median for the sample of truck drivers is, then, $1.80.

The value of the median happens here to be the same as the value that we obtained for the mode of the sample. The median and the mode are not necessarily exactly equal to each other in all samples. Usually, however, they are fairly close to each other in value.

In Table 3.2 the number of elementary units in the sample is an even number, twelve. If the number of values of the variable in the sample is an odd number, one value of the variable will be in the middle of the distribution when the values of the variable in the sample are arranged in order of magnitude. There will be the same number of values of the variable above this middle value as there are below it in the list. In such situations, the horizontal line across the table splitting the sample into halves passes through one of the values of the variable. That value is the median of the sample.

Table 3.3 may be used to illustrate the determination of the median when the number of values of the variable is an odd number, in this case, nine. Consequently, the median value of the variable, that is, the median score, for the sample is 52.

If it happens that several values of the variable in the list are the same as the median, one need not worry about which particular one of the individuals is to be designated as the median individual in the sample. Any one or all of the individuals with that value may be considered to be at the median of the sample.

TABLE 3.3

Scores of Nine Students on a Mathematics Test*

Student	Score
A	60
B	60
C	59
D	54
E	52
F	52
G	45
H	39
I	37

*The maximum possible score on the test was 60.

3.3.4 The arithmetic mean. A fourth method for estimating the central value of the variable in a universe is the method of the *average*, or more correctly, the method of the *arithmetic mean*. In other books on mathematics you will find reference to other kinds of means, such as geometric means and harmonic means. Statisticians rarely need to use any mean except the arithmetic mean, and so there need be no confusion in this book if we agree now that when we use the term "the mean" we shall be using an abbreviated title for "the arithmetic mean." We shall reserve the English letter m to stand for the mean of a sample. And we shall reserve the Greek letter μ to stand for the mean of a universe. (See section A.1.)

The mean usually is called the average in ordinary conversation. But it is almost standard practice among statisticians to use the word "mean" rather than the word "average."

You know already how to find the mean or average of a set of numbers. To find the mean of the pay rates of the twelve truck drivers in Table 3.1, we add all the pay rates and divide the total by 12. The total of the pay rates is \$21.48. Dividing this total by 12, we find that the mean is \$1.79. Notice that the value of the mean here is not exactly the same as the values of the midpoint of the range, the mode, and the median. However, all four of these estimates are fairly close.

In Table 3.3, the total of the pupils' scores is 458. Dividing this total by 9 gives $m = 50.9$ (51 when rounded off to the nearest whole number) as the mean score.

Very often, statisticians write the value of the mean with one decimal place or one significant digit more than were in the original data. The reason for this is that, as we shall see later in this book, the sampling errors in the distribution of means tend to be much smaller than the deviations in the distribution of the original data. In other words, the mean of a sample is considered to be more reliable than any individual observed value in the sample.

3.3.5 Choosing an appropriate estimate of the central value. An investigator rarely uses more than one of the four methods given above for estimating the central value of the variable in a universe. Usually, he selects the method that he has reason to believe is the most appropriate for the situation that he is investigating and then uses only that one. Here are some of the arguments for and against each of the estimates of central value.

For many purposes, it would be easy to find the mode and the mode would be a useful and simple estimate of the central value of the variable in the universe. For example, suppose that you asked ten people to carry out an arithmetical computation for you and six of the people obtained the same answer, two obtained another answer, one obtained a third answer, and one obtained a fourth answer. If you had to accept one of the answers, you probably would accept the answer that was obtained by six people. That is, in such a situation you might accept the mode as the most appropriate answer. You would not likely choose the average of all the answers or the median of the answers or the midpoint of the range of the answers.

Unfortunately, in many small samples the mode is not clearly defined by the data and therefore cannot be used as an estimate of the central value in the universe in such situations. On the other hand, the midpoint of the range, the median, and the mean can always be computed from the values in a truly quantitative classification, even in small samples.

Generally, the mean is the most useful estimate of the central value in the universe in analyzing those types of data for which it is a meaningful value. This is true for several fundamental theoretical reasons. In the first place, it often is easier to specify confidence limits for the mean than for the midpoint of the range or the mode or the median, and usually, though not always, the sampling error is smaller in an estimate based on the mean than it is for an estimate based on the midpoint of the range or the mode or the median.

The second reason for preferring the mean in most of the statistical situations to which it is applicable is the very basic relation of the mean to the normal probability law. We shall study the normal probability law in Chapter 6.

A third reason for the importance of the mean to statisticians is its theoretical relation to "the method of maximum likelihood" in connection with the estimation of the parameters of such important universe distributions as the normal distribution, the binomial distribution, and others. We will not study "the method of maximum likelihood" in this book because it involves a knowledge of the calculus and because we can manage to get along satisfactorily without using it in this course in elementary statistics.[1]

[1]The estimate obtained for a parameter of a universe by the method of maximum likelihood is that value computed from the sample which, if it were the true value of the parameter, would maximize the probability of the occurrence of the observed sample.

In section A.6.18 an example is given for which, because of the way in which the sample values of a truly quantitative variable have been arranged in groups, it is impossible to compute the mean. However, the median may be computed easily.

For partly quantitative data, such as, for example, examination marks that are not truly quantitative measurements or the ranks assigned by students to indicate their preferences for types of motion pictures or types of books, the median ordinarily is a more appropriate estimate of central tendency than the mean.

The mean of a sample has the important property that the sum of the squares of the deviations of the values of the variable in the sample is a minimum if the deviations are measured from the mean of the sample. One reason why this property of the mean is important is that the sum of the squares of the deviations from the mean is involved in the definition of the standard deviation of a sample. In other words, this property of the mean is an important link between it and the standard deviation. These are very important and useful quantities in statistics.

The median of a sample has the property that the sum of the absolute values of the deviations of the values of the variable in the sample is a minimum if the deviations are measured from the median of the sample. However, we rarely make use of this property of the median in practical statistics, and, consequently, it is not an important property of the median.

It is not as easy to find the median of several combined groups as it is to find the mean of several combined groups. To find the mean of combined groups, all that we need to do is combine the means of the groups, using the numbers of values of the variable in the groups as weights. For example, if the average earnings in a week of a sample containing 25 salesmen is $100 and the average earnings during the same week of another sample containing 40 salesmen is $96, the average earnings for the combined groups, that is, for the 65 salesmen, during that week is found easily by computing $[25(100) + 40(96)]/(25 + 65) = 6340/65 = \$97.54$. To find the median of combined groups, it is necessary to return to the original values of the variable and start from there.

3.4 SOME ALGEBRAIC SYMBOLS

Before we can work out general methods and general formulas for the statistical methods of analyzing data, we must choose some simple algebraic symbols. Keyser says that only a god could get along without good symbols, and that only a divine fool would do so.[2] Some frequently used mathematical symbols are given on page 345. Formulas are symbols. A formula saves

[2] C. J. Keyser, *The Pastures of Wonder*. New York: Columbia University Press, 1929, p. 45.

time and labor by organizing the computations and omitting all unnecessary steps. It generalizes a process so that it can be applied to any appropriate situation.

You have used the letter x as a variable many times in algebra. We shall use x to stand for the variable in many of the distributions in this book. The data in a sample often will be, then, values of the variable x. We need a method by which we can put a label or a tag on each of the values of x in a distribution. For example, in Table 3.1 there are twelve values of the variable, that is, there are twelve pay rates in the distribution. How can we invent a symbol that will indicate each of the values of the variable?

3.4.1 Subscripts. Subscripts are very convenient for labeling or tagging the different items in a distribution. For example, to distinguish the twelve values of x if x stands for the pay rate of the truck driver, we write the twelve symbols x_1, x_2, x_3, x_4, x_5, x_6, x_7, x_8, x_9, x_{10}, x_{11}, and x_{12}. The symbol x_1 stands for the first value of the variable in the list, the symbol x_3 stands for the third value of the variable in the list, and so on. The numbers that are placed to the right of and below the letter are called *subscripts*. The symbol x_1 is read "x-sub-one," meaning x with the subscript one.

TABLE 3.4

Generalized Table for Serial Distributions

Elementary Unit Number	Value of the Variable
1	x_1
2	x_2
3	x_3
.	.
.	.
.	.
n	x_n

Table 3.4 is a generalized table that represents any or all serial distributions. The variable in the distribution is designated by x. The number of values of the variable in the sample is designated by n. The n values of the variable in the sample are designated by x_1, x_2, x_3, $\cdots$, x_n. The dots between x_3 and x_n indicate that we have abbreviated the list.

3.4.2 The mean. In order to compute the mean of a sample we add all the values of the variable in the sample and divide the total by the number of values of the variable in the sample. One way to express the process of adding all the values of the variable in the sample is to write $x_1 + x_2 + x_3 + \cdots + x_n$. Then, because the number of values of the variable in the sample is n, we would divide the sum by n. A formula for the mean m of the sample in Table 3.4 is $m = (x_1 + x_2 + x_3 + \cdots + x_n)/n$.

3.4.3 The sigma notation for summation. Statisticians and most other scientists in the world today have agreed upon a symbol with which you probably are not familiar. Consequently, the symbol may seem a bit strange to you at first. Do not expect to appreciate all its advantages the first time that you see it. Your understanding of the new symbol and your appreciation of its usefulness will increase gradually as you use it and as you learn how awkward statistical calculation would be if we did not have such a good symbol.

The new symbol is the Greek capital letter sigma, Σ. It is used to indicate the arithmetic or algebraic operation of summing, that is, addition. For example, when statisticians wish to indicate the summing or adding of all the values of the variable in a sample, they write merely Σx. In other words, Σx is a brief symbol which means the sum of all the values of x in which one is interested at a particular moment. Consequently, by definition, the sum of n values of x in a sample may be written as $\Sigma x = x_1 + x_2 + x_3 + \cdots + x_n$.

Our formula for the mean of a sample containing n values of x can be written now in the form

$$m_x = \frac{\Sigma x}{n} \tag{3.1}$$

The mean μ_x of a universe that consists of N values of the variable x is given by the formula $\mu_x = \Sigma x / N$.

If in some problem you have two variables and you designate one variable by x and the other by y, then Σx will stand for the sum of all the values of x in the sample, and Σy will stand for the sum of all the values of y in the sample.

This symbolism is called the sigma notation for summation. The symbol Σ is one of the most time-saving, labor-saving, and mind-saving symbols in all mathematics. It is rapidly becoming a familiar symbol in all the sciences.

Be sure to remember that the symbol Σ is not a symbol for a variable or for an unknown. We use the letters x, y, z, p, and other letters to stand for variables and unknowns. Consequently, when you write xy in algebra you mean x multiplied by y and you may read xy as "x times y." But Σ is not a symbol for a variable or an unknown and, therefore, never make the mistake of thinking that Σx means Σ multiplied by x and never read Σx as "Σ times x." Remind yourself frequently for a while that Σ stands for the following words: "the sum of all the values of." The symbol that is written after Σ tells you the variable whose values are summed.[3]

---

[3]Whenever there is danger of confusion or ambiguity about what values of x are to be summed, more specific symbolism such as the following should be used:

$$\sum_{i=5}^{10} x_i = x_5 + x_6 + x_7 + x_8 + x_9 + x_{10}$$

3.5 ESTIMATES OF VARIABILITY OR DISPERSION

In order to answer all the questions that were asked near the beginning of this chapter, we need to know more than an estimate of the central value of the variable in the universe. We need to know something about how the values of the variable in the universe spread out from the center of the universe. In other words, we need to be able to describe the dispersion or variability of the values of the variable in the universe.

The very fact that statisticians call the characteristic they are investigating a variable is enough to indicate that the main thing in which they are interested is the variability in data. If there were no variability in data, that is, if there were perfect uniformity in all human affairs, there could be no statistical analysis or statistical theory, and there would be no need for statistics.

But what a drab world it would be if there were no differences, that is, no variability, among people or things or events! In *R. U. R.*, Karel Capek's modern satiristic drama about automation, which has been staged successfully many times in recent years, the mood of the play is set by the line "We made the robots' faces too much alike. A hundred thousand faces all alike . . . . A hundred thousand expressionless bubbles. It's like a nightmare!"[4]

It will help you to learn to be a good statistician, then, if you bear in mind that variations are the main interest of statistics. It is mainly by analyzing the variations in samples that the statistician is able to learn how to determine the reliability of estimates of the parameters of universes. This brings us to another aspect of the nature of statistical samples that is equally as important as the fact that without variations in data there could not be any statistical theory. The second fact is that if there were no *regularity of form* in the variations in statistical samples, then there could be no statistical theory that would be of practical usefulness.

If the variations in statistical samples did not conform to certain mathematical laws, namely, the laws of the mathematical theory of probability, then statistical analysis of the data in a sample would give us no worthwhile information about the parameters in the universe from which the sample was drawn; any inferences that one drew from a sample and applied to the universe would be no better than guesswork.

Consequently, there are two things to remember if you wish to be a good statistical thinker: (1) you must focus your attention on the variations in the samples with which you are working; (2) you must, at the same time, focus your attention on the form or pattern, the regularity, the systematic properties of the variations. Frequency tables, histograms, and frequency curves are devices often used to show these properties of distributions.

[4]Karel Capek, *R. U. R.* (*Rossum's Universal Robots*), translated by Paul Selver. Garden City, New York: Doubleday, Page & Co., 1923.

There are many methods for describing numerically the variability in a sample. Two methods, finding the range and finding the standard deviation, will now be described.

3.5.1 The range. We concluded earlier in this chapter that ordinarily the range of the values of the variable in a sample gives us only a small amount of useful information about the universe from which the sample was drawn. But the range is much easier to determine for a small sample than is the standard deviation. For this reason, statisticians have found a way to use the range rather than the standard deviation in connection with many statistical quality control procedures.

3.5.2 The variance and the standard deviation. We obtain the value of the variance and the value of the standard deviation of a distribution by studying the differences between the values of the variable and their mean. For example, the mean of the twelve values of the variable in Table 3.2 is $1.79. Let us study now the differences between the individual pay rates of the twelve truck drivers and the average pay rate.

For this analysis, it is convenient to construct Table 3.5. The individual pay rates are listed in the second column. Then the mean, $1.79, was subtracted from each of the individual pay rates and the differences were recorded in column three. Notice that the deviations of the low pay rates from the mean are negative numbers and that the deviations of the high pay rates from the mean are positive numbers.

TABLE 3.5

Tabulation for Finding Variance and Standard Deviation

Driver's Initials	Hourly Rate of Pay x (Dollars)	Deviations $x - 1.79$ (Dollars)	$(x - 1.79)^2$ (Dollars Squared)
E. K.	1.35	−0.44	0.1936
H. B.	1.50	−0.29	0.0841
M. C.	1.55	−0.24	0.0576
H. V.	1.55	−0.24	0.0576
L. W.	1.65	−0.14	0.0196
T. A.	1.80	0.01	0.0001
A. J.	1.80	0.01	0.0001
J. S.	1.80	0.01	0.0001
R. Q.	1.95	0.16	0.0256
B. M.	2.05	0.26	0.0676
F. T.	2.15	0.36	0.1296
V. P.	2.33	0.54	0.2916
Total		0.00	0.9272

The fourth column was formed by squaring each of the numbers in the third column. Consequently, the fourth column contains the squares of the deviations, or variations, from the mean of the sample. All of the numbers in the fourth column are positive because, as you know already, when a negative number is multiplied by a negative number the product is a positive number. The original data of the sample are in dollars. The numbers in the fourth column of Table 3.5 are dollars squared.

Then the total of the numbers in the fourth column was found. Be sure that you know how to describe this total. It is the sum of the squares of all the deviations from the mean of the sample. Now, divide the total of column four by the number of values of the variable in the sample, that is, by 12: $0.9272/12 = 0.0773$ dollars squared. This quantity is called the variance of the sample of pay rates.

"Dollars squared" is an awkward term and an awkward idea. Obviously, if we wish to describe the variations of the pay rates in the sample, we would prefer to use a quantity that is expressed in the same unit of measurement as the original data, namely, dollars. Fortunately, all that we need to do to obtain such a quantity is take the positive square root of the quantity 0.0773 dollars squared and the result will be in dollars. $\sqrt{0.0773}$ dollars squared $=$ $\$0.28$.

This quantity, $\$0.28$, is called the *standard deviation* of the sample. In this book, we shall use the small letter s to stand for the standard deviation of a sample, and we shall use the small Greek letter σ to stand for the standard deviation of the universe from which the sample was drawn. Consequently, the proper symbol for the variance of a sample is s^2, and the proper symbol for the variance of a universe is σ^2. For the sample from which Table 3.5 was constructed, we can now write $s^2 = 0.0773$ dollars squared and $s = \$0.28$.

The value of the variance s^2 that is computed from a sample is interpreted as an estimate of the value of the variance σ^2 in the universe from which the sample was drawn. Similarly, the value of the standard deviation s that is obtained by taking the positive square root of the variance of a sample is interpreted as an estimate of the standard deviation σ of the universe from which the sample was drawn.

It is easy now to write formulas for the variance and the standard deviation of any serial distribution in a sample. All that we need to do is generalize, summarize, and organize the work involved in building Table 3.5 and in the computations that followed. First, we need to notice that a generalized symbol for the heading of a column such as the fourth column of Table 3.5 is $(x - m)^2$. Now that we use Σ for "summation," a good symbol for the total of such a column is $\Sigma(x - m)^2$. We use n to stand for the number of values of the variable in the sample. Consequently, the *variance* of a sample may be defined as

$$s_x^2 = \frac{\Sigma(x - m)^2}{n} \tag{3.2}$$

and the *standard deviation* of a sample may be defined as

$$s_x = \sqrt{\frac{\Sigma(x - m)^2}{n}} \tag{3.3}$$

However, if you study section A.5.2, you will learn that there are much easier methods than formula 3.3 for computing the standard deviation of a sample. One of them is

$$s_x = \frac{1}{n}\sqrt{n\Sigma x^2 - (\Sigma x)^2} \tag{3.4}$$

The variance and the standard deviation of a universe that consists of N values of the variable x are defined by the formulas

$$\sigma_x^2 = \frac{\Sigma(x - \mu)^2}{N} \quad \text{and} \quad \sigma_x = \sqrt{\frac{\Sigma(x - \mu)^2}{N}}$$

You may be wondering why we did not use the total of the third column of Table 3.5 rather than the total of the fourth column to describe the variability of the sample. The reason is that the total of the third column, that is, $\Sigma(x - m)$, is always zero if you subtract the exact value of m from every value of the variable in the sample. To prove that, proceed as follows:

$$\Sigma(x - m) = \Sigma x - \Sigma m = \Sigma x - nm = \Sigma x - n\frac{\Sigma x}{n} = \Sigma x - \Sigma x = 0$$

Similarly, in any universe of values of x for which the mean is μ, it is true that $\Sigma(x - \mu) = 0$.

On the other hand, $\Sigma(x - m)^2$, the total of column four of a table such as Table 3.5, never is zero except in the rare instance in which all the values of the variable in a sample are identical; in this trivial case there would be no variations or deviations from the mean and the variance and the standard deviation of the sample would be zero.

3.6 STANDARD UNITS AND STANDARD SCORES

The mean and the standard deviation enable us to develop a method for making comparisons. The method involves the process of reducing data to standard units. We have already determined that in the example of the pay rates of twelve truck drivers $m = \$1.79$ and $s = \$0.28$. To reduce the pay rate of V. P. to standard units, subtract the mean and then divide by the standard deviation. $\$2.33 - \$1.79 = \$0.54$ and $\$0.54/\$0.28 = +1.93$ standard units.

We interpret this result by stating that the hourly pay rate of V. P. is 1.93 standard units greater than the mean or average hourly pay rate of the twelve persons in the sample. The number $+1.93$ provides a scientific description of the position of V. P. in the distribution in the sample. Simi-larly, for F. T., \$2.15 − \$1.79 = \$0.36 and \$0.36/\$0.28 = +1.29 standard units.

In general, if x is a variable and if we know the values of the mean μ_x and the standard deviation σ_x of the variable in the universe, then the formula for reducing a specific value of x to standard units is

$$z = \frac{x - \mu_x}{\sigma_x}, \text{ that is, } z = \frac{\text{a value of a variable} - \text{the mean of the variable}}{\text{the standard deviation of the variable}}$$

$$(3.5)$$

If we do not know the values of μ_x and σ_x for the universe, we usually sub-stitute the values of m_x and s_x computed from a random sample drawn from the universe as the best available estimates of μ_x and σ_x.

The standard unit equivalents to the hourly rate of pay of each of the twelve truck drivers are shown in Table 3.6.

TABLE 3.6

Hourly Rates of Pay of Twelve Truck Drivers in Dollar Units and in Standard Units

Driver's Initials	Hourly Rate of Pay in Dollar Units x	Hourly Rate of Pay in Standard Units z
E. K.	\$1.35	−1.57
H. B.	1.50	−1.04
M. C.	1.55	−0.86
H. V.	1.55	−0.86
L. W.	1.65	−0.50
T. A.	1.80	+0.04
A. J.	1.80	+0.04
J. S.	1.80	+0.04
R. Q.	1.95	+0.57
B. M.	2.05	+0.93
F. T.	2.15	+1.29
V. P.	2.33	+1.93

First of all, notice that the individuals whose pay rates are above the mean have positive numbers of standard units, and the individuals whose pay rates are below the mean have negative numbers of standard units. An individual's number of standard units must always have either a plus sign or a minus sign, except when it is zero. The sign tells us whether the person is above average or below average in the distribution. The numer-ical part tells how far the person is from the mean, but it does not tell us the direction.

We shall reserve the letter z in this book for values of a variable that are expressed in standard units. The unit of measurement of z is called a *standard unit*. The value of z for a person in a distribution often is called his or her z-score or *standard score*.

Why do we call these numbers "standard units"? The reason is that when you divide a number of dollars by a number of dollars, the quotient is not a number of dollars; it is an abstract or pure number. The numbers in the third column of Table 3.6 could be described as pure units or abstract units but, because they are obtained by dividing by the standard deviation, we shall call them standard units.

No doubt you do not see clearly at this time how much importance to attach to a person's z-score or standard score, that is, to his position in a distribution expressed in standard units. To interpret a z-score, it is necessary to think in terms of probability and the laws of probability. As you gain experience with statistical distributions, you will learn that in most good samples of reasonable size there are few, if any, elementary units more than three standard units above the mean or more than three standard units below the mean of the sample. Consequently, when we find an individual with a z-score that is greater than $+3$ or less than -3, we are likely to think of that individual as a very unusual member of the distribution.

For example, the average intelligence quotient of people in the United States is 100, and the standard deviation of that variable is about 17. Consequently, a person whose I.Q. is 49 has a z-score of -3.00. There would be no point in letting such a person attend a regular high school or college. Similarly, a person whose I.Q. is 151 has a z-score of $+3.00$. In other words, a person whose I.Q. is 151 is three standard units above the mean or average intelligence of Americans. Psychologists call such a person a "near genius." Only about one person in a thousand has I.Q. three standard units or more above the mean. In fact, only about two or three people out of a hundred randomly selected people have I.Q. more than two standard units above the mean, that is, above 134.

Here is another example. According to Watson and Lowery, the average weight of seventeen-year-old girls in America is about 133.5 pounds. The standard deviation of their weights is about 18 pounds.[5] Consider a girl who is seventeen years old and who weighs 151.5 pounds. How many standard units is she above the average weight of American girls at her age? To obtain the answer, subtract 133.5 pounds from 151.5 pounds and divide the difference by 18 pounds. The result is a z-score of $+1.00$.

Would you say that this indicates that she is unusually or significantly overweight? Probably, the correct answer is "No," because if you knew how to use Table 6.7 you would find that about 16 per cent of American girls are more than one standard unit above the mean in weight.

[5]E. H. Watson, M. D., and G. H. Lowery, M. D., *Growth and Development of Children.* Chicago: The Year Book Publishers, Inc., 1954, p. 50.

3.7 A WAY TO COMPARE TWO GROUPS

You can use the mean and the standard deviation to make comparisons between two or more groups. For example, suppose that you have picked two groups of male students with approximately the same number of students in each group. Suppose that the height of each student in the two groups is measured and recorded to the nearest inch. Suppose further that the mean height of the students in the first group is 69.0 inches and that the mean height of the students in the second group is 69.4 inches. Suppose also that the standard deviation of the heights of the students in the first group is 2 inches and that the standard deviation of the heights of the students in the second group is 3 inches.

Then you can conclude from this information that there is a difference of only about four tenths of an inch in the average heights of the two groups. But because there is a relatively great difference in the two standard deviations, you know that the group with the larger standard deviation probably contains more unusually tall students and more unusually short students than the other group. The group with standard deviation of 2 inches is much more nearly uniform or homogeneous in height than the other group. If you were choosing a group for certain purposes, you might wish to have them as nearly uniform or homogeneous in height as possible. For other purposes, you might wish to have as much variation as possible in height.

Here is another illustration that may help you to grasp the idea of the importance of the mean and the standard deviation in making comparisons. Suppose that two students, A and B, take the same three examinations in three courses. Let the scores of A be 35, 65, and 95, and let the scores of B be 63, 65, and 67. You will notice that both students obtained an average score of 65 on the three examinations. However, you will probably agree that A and B are not equally capable students insofar as these three examinations are concerned. There is great variation between the three scores for A, and there is very little variation between the three scores for B. If you compute the standard deviation for the three scores of each student, you will find that the standard deviation of A's scores is about fifteen times as great as the standard deviation of B's scores.

Here are two samples of measurements made by two chemical analysts in repeated determinations of the per cent of carbon in a container of ephedrine hydrochloride.[6] Analyst A: 59.09, 59.17, 59.27, 59.13, 59.10, 59.14. Analyst B: 59.51, 59.75, 59.61, 59.60. First, we must realize that it is not reasonable to expect people always to make perfectly accurate measurements. The most that can be expected is that (1) the average of all a person's measurements should be very close to the true value, and (2) the variation in a person's measurements should be small.

[6]Francis W. Power, "Accuracy and Precision of Microanalytical Determination of Carbon and Hydrogen. A Statistical Study." *Analytical Chemistry* 11, No. 12 (December 15, 1939), pp. 660–673. See p. 660.

For A, $m = 59.15$ per cent. For B, $m = 59.62$ per cent. For A, $s = 0.060$ per cent. For B, $s = 0.086$ per cent. The indication is that A's measurements have less variation, that is, greater *precision* than B's.

Although the two analysts did not know the true per cent of carbon in the chemical while they were making their analyses, the chemical had been prepared by the National Bureau of Standards in Washington and was certified to be 59.55 per cent carbon. Consequently, we can find the actual error in each measurement and the mean error for each analyst. For A, the mean error is -0.40. For B, the mean error is $+0.07$. Consequently, B's measurements have very much greater *accuracy* than A's.

3.8 THE COEFFICIENT OF VARIATION

Is there greater variation or variability in height than there is in weight among students? Because heights are measured in inches and weights are measured in pounds, it is impossible to make a direct comparison between the mean and the standard deviation of the heights of a sample of students and the mean and the standard deviation of the weights of the same sample.

For example, for a representative sample of the seventeen-year-old boys in the United States, the mean height probably would be about 69.5 inches and the standard deviation of their heights probably would be about 2.3 inches. The mean weight of the boys in this sample probably would be about 148 pounds and the standard deviation of their weights probably would be about 20 pounds.[7]

One way to compare these two kinds of variability, namely, variability in height and variability in weight, is to compute what statisticians call the *coefficient of variation* for each distribution. It is defined by the formula

$$\text{coefficient of variation} = \frac{\text{standard deviation}}{\text{mean}} \times 100 \text{ per cent} \qquad (3.6)$$

For the distribution of height of the seventeen-year-old boys, the coefficient of variation is $[(2.3 \text{ inches})/(69.5 \text{ inches})](100) = 3.31$ per cent. For the distribution of weight of the seventeen-year-old boys, the coefficient of variation is $[(20 \text{ pounds})/(148 \text{ pounds})](100) = 13.51$ per cent.

Consequently, if we use the two coefficients of variation to indicate the relative amounts of variability in the two distributions, height and weight, we conclude that normally there is much greater variability in weight among American seventeen-year-old boys than there is in height.

EXERCISES

1. Table 3.7 shows the scores on a mathematical proficiency test of a sample

[7]E. H. Watson, M. D., and G. H. Lowery, M. D., *Growth and Development of Children.* Chicago: Year Book Publishers, Inc., 1954, pp. 50, 52.

of 65 students chosen at random from the 653 high school graduates who applied for admission to a college a few years ago. The scores are listed in the alphabetical order of the surnames of the students. Rearrange the serial distribution to form an ordered array.

TABLE 3.7

Scores of 65 High School Graduates on an American Council on Education Test of General Proficiency in Mathematics

Student's Initials	Score	Student's Initials	Score	Student's Initials	Score	Student's Initials	Score
E. A.	49	C. E.	64	B. L.	51	R. O.	66
G. A.	49	B. E.	69	V. L.	53	N. O.	66
O. A.	50	T. F.	56	H. L.	54	O. P.	44
B. A.	53	W. F.	66	K. L.	54	G. P.	50
R. A.	60	M. G.	51	R. L.	57	H. P.	62
L. A.	69	B. G.	55	B. M.	46	V. P.	63
L. B.	72	W. G.	59	S. M.	48	C. P.	65
E. C.	50	F. G.	72	H. M.	48	G. R.	51
S. C.	53	P. H.	49	G. M.	49	B. R.	56
K. C.	64	E. H.	61	C. M.	50	C. R.	61
W. D.	48	B. J.	48	M. M.	54	D. S.	40
B. D.	51	E. J.	49	S. N.	54	G. T.	63
D. E.	47	F. J.	52	J. N.	55	R. V.	49
B. E.	48	M. J.	54	H. O.	57	B. V.	49
M. E.	53	J. K.	60	W. O.	61	M. V.	57
F. E.	54	S. L.	50	C. O.	62	C. W.	61
R. E.	58						

2. Find the range, the midpoint of the range, the mode, and the median for the sample in Table 3.7. (*Ans.* range is 40 to 72, midpoint of range = 56, mode = 49, median = 54)

3. Assuming that the mathematical proficiency test on which the scores in Table 3.7 are based was constructed and standardized so carefully that the classification may be considered to be a truly quantitative classification, compute the mean and the standard deviation for the sample. (*Ans. m* = 55.4, *s* = 7.07)

4. Compute the mean and the standard deviation for the proportion p of red beads in the 25 samples that you drew from the universe of 2000 red beads and 3000 white beads.

5. Find the range, the midpoint of the range, the mode, and the median for the proportion p of red beads in the distribution of p formed from the 25 samples that you drew from the universe of 2000 red beads and 3000 white beads.

6. Determine the range, the midpoint of the range, the mode, and the median for the proportion p of red beads in the distribution of p formed from the large number of samples that the whole class drew from the universe of 2000 red beads and 3000 white beads. (*Ans.* The midpoint of the range, the mode, and the median ought to be approximately 0.400.)

7. Compute the z-score (standard score) for each of the 65 students in Table 3.7. What per cent of the z-scores are not in the interval *between* $z = -1.96$ and $z = +1.96$? (*Ans.* 4.6 per cent) What per cent of the z-scores are not in the interval *between* $z = -2.58$ and $z = +2.58$? (*Ans.* zero per cent)

8. Compute the coefficient of variation for the samples in Tables 3.2, 3.3, and 3.7. (*Ans.* 15.6 per cent, 16.2 per cent, 12.8 per cent)

9. Two subcontractors, A and B, supply a part for an aircraft assembly. The specification is that the part shall be 8.5 in. long, with a tolerance of ± 0.10 in. error. A random sample of five parts from a large lot delivered by A measured 8.48, 8.53, 8.51, 8.46, and 8.52. A random sample of five parts from a large lot delivered by B measured 8.45, 8.60, 8.41, 8.45, and 8.59. Determine the accuracy and the precision of the two samples. (*Ans.* $m_A = 0.00$, $s_A = 0.026$, $m_B = 0.00$, $s_B = 0.079$)

3.9 FREQUENCY DISTRIBUTIONS

If the data collected in an investigation involve a very large number of elementary units, it often is too laborious to try to deal with the data by the methods illustrated for serial distributions, unless the data are to be machine tabulated by punch card methods or some other mechanical process. In nonmechanical tabulation and analysis of statistical data, it usually is much easier to analyze and interpret the data in a large sample if the data have been organized and presented in the form of a frequency distribution, that is, in a frequency table.

Table 3.8 is a *frequency distribution*. The variable is the age at which an American-born man listed in *Who's Who in America* was first married. Age is a continuous variable. The data were obtained by drawing a probability sample of the men born in the United States, who have been married, and who are listed in the 1952–1953 edition of *Who's Who in America*.

TABLE 3.8

Age at Time of First Marriage of American-Born Men Listed in Who's Who*

Age at Marriage (Years)	Number of Men f
18–20	5
21–23	38
24–26	73
27–29	56
30–32	26
33–35	19
36–38	11
39–41	3
42–44	3
45–47	3
48–50	1
51–53	2
54–56	1
57–59	1
Total	242

*Source: A sample drawn from *Who's Who in America*, Vol. 27, 1952–1953.

What are some of the questions that one might ask about the sample in Table 3.8 and about the universe from which the sample was drawn? How many men are there in the sample? What are the earliest and the latest ages at which any of the men in the sample married? How many men in the sample married before reaching a particular age, say, 24 years? What is the age at which the largest number of the men in the sample married?

Someone might wish to know the median age at time of first marriage for the sample, so that he could say that 50 per cent of the men married before reaching that age and the other 50 per cent married after reaching that age. Most of us would want to know the mean or average age at which the men in the sample married, so that we would have an estimate of the average age at which the men in the universe from which the sample was drawn married. Then we would want to have some confidence limits for the mean in order to indicate the reliability of the estimated mean obtained from the sample.

In order to determine confidence limits for the average age at time of first marriage for all the American-born men in *Who's Who*, we would need to know the standard deviation of the variable in the sample. The standard deviation would give us important information about the variability of the ages of the men in the sample at the time of marriage as well as about the reliability of the estimate of the mean age at time of marriage of the men in the universe. Furthermore, the standard deviation would be useful if we wished to compare individuals in the sample, by enabling us to express the ages of the individuals in standard units.

The mean and the standard deviation of the sample would be the most useful tools for comparing this distribution of men with another distribution of men or with a similar distribution of women. For example, we might wish to make comparisons between the universe represented by the sample in Table 3.8 and the universes represented by the distributions in Table 3.17.

Finally, all of us likely would want to see a good graph for the distribution in Table 3.8. A graph of the form shown in Figure 3.1 for a frequency distribution is called a *histogram*.

A frequency table is one of the most important of all the ways for organizing and presenting data. Why? First, because without the idea of frequency distributions it would be difficult, if not impossible, to develop the mathematical theory of sampling errors and other parts of statistical theory. Second, because often it makes good graphical presentation of the data very simple and easy. Third, because of its compactness and its simplicity. Fourth, because sometimes it enables us to avoid a great deal of tedious arithmetic computation. It is of the utmost importance that you learn to read a frequency table, that is, that you learn how to extract the information contained in a frequency table.

For example, Table 3.8 shows that five of the men married between the

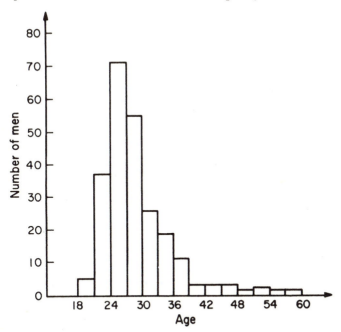

Fig. 3.1. Age at Time of First Marriage of a Sample of the American-Born Men Listed in *Who's Who*

beginning of their eighteenth year and the end of their twentieth year. Thirty-eight of the men married between the beginning of their twenty-first year and the end of their twenty-third year. One man married sometime between the beginning of his fifty-seventh year and the end of his fifty-ninth year. There were 242 men in the sample.

3.9.1 Frequency. The numbers in the second column of Table 3.8 are called *frequencies*. They tell us how frequently a man was found in the sample for each of the age classes in column one. It is convenient to use the letter f as a symbol for frequency. The total of the second column, that is, the total frequency, is 242. Consequently, we can write $\Sigma f = 242$ for Table 3.8.

In the earlier chapters of this book, we have used the letter n as a symbol for the total number of elementary units in the sample. We shall continue to use n for that purpose. Therefore, for frequency distributions the symbols Σf and n stand for the same thing, and you may use whichever symbol seems the more convenient at the time.

3.9.2 Cumulated frequency. In Table 3.9, three new columns have been added to Table 3.8. The new columns help us to answer some of the questions that were asked earlier about the distribution. For example, column three shows that 43 (5 + 38) of the men married before they reached twenty-four years of age. That column shows also that 116 (5 + 38 + 73) of the men married before they reached twenty-seven years of age. The numbers in the third column are called *cumulated frequencies*. The last cumulated frequency is equal to the total frequency, in this case, 242.

TABLE 3.9

Frequency, Cumulated Frequency, Relative Frequency, and Cumulated Relative Frequency for the Distribution in Table 3.8

Class (Years)	Frequency f	Cumulated Frequency	Relative Frequency	Cumulated Relative Frequency
18–20	5	5	0.021	0.021
21–23	38	43	0.157	0.178
24–26	73	116	0.302	0.480
27–29	56	172	0.231	0.711
30–32	26	198	0.108	0.819
33–35	19	217	0.079	0.898
36–38	11	228	0.046	0.944
39–41	3	231	0.012	0.956
42–44	3	234	0.012	0.968
45–47	3	237	0.012	0.980
48–50	1	238	0.004	0.984
51–53	2	240	0.008	0.992
54–56	1	241	0.004	0.996
57–59	1	242	0.004	1.000
Total	242		1.000	

3.9.3 Relative frequency. The fourth column in Table 3.9 shows the percentage of men in the sample in each age class. For example, 0.021, that is, 2.1 per cent of the men in the sample married after reaching age eighteen but before reaching age twenty-one. And 0.157, that is, 15.7 per cent of the men married after reaching age twenty-one but before reaching age twenty-four. The numbers in the fourth column are called *relative frequencies*. This is an important term in statistics, and we shall use it often in the remainder of this book.

The relative frequencies are computed by dividing the actually observed frequencies by the total frequency. For example, the first relative frequency is 5/242 = 0.021 and the second relative frequency is 38/242 = 0.157.

The total of the relative frequencies is 1.000 or 100.0 per cent, which means merely that 100.0 per cent of the men in the sample have been taken into consideration in computing the relative frequencies for the respective age classes. Occasionally, the sum of the relative frequencies does not equal exactly 1.000. This may be caused by the rounding off of the decimals in the relative frequency computations. Usually, it is a simple matter to adjust the relative frequencies so that the total is exactly 1.000.

For example, when the relative frequencies in the fourth column of Table 3.9 were computed, their sum turned out at first to be 0.999 instead of 1.000. A check was made to see if a mistake had been made in the computation of the relative frequencies. Then an examination was made of the relative frequencies before they were rounded off to one decimal place to find the one for which the largest decimal fraction had been discarded in the rounding process. It was found that for the age 30-32 class the computation gave 0.10743 as the relative frequency before the rounding off process, and the fraction 0.00043 was the largest fraction discarded in the rounding process for all the relative frequencies. Consequently, the relative frequency was changed from 0.107 to 0.108 so that the total of the relative frequencies would be exactly 1.000.

Of course, it is just as likely to happen that the total of the relative frequencies may turn out to be slightly greater than 1.000 as a result of the rounding process. In that case, you can make a downward adjustment in the relative frequency that was increased in the rounding process but that contained the smallest decimal fraction in the discarded digits.

3.9.4 Cumulated relative frequency. The numbers in the fifth column of Table 3.9 are *cumulated relative frequencies*. For example, the number 0.178 tells us that 17.8 per cent of the men in the sample married before they reached age twenty-four. Similarly, 48.0 per cent of the men married before they reached age twenty-seven. The cumulated relative frequencies are obtained by cumulating, that is, by adding the relative frequencies. Obviously, the last cumulated relative frequency must be exactly 1.000. Figure 3.2 is a graph of the cumulated relative frequencies for the distribution in Table 3.9.

Relative frequencies and cumulated relative frequencies are useful for comparison of the shapes of two frequency distributions in which the class limits are the same for both distributions but the total frequency is not. The reason for this is that if you plot the histograms for two frequency distributions in which the total frequency is not the same for both distributions, the two histograms will not have equal areas. If, however, you use relative frequencies to plot the histograms, the two histograms will have exactly the same total area, namely, 1.00. For example, you might wish to compare the shapes of the two distributions of men or the two distributions of women in Table 3.17, using five-year age intervals for all the ages.

If there are between 100 and 1000 elementary units in the distribution, the figures recorded in the relative frequency column and the cumulated relative frequency column usually are carried to three decimal places. If there are less than 100 elementary units in the distribution, it is customary to round off the relative frequencies and the cumulated relative frequencies to two decimal places. If there are more than 1000 elementary units in the

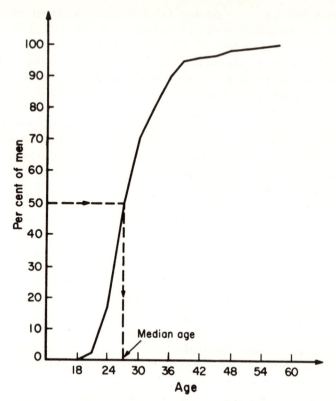

Fig. 3.2. Per Cent of Men Already Married before Reaching Specified Ages in a Sample of the American-Born Men Listed in *Who's Who*

distribution, you might wish to retain four decimal places in the relative frequencies and the cumulated relative frequencies.

It will be demonstrated for you now that there are two different types of frequency distribution of a variable, namely, an ungrouped frequency distribution of a discrete variable and a grouped frequency distribution of any variable, discrete or continuous.

3.10 UNGROUPED FREQUENCY DISTRIBUTION OF A DISCRETE VARIABLE

Table 3.10 is an illustration of an ungrouped frequency distribution of a discrete variable. It was constructed by using the data in Table 1.3, which shows the number of red beads in each of 200 samples of beads. Notice that in Table 3.10 each different possible value of the variable is kept separated from all other possible values of the variable. Other illustrations of ungrouped frequency distributions of discrete variables may be seen in Tables 4.2, 6.1, 6.20, 6.21, 6.22, 6.23, and 7.1.

TABLE 3.10

**Ungrouped Frequency Distribution of the Number of
Red Beads in a Sample of 250 Beads***

Observed Value of the Variable (Number of Red Beads in a Sample)	Tallies	Frequency (Number of Samples)
69	/	1
70		—
71		—
72	/	1
73	/	1
74	/	1
75	/	1
76	//	2
77	//	2
78	////	4
79	///	3
80	////	4
81	//// //	7
82	////	5
83	//// //	7
84	//// ///	8
85	//// ///	8
86	//// ////	10
87	//// ////	9
88	//// ////	10
89	//// ////	10
90	//// //// /	11
91	//// ////	10
92	//// ////	10
93	//// ////	10
94	//// ////	9
95	//// ////	9
96	//// ///	8
97	//// //	7
98	////	5
99	//// /	6
100	////	4
101	////	4
102	///	3
103	///	3
104	//	2
105		—
106	//	2
107	/	1
108		—
109	/	1
110		—
111	/	1
	Total	200

**Source:* Table 1.3.

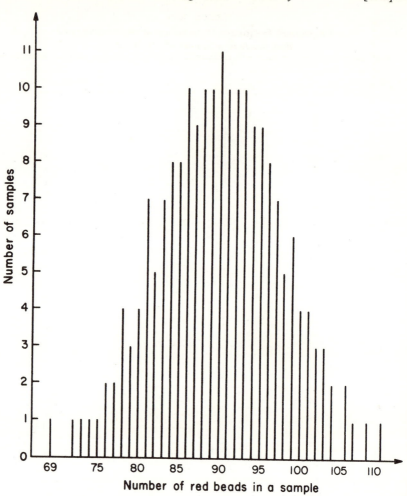

Fig. 3.3. The Number of Red Beads in Each of 200 Samples of 250 Beads from a Universe Containing 1800 Red Beads and 3200 White Beads

3.10.1 The constant step k. If you examine the successive different possible values of the variable in an ungrouped frequency distribution, you will notice that the differences or steps are constant. In Table 3.10, the step or difference between any pair of successive different possible values of the variable is 1. In Table 6.20 the difference or step between any pair of successive different values of the variable is a half-inch. If we let k stand for the constant step between any two successive different possible values of the variable in an ungrouped frequency distribution of a discrete variable, then for Table 4.2, $k = 1$. For Table 6.20, $k = \frac{1}{2}$. In Table 6.21, $k = 1$. If you

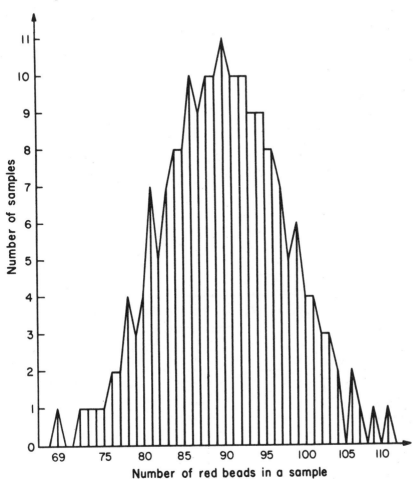

Fig. 3.4. The Number of Red Beads in Each of 200 Samples of 250 Beads from a Universe Containing 1800 Red Beads and 3200 White Beads

construct an ungrouped frequency distribution for the data in Table 1.4 or Table 1.5, you will find that $k = 0.004$.

3.10.2 Graphs for an ungrouped frequency distribution. Figure 3.3 is a graph of the ungrouped frequency distribution in Table 3.10. This graph consists of a *set of ordinates*. Each ordinate represents the frequency with which a particular value of the variable appears in Table 3.10. Consequently, there are as many ordinates in the graph as there are different possible values of the variable in Table 3.10. Sometimes, the tops of the ordinates are joined by line segments to form a graph such as Figure 3.4 which is called a *frequency polygon*.

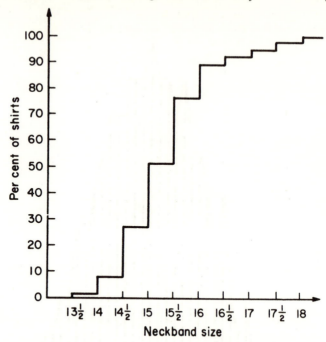

Fig. 3.5. Per Cent of Men's White Shirts That Have Neckband Less Than a Specified Size

If you construct a graph for the cumulated frequencies or the cumulated relative frequencies in an ungrouped frequency distribution, it will be a somewhat irregular "stair-step" type of figure. Figure 3.5 is a graph of the cumulated relative frequency for the distribution in Table 6.20.

3.11 GROUPED FREQUENCY DISTRIBUTIONS

Table 3.8 is an illustration of a grouped frequency distribution. The variable is the age of a man, and age is a continuous variable. In a grouped frequency distribution, it is permissible for several different possible values of the variable to occur within each class or group in the table. For example, in Table 3.8, any integral or fractional number of years from eighteen up to, but not including, twenty-one is a possible age at time of marriage for men in the 18–20 class or group, and all those possible values of the variable have been grouped or lumped together as the first class in the distribution. You can find other examples of grouped frequency distributions in Tables 3.15, 3.16, 6.3, 6.4, 6.9, 6.12, and 6.16.

TABLE 3.11

Number of Red Beads in Each of 200 Samples of 250 Beads from a Universe Containing 1800 Red Beads and 3200 White Beads

Observed Number of Red Beads in Sample	Tallies	Observed Number of Samples (Frequency)
68–70	/	1
71–73	//	2
74–76	////	4
77–79	///// ////	9
80–82	///// ///// ///// /	16
83–85	///// ///// ///// ///// ///	23
86–88	///// ///// ///// ///// ///// ////	29
89–91	///// ///// ///// ///// ///// ///// /	31
92–94	///// ///// ///// ///// ///// ////	29
95–97	///// ///// ///// ///// ////	24
98–100	///// ///// /////	15
101–103	///// /////	10
104–106	////	4
107–109	//	2
110–112	/	1
Total		200

Source: Table 1.3.

A table is said to be a grouped frequency distribution if more than one *possible* value of the variable are grouped together in a single class. Notice that it is the grouping of the different possible values of the variable that produces a grouped frequency distribution.

TABLE 3.12

Proportion of Red Beads in Each of 200 Samples of 250 Beads from a Universe Containing 1800 Red Beads and 3200 White Beads*

Observed Proportion of Red Beads in a Sample	Observed Number of Samples (Frequency)
.272–.280	1
.284–.292	2
.296–.304	4
.308–.316	9
.320–.328	16
.332–.340	23
.344–.352	29
.356–.364	31
.368–.376	29
.380–.388	24
.392–.400	15
.404–.412	10
.416–.424	4
.428–.436	2
.440–.448	1
	Total 200

Source: Table 1.4.

TABLE 3.13

Sampling Errors in Proportion of Red Beads in 200 Samples of 250 Beads from Universe of 1800 Red Beads and 3200 White Beads*

Observed Sampling Error	*Observed Number of Samples (Frequency)*
−.088 to −.080	1
−.076 to −.068	2
−.064 to −.056	4
−.052 to −.044	9
−.040 to −.032	16
−.028 to −.020	23
−.016 to −.008	29
−.004 to .004	31
.008 to .016	29
.020 to .028	24
.032 to .040	15
.044 to .052	10
.056 to .064	4
.068 to .076	2
.080 to .088	1
Total 200	

Source: Table 1.5.

It is important to realize that every frequency distribution of a continuous variable is a grouped frequency distribution. At first glance, you might be tempted to think that Table 3.17, is an ungrouped frequency distribution for ages sixteen to thirty. That is not so, however. Age is a continuous variable and the first class contains all possible values of the variable from sixteen years to 16.999 . . . years, that is, from the first moment of the sixteenth year to the last moment of the sixteenth year. All possible ages from sixteen years up to but not including seventeen years are grouped together in the sixteen-year class.

On the other hand, Table 6.20 is not a grouped frequency distribution. The variable is discrete. There is only one possible value of the neckband size of a shirt in each class. There are several shirts in each class, but all the shirts in any particular class have the same neck size. Tables 3.11, 3.12, and 3.13 are examples of grouped frequency distributions in which the variables are discrete.

3.11.1 Graphs for a grouped frequency distribution. The most common type of graph for a grouped frequency distribution is the *histogram*. Figures 3.6 and 3.7 are histograms for the grouped frequency distributions in Tables 3.12 and 3.13. There are other examples of histograms in this book.

The horizontal width of each rectangle in a histogram is one *class interval*.

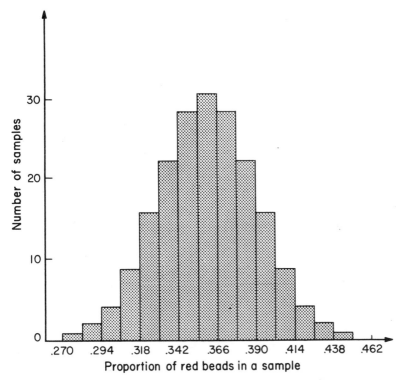

Fig. 3.6. The Proportion of Red Beads in Each of 200 Samples of 250 Beads from a Universe Containing 1800 Red Beads and 3200 White Beads

The vertical height of each rectangle in a histogram is the same as the frequency in the class interval on which the rectangle stands. Consequently, the area of each rectangle in a histogram may be considered to be proportional to the frequency in the class interval on which the rectangle stands. It follows from this that the area of all the rectangles combined is proportional to the total frequency in the distribution.

Sometimes we draw a smooth curve through the tops of the rectangles in a histogram. The smooth curve should be drawn so that the total area under the curve is the same as the total area in the whole histogram. A smooth curve of this type is called a *frequency curve*. Figure 3.8 is an example.

3.12 MEAN AND STANDARD DEVIATION OF A FREQUENCY DISTRIBUTION

Let us find the mean and the standard deviation of the frequency dis-

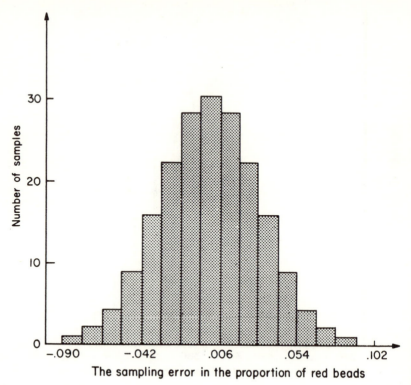

The sampling error in the proportion of red beads

Fig. 3.7. Sampling Errors in Proportion of Red Beads in 200 Samples of 250 Beads from Universe of 1800 Red Beads and 3200 White Beads

tribution in Table 3.8. We can do this by building Table 3.14 which is somewhat similar to Table 3.5, which was used for finding the standard deviation of a serial distribution. First, find the midpoint of each age-class interval and write it in the second column. Use these midpoints as the values of x. Detailed illustrations to guide you in finding midpoints will be found in section A.6.5. Multiply each value of x by the corresponding value of f to obtain the column under xf. Using the total $\Sigma xf = 6957.0$ and $n = \Sigma f = 242$, we find that the mean is 28.75 years.

Now subtract this value of m from each value of x to obtain the numbers in the fifth column of Table 3.14. Multiply each of these numbers by itself and write the results in the sixth column. Then multiply each number in the sixth column by the corresponding number in the frequency column and write the results in column seven. We need the total of column seven to find the variance and the standard deviation. The computations below Table 3.14 show that for the sample in Table 3.8 the variance is 39.3600 (years squared) and the standard deviation is 6.27 years.

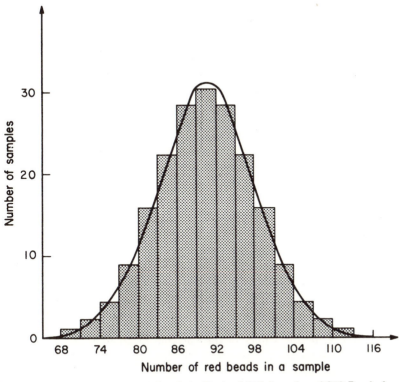

The Number of Red Beads in Each of 200 Samples of 250 Beads from a Universe Containing 1800 Red Beads and 3200 White Beads

TABLE 3.14
Tabulation for Finding the Mean and the Standard
Deviation of the Sample in Table 3.8

Age at Marriage	x	f	xf	$(x-28.75)$	$(x-28.75)^2$	$(x-28.75)^2 f$
18–20	19.5	5	97.5	−9.25	85.5625	427.8125
21–23	22.5	38	855.0	−6.25	39.0625	1484.3750
24–26	25.5	73	1861.5	−3.25	10.5625	771.0625
27–29	28.5	56	1596.0	−0.25	0.0625	3.5000
30–32	31.5	26	819.0	2.75	7.5625	196.6250
33–35	34.5	19	655.5	5.75	33.0625	628.1875
36–38	37.5	11	412.5	8.75	76.5625	842.1875
39–41	40.5	3	121.5	11.75	138.0625	414.1875
42–44	43.5	3	130.5	14.75	217.5625	652.6875
45–47	46.5	3	139.5	17.75	315.0625	945.1875
48–50	49.5	1	49.5	20.75	430.5625	430.5625
51–53	52.5	2	105.0	23.75	564.0625	1128.1250
54–56	55.5	1	55.5	26.75	715.5625	715.5625
57–59	58.5	1	58.5	29.75	885.0625	885.0625
Total		242	6957.0			9525.1250

$$m = \frac{6957.0}{242} = 28.748 = 28.75$$

$$s^2 = \frac{9525.1250}{242} = 39.3600$$

$$s = \sqrt{39.3600} = 6.274 = 6.27$$

If you were asked to look now at formulas 3.1, 3.2, and 3.4 for the mean, the variance, and the standard deviation of any serial distribution and to suggest somewhat similar formulas for the mean, the variance, and the standard deviation of any frequency distribution in a sample, it would not take you long to produce the formulas

$$m_x = \frac{\Sigma xf}{n}, \quad n = \Sigma f \quad \text{(mean)} \tag{3.7}$$

$$s_x^2 = \frac{\Sigma(x - m)^2 f}{n} \quad \text{(variance)}$$

$$s_x = \frac{1}{n} \sqrt{n\Sigma x^2 f - (\Sigma xf)^2} \quad \text{(standard deviation)} \tag{3.8}$$

where x stands for the values of the variable in an ungrouped frequency distribution and for the midpoints of the class intervals in a grouped frequency distribution.

Short cuts that reduce the amount of computation required in determining the mean and the standard deviation of practically every kind of distribution likely to be encountered may be found in sections A.5 and A.6.

3.13 REMARKS

People are always interested in averages. The fact that they are also interested in such things as the range and the rank order in human affairs is evidence that they are groping for knowledge about the variability of the data. The standard deviation gives us an excellent method for describing the variability of distributions. It also furnishes the key to the method for determining the reliability of estimates made from samples. Consequently, the standard deviation is of strategic importance in statistics.

It is a fact that for the most important of all the types of distribution, namely, normal distributions, the mean and the standard deviation of a random sample drawn from the universe contain all the relevant information that the sample can disclose with respect to the distribution in the universe.

Can you imagine anything more neat and powerful than being able to compute two numbers that indicate all the properties of a normal distribution in a universe represented by a random sample?

It is exceedingly important that you make a lasting mental note of the general form of the frequency distributions that you have seen and of the general shape of the histograms, frequency curves, and frequency polygons that you have seen. You will meet the same form of frequency distribution and the same shape of histogram and the same shape of frequency curve so often as you work with statistics that they ought to become as familiar to you as your own right hand.

In this chapter we have taken a big stride forward in understanding and appreciation of the ideas of distribution and probability. From now on, these two ideas should assume an important place in your thinking about situations and events and human problems.

In other words, a transition ought to be taking place now in your way of thinking about the world in which you live, with its infinite variety and continual change that often may have seemed chaotic. You probably were in the habit of thinking in terms of isolated individual persons and things. Now you ought to be forming the habit of seeing patterns, namely, frequency distributions, in the characteristics of people and of events that are of interest to you.

Remember that if there were no variations in things, there could be no statistical science. But remember also that if there were no structural principles—that is, no laws of form, shape, or distribution—among the variations in this world, there could be no statistical science to enable us to draw inferences, with known reliability, about universes by analyzing samples drawn from the universes.

The study of frequency distributions does not end with Chapter 3. It never ends so long as one is studying statistics. Frequency distributions will be your constant companions throughout most of the remainder of this book. It is hoped that the idea of frequency distributions not only will be a good companion for you during the remainder of this course, but that it will be a frequently used idea during the remainder of your life. It should become a part of your very self so that it influences your habitual way of thinking. You will be better off if it does.

EXERCISES

1. How many stenographers are there in the sample in Table 3.15? How many of them had weekly earnings in the range from $59.75 to $64.75? What is the approximate range in the weekly earnings of the stenographers in the distribution?

TABLE 3.15

**Weekly Earnings of General Stenographers
in Chicago in April, 1962***

Weekly Earnings†	Number of Stenographers
$55 and under $60	35
60 and under 65	315
65 and under 70	958
70 and under 75	1,115
75 and under 80	1,565
80 and under 85	1,159
85 and under 90	1,005
90 and under 95	806
95 and under 100	515
100 and under 105	363
105 and under 110	183
110 and under 115	60
115 and under 120	16
120 and under 125	4
125 and under 130	2
Total	8,101

*Source: U. S. Department of Labor, Bureau of Labor
Statistics, Bulletin No. 1303-64, "Occupational Wage
Survey, Chicago, Illinois, April, 1962."

†The weekly earnings of each individual in the sample
were recorded to the nearest half dollar.

2. Construct a histogram for the distribution in Table 3.15 or Table 3.16. To
what number in the table is the total area of the histogram proportional?

TABLE 3.16

**Hourly Wages of Drivers of Light Trucks
in Boston in October, 1962***

Hourly Wages	Number of Truck Drivers	Hourly Wages	Number of Truck Drivers
$1.20 and under $1.30	9	$2.20 and under $2.30	34
1.30 and under 1.40	46	2.30 and under 2.40	12
1.40 and under 1.50	1	2.40 and under 2.50	3
1.50 and under 1.60	22	2.50 and under 2.60	19
1.60 and under 1.70	21	2.60 and under 2.70	5
1.70 and under 1.80	34	2.70 and under 2.80	4
1.80 and under 1.90	38	2.80 and under 2.90	41
1.90 and under 2.00	82	2.90 and under 3.00	—
2.00 and under 2.10	51	3.00 and under 3.10	—
2.10 and under 2.20	79	3.10 and under 3.20	50
		Total	551

*Source: U. S. Department of Labor, Bureau of Labor Statistics, Bulletin No.
1345-15, "Occupational Wage Survey, Boston, Massachusetts, October, 1962." Light
trucks are trucks with less than one and one-half tons capacity.

3. Construct a table showing the relative frequencies and the cumulated relative frequencies for the distribution in Table 3.15 or Table 3.16.

4. Construct a graph of the cumulated relative frequencies that you computed in exercise 3.

5. Draw a smooth frequency curve over the histogram that you constructed in exercise 2.

6. Compute and tabulate the relative frequencies and the cumulated relative frequencies for the distributions in Tables 6.3 and 6.4.

7. Compute and tabulate the relative frequencies and the cumulated relative frequencies for the distributions in Tables 6.20, 6.21, and 6.22.

8. Construct graphs for the relative frequencies and for the cumulated relative frequencies of the distributions in Tables 6.3 and 6.4.

9. Construct graphs for the relative frequencies and for the cumulated relative frequencies of the distributions in Tables 6.20, 6.21, and 6.22.

10. Draw smooth curves through the graphs that you constructed in problems 4 and 8. The curves for most cumulated relative frequency distributions are of the shape called an *ogive*.

11. What is the value of the constant step k in each of the distributions in Tables 3.10, 4.2, 6.20, 6.21, 6.22, and 7.1. (*Ans.* 1, 1, ½, 1, ½, 1)

12. Compute the mean and the standard deviation for the distribution in Table 3.15 or 3.16. Use the short method described in section A.6.16 (p. 381). (*Ans.* Table 3.15, $m = \$81.60$, $s = \$11.52$; Table 3.16, $m = \$2.12$, $s = \$0.50$) Use the coefficients of variation to compare the variabilities of the samples in Tables 3.15 and 3.16. (*Ans.* 14.1 per cent and 23.6 per cent) What is your interpretation of the results?

13. Rewrite Table 3.17 using five-year class intervals for all the ages. Then compute the mean and the standard deviation for each of the four distributions. (*Ans.* $m_1 = 29.7$, $m_2 = 27.3$, $m_3 = 27.2$, $m_4 = 24.7$, $s_1 = 7.99$, $s_2 = 6.91$, $s_3 = 6.84$, $s_4 = 6.49$) Use the coefficients of variation to compare the four distributions in Table 3.17. What is your interpretation of these results?

14. Use the means, standard deviations, and coefficients of variation to compare the distributions of men in Table 3.17 with the distribution of men in Table 3.8.

15. Compute the standard score for a man in Table 3.8 who married on his eighteenth birthday. (*Ans.* $z = -1.71$) Compute the standard score for a man in each of the two distributions of men in Table 3.17 who married on his eighteenth birthday. (*Ans.* skilled man, $z = -1.46$; unskilled man, $z = -1.35$) Is it more unusual for a man in Table 3.8 to marry at exactly eighteen years of age than it is for the man in Table 3.17 to marry at exactly age eighteen? (*Ans.* Yes, because $z = -1.71$ represents a greater deviation from the mean than $z = -1.46$ or $z = -1.35$)

16. If all the values of the variable in a sample that has a mean of 65 and standard deviation of 5 are changed into standard units, the mean of the distribution of standard scores will be _____ and the standard deviation of the distribution of standard scores will be _____.

17. If the mean of a distribution is 80 and the standard deviation of the distribution is 15, which of the following is the standard score corresponding to the value 65 of the variable? (a) $z = -1.96$ ☐   (b) $z = -2.58$ ☐   (c) $z = -1.00$ ☐ (d) $z = +1.00$ ☐   (e) $z = +2.58$ ☐   (f) $z = +1.96$ ☐.

TABLE 3.17

**Age at Marriage of Skilled and Unskilled Workers
Who Married in York, England in 1936***

Age at Marriage†	Skilled Workers		Unskilled Workers	
	Males	*Females*	*Males*	*Females*
16	—	1	—	5
17	—	2	1	9
18	—	4	3	18
19	4	14	7	36
20	4	15	20	46
21	6	26	27	49
22	19	36	33	49
23	33	41	50	43
24	43	47	45	28
25	35	45	38	24
26	56	35	39	23
27	51	36	25	20
28	37	31	21	14
29	20	13	18	6
30	20	14	20	5
31–35	51	31	42	19
36–40	17	19	14	10
41–45	17	9	6	9
46–50	3	9	5	4
51–55	3	1	—	2
56–60	9	3	2	—
61–65	5	1	4	1
Total	433	433	420	420

**Source:* B. Seebohm, Rowntree, *Poverty and Progress.* London: Longmans, Green and Co., 1941, Appendix U, p. 527. Reproduced by permission of the author.

†That is, age on last birthday.

Chapter 4

BINOMIAL DISTRIBUTIONS AND BINOMIAL PROBABILITIES

4.1 PROBABILITY

Usually, the ends of the field to be taken by the teams playing in a football game are determined by the toss of a coin. Why is this considered a fair way to decide such a matter? It is fair because each captain has an equal chance of winning the toss. It is just as likely that the coin will turn up "heads" as that "tails" will show.

In other words, the probability or the chance of the coin showing "heads" is the same as the probability or the chance of the coin showing "tails"; the odds are one-to-one. To state this another way, if an unbiased coin is tossed a large number of times, we expect that it will turn up "heads" approximately half of the time and "tails" approximately half of the time.

However, if someone produced a tossing coin and you and he tossed the coin one hundred times and it turned up "heads" only four times, most likely you would consider that coin biased or unfair for use in a tossing game. You might feel that if you were to continue to play with that coin, you would win on "heads" only about 4 per cent of the time and your opponent would win on "tails" about 96 per cent of the time. Notice that $0.04 + 0.96 = 1$.

In other words, we tend to interpret the relative frequency with which many events have occurred in the past as the probability that they will occur under similar circumstances in the future.

A die is another common playing device. It is a cube and, therefore, has six sides or faces. On the faces are one, two, three, four, five, and six dots. The probability that a chosen face will turn up if an unbiased die is tossed is $\frac{1}{6}$. The probability that the chosen face will not turn up is $\frac{5}{6}$. Notice that $\frac{1}{6} + \frac{5}{6} = 1$.

Several definitions of probability are available, and some of them involve quite sophisticated mathematics. All we need here is a simple, working definition such as many good statisticians use in their practical work. *The probability of an event's occurring is the expected proportion of times on which the event will happen, as indicated by all the available relevant information.*

Probabilities must be expressed in the form of numbers if they are to be suitable for statistical application. Relative frequencies accomplish this for us. If in n observations of an event it turns out favorably on a occasions and unfavorably on the remaining $n - a$ occasions, then the relative frequency of success is a/n and the relative frequency of failure is $(n - a)/n$. The statistical probability of success in the event is the same as the relative frequency of success. The statistical probability of failure in the event is the same as the relative frequency of failure.

Because the relative frequency, that is, the proportion of the times that the event turns out in a certain way, cannot be less than zero or more than 1.00 (that is, 100 per cent), a probability cannot be less than zero or more than 1.00. We usually think of zero probability as representing those results we expect never to happen. And we associate the probability 1.00 with results we expect to happen every time there is a trial or opportunity. All other probabilities have numerical values between 0 and 1.

Notice that, according to our definition, the probability associated with an event might change from time to time as additional relevant information about the relative frequency of occurrence of the event becomes available. That is not a serious practical shortcoming of our definition. For example, the United States Government revises the data in Table 4.1 periodically as more recent information becomes available. Furthermore, separate tabulations are available for white males, white females, Negro males, and Negro females because the relevant information indicates that no two of these four groups have the same life expectancies.

On the basis of life expectancy tables life insurance companies calculate how much to charge as premium on each policy. They tabulate the birth and the death dates for a large group of people. For example, Table 4.1 shows the life history of 100,000 people. Notice that out of the 100,000 people born alive, 97,024 were alive at the end of the first year or (the same thing) at the beginning of the second year.

In other words, 2976 of the original 100,000 people died during the first year of life. Using this information, we may say that the statistical probability that a randomly selected baby born alive will die before reaching one year of age is $2976/100,000 = 0.02976$. The statistical probability that a randomly selected baby born alive will be alive one year after birth is $97,024/100,000 = 0.97024$.

TABLE 4.1

Life Table for the Total Population of
the United States, 1949–1951*

Year of Age	No. of Survivors at Beginning of Year of Age Out of 100,000 Born Alive	Year of Age	No. of Survivors at Beginning of Year of Age Out of 100,000 Born Alive
0	100,000	22	95,103
1	97,024	23	94,963
2	96,801	24	94,820
3	96,667	25	94,676
4	96,565	30	93,919
5	96,482	35	92,976
6	96,408	40	91,648
7	96,342	45	89,634
8	96,283	50	86,591
9	96,229	55	82,176
10	96,177	60	75,921
11	96,127	65	67,555
12	96,075	70	56,987
13	96,019	75	43,903
14	95,957	80	29,313
15	95,885	85	15,785
16	95,801	90	6,144
17	95,706	95	1,511
18	95,601	100	199
19	95,487	105	12
20	95,366	109	1
21	95,238	110	0

Source: U. S. Department of Health, Education and Welfare, National Office of Vital Statistics, *Vital Statistics — Special Reports*, 41, No. 1 (November 23, 1954).

Notice that the sum of the probability of a person's dying during the first year of life and the probability of a person's being alive at the end of the first year of life is 1. That is, $0.02976 + 0.97024 = 1$.

Suppose that today is your seventeenth birthday. From the point of view of an insurance company, what is the statistical probability that you will live at least to age seventy? According to Table 4.1, the statistical probability is $56,987/95,706 = 0.59544$ that a randomly selected person of age seventeen will be alive at age seventy.

The statistical probability that a randomly selected person of age seventeen will not be alive at age seventy may be found either (1) by adding the number of deaths during all the years from seventeen through sixty-nine and dividing by 95,706, or (2) by subtracting the number of survivors at age seventy from the number alive at age seventeen and dividing by 95,706, or (3) most easily, by subtracting 0.59544 from 1. The probability that a randomly selected person of age seventeen will die before reaching the age of seventy is $1 - 0.59544 = 0.40456$.

Do you see now why your attention has been invited several times to the fact that, whenever an event can happen only in either of two ways, the two probabilities together amount to 1? If one of the two probabilities of this kind is known, the easiest way to find the other probability is by subtracting the known probability from 1.

4.1.1 First rule of probability. If P is the probability that an event will happen in a single trial and if Q is the probability that the event will not happen in a single trial, then $P + Q = 1$. This means simply that, in a single trial, the event is certain to happen in one of the two ways. For example, if we toss a coin we are certain that it will fall in one of two ways, namely, "heads" or "tails," but we are not certain which of the two ways will happen.

The equation $P + Q = 1$ is called the first rule of probability. Frequently, it is convenient to write the equation in the form $P = 1 - Q$ or in the form $Q = 1 - P$.

4.1.2 Second rule of probability. If we toss a coin, it cannot turn up "heads" and "tails" at the same time. Similarly, a die cannot show two numbers facing upward at the same time. Such events are said to be *mutually exclusive*. The probability of the happening of one or another of two or more mutually exclusive events is the sum of the probabilities of the separate events. This is the second rule of probability.

For example, the probability that a tossed coin will show "heads" is ½; the probability that it will show "tails" is ½; and these are mutually exclusive events. Therefore, the probability that the tossed coin will show either "heads" or "tails" is $½ + ½ = 1$.

Likewise, the probability that a die will show "one" in a single trial is ⅙. The probability that the die will show "two" in the same trial is ⅙. These are mutually exclusive events. Consequently, the probability that the die will show either "one" or "two" in a single trial is $⅙ + ⅙ = ⅓$. Similarly, the probability that a die will show either "one" or "two" or "five" in a single trial is $⅙ + ⅙ + ⅙ = ½$.

4.1.3 Third rule of probability. The way in which a "true" coin falls in one fair trial has no effect on the way in which it falls in subsequent tosses of the coin. Similarly, the fact that a "true" die turns up "six" in one fair trial does not affect the next trial. Such events are said to be *independent*. The probability of the happening of two or more independent events is the product of the separate probabilities of the events. This is the third rule of probability.

For example, the probability that a "true" coin will turn up "heads" in the first toss and "heads" in the second toss is $(\frac{1}{2})(\frac{1}{2}) = \frac{1}{4}$. Likewise, the probability that a "true" die will turn up "one" in the first trial, "two" in the second trial, and "six" in the third trial is $(\frac{1}{6})(\frac{1}{6})(\frac{1}{6}) = 1/216$.

Similarly, in a single trial consisting of tossing a "true" coin and a "true" die, the probability that the coin will show "heads" and that the die will show "four" is $(\frac{1}{2})(\frac{1}{6}) = 1/12$. That this is the correct probability is easily seen if you consider that there are twelve possible ways in which a coin and a die can be arranged; any one of the six faces of the die may accompany "heads" of the coin, and any of the same six faces of the die may accompany "tails." Only one of the twelve possible ways is favorable to the demand that the coin show "heads" and the die show "four." Therefore, $P = 1/12$.

Do not be too quick to change your mind about the odds connected with an event because it happened in a certain way several times. For example, history shows that nearly the same number of boys as girls are born. Suppose that in your community, ten of the first eleven babies born in the month of March are girls. What is the probability that the next baby born in that community will be a girl? The correct answer is approximately $\frac{1}{2}$. Likewise, if a "true" die is rolled four times fairly and turns up "one" all four times, the probability that it will show "one" in the next fair trial is still $\frac{1}{6}$, neither more nor less.

4.1.4 The law of large numbers. People frequently speak of the law of averages as operating in connection with chance events. A better name, perhaps, is the law of large numbers. The law of large numbers refers to the relative frequency of the occurrence of an event in a large number of trials. The law of large numbers means that the relative frequency will in the long run come close to and remain close to the actual probability that should be associated with the event. For example, if an unbiased— that is, perfectly symmetrical—coin is tossed fairly a very, very large number of times we can expect that the relative frequency with which it turns up "heads" will be very, very close to $\frac{1}{2}$. Furthermore, by making the number of trials larger and larger, we can make the relative frequency come close to and remain close to $\frac{1}{2}$.

Try to understand clearly why it is safe for a life insurance company to use a life table, which is equivalent to a probability table, as the basis for its business, whereas it would be foolish for you or any other specific individual to depend upon the same life table for an accurate indication of exactly how long a specific person will live.

The reason is, of course, because the life table indicates only the average length of life for a very large group of people at each year of age. Because a life insurance company insures a large number of people and bases its premiums upon a life table, that is, upon the averages of very large groups, it can depend safely upon the assumption that the average lengths of life for its policyholders will not depart very far from the averages indicated by the life table.

Thus you see that a life insurance company is protected by the law of large numbers. In other words, the life insurance company is safe in basing its business on the life table because of the high degree of reliability that is a property of the means of large random samples. Some of the policyholders may die before reaching the average ages expected, but others will live longer than the average ages expected; the effect of those who live longer than the expected averages is to cancel the effect of those who die before reaching the expected averages.

On the other hand, because a specific person is only one elementary unit in a large distribution, we know from all the distributions that have been used as illustrations in this book that it is not surprising if one elementary unit selected at random from a large distribution deviates very far from the mean of the distribution.

In technical language of statistics, we can state the above principle as follows: Confidence limits for the mean of a large random sample can be determined by using the mean of the universe and allowing for a small amount of sampling error or error of estimation in the sample mean. If we try to make a similar estimate for one elementary unit selected at random from the same universe, we must allow for a relatively large amount of sampling error or error of estimation. Formula 7.2 and the discussion in section 7.2.1 show that the allowance for sampling error or error of estimation for one elementary unit in a distribution is $\sqrt{n}$ times larger than the allowance for sampling error or error of estimation in the mean of a random sample of n elementary units drawn from the distribution.

EXERCISES

1. Use Table 4.1 to find the probability that a randomly selected person whose age is nineteen will be alive at age forty-five. (*Ans.* 0.93870) What is the probability that that person will die before reaching age forty-five? (*Ans.* 0.06130)

2. What is the probability that two randomly selected people, both of whom are twenty years of age, will be alive at age forty-five? (*Ans.* 0.88340) What is the probability that neither of the two persons will be alive at age forty-five? (*Ans.* 0.00361)

3. What is the probability that one or the other but not both of the two twenty-year-old people in exercise 2 will be alive at age forty-five? (*Ans.* 0.11299)

4. Assuming that Table A.15 is a perfect table of random digits, what is the probability that a digit drawn at random from that table will be 6? (*Ans.* 1/10)

5. What is the probability that a digit drawn at random from Table A.15 will be 3 and that the digit immediately next on the right (or the left) also will be 3? (*Ans.* 1/100)

6. What is the probability that a digit drawn at random from Table A.15 will be 7 and that the digit immediately next on the right (or the left) will not be 7? (*Ans.* 9/100)

7. What is the probability that a digit drawn at random from Table A.15 will be either 5 or 9? (*Ans.* 1/5)

8. What is the probability that a digit drawn at random from Table A.15 will be neither 5 nor 9? (*Ans.* 4/5)

9. Suppose that during a war there is one submarine stationed in a North Atlantic shipping lane watching for merchant ships. The submarine will not sight every ship that passes through the shipping lane. Moreover, some torpedoes miss their targets. Furthermore, frequently a torpedo hit damages a ship but does not sink it.

If the submarine sights 63 per cent of the ships that use the shipping lane, and if 13 per cent of the torpedoes fired by the submarine at sighted ships hit the target, and if one-third of the ships hit by a single torpedo sink, what is the probability that a particular unescorted merchant ship passing through the shipping lane will not be sunk if the submarine fires one torpedo at each ship that it sights. (*Ans.* $P = 0.973$)

10. Suppose that there are six black beads and four white beads in a bag. Before any beads are drawn from the bag, state the following probabilities: (1) That the first bead drawn will be black. (*Ans.* 6/10) (2) That the second bead drawn will be black if the first bead drawn is not replaced before the second bead is drawn. (*Ans.* 6/10) (3) That the third bead drawn will be black if neither of the first two beads drawn is replaced before the third bead is drawn. (*Ans.* 6/10)

4.2 ILLUSTRATION OF THE BINOMIAL PROBABILITY LAW

Let us consider the probability of winning in a game called "twenty-six." In this game, the player places his bet with the banker, chooses one of the faces, say "five," of a die, shakes ten dice in a box, and rolls them onto the table. The number of successes, that is, the number of dice that show the face that he chose (in this case, "five") is counted and recorded on a sheet of paper. Then the player repeats the rolling twelve times, making, in all, thirteen trials.

If the total number of successes in the thirteen trials is twenty-six or more, the banker pays the player four times the amount that he deposited before the game started. If the total number of successes is less than twenty-six but more than eleven, the player loses his deposit. If the total number of successes in the thirteen trials is thirty-three or more, the player wins eight times his deposit. If the total number of successes is thirty-six or more, the banker pays the player twelve times as much as he deposited. If the total number of successes is eleven or less, the banker pays the player four times as much as he deposited.

The probability that in a single trial with one die the "five" will turn up is ⅙. Consequently, the expected or average number of "fives" turning up in 13 throws of 10 dice (or, the same thing, in 130 throws of one die) is

$130(\frac{1}{6}) = 21\frac{2}{3}$. This means that if you play the game of "twenty-six" a large number of times you will average about $21\frac{2}{3}$ successes per game.

All of the probabilities and odds involved in the game of "twenty-six" can be computed exactly by using the binomial expansion $(\frac{5}{6} + \frac{1}{6})^{130}$. It would be necessary to write out the result of multiplying $(\frac{5}{6} + \frac{1}{6})$ by itself 130 times. That would be a very laborious task. You will be shown later in this chapter how to find excellent approximations to these binomial probabilities and odds with very little labor by using the normal probability table.

4.2.1 The binomial distribution $(\frac{5}{6} + \frac{1}{6})^{10}$.
If you multiply $(\frac{5}{6} + \frac{1}{6})$ by itself ten times, the result is

$$(\tfrac{5}{6} + \tfrac{1}{6})^{10} = (\tfrac{5}{6})^{10} + \frac{10}{1}\,(\tfrac{5}{6})^9(\tfrac{1}{6})^1 + \frac{10(9)}{1(2)}\,(\tfrac{5}{6})^8(\tfrac{1}{6})^2$$

$$+ \frac{10(9)(8)}{1(2)(3)}\,(\tfrac{5}{6})^7(\tfrac{1}{6})^3 + \frac{10(9)(8)(7)}{1(2)(3)(4)}\,(\tfrac{5}{6})^6(\tfrac{1}{6})^4$$

$$+ \frac{10(9)(8)(7)(6)}{1(2)(3)(4)(5)}\,(\tfrac{5}{6})^5(\tfrac{1}{6})^5$$

$$+ \frac{10(9)(8)(7)}{1(2)(3)(4)}\,(\tfrac{5}{6})^4(\tfrac{1}{6})^6 + \frac{10(9)(8)}{1(2)(3)}\,(\tfrac{5}{6})^3(\tfrac{1}{6})^7$$

$$+ \frac{10(9)}{1(2)}\,(\tfrac{5}{6})^2(\tfrac{1}{6})^8 + \frac{10}{1}\,(\tfrac{5}{6})^1(\tfrac{1}{6})^9 + (\tfrac{1}{6})^{10}$$

There are eleven terms in the expansion of $(\frac{5}{6} + \frac{1}{6})^{10}$. Each term tells us the probability of obtaining a specific number of successes in a single throw of ten dice. For example, the first term tells us that the probability of obtaining zero successes in a single throw of ten dice is $(\frac{5}{6})^{10}$, which is exactly the answer that we obtain by applying the third rule of probability to this situation.

The second term of the binomial expansion tells us that the probability of obtaining 9 failures and 1 success in a single throw of ten dice is $(\frac{10}{1})(\frac{5}{6})^9(\frac{1}{6})^1$. Similarly, the third term tells us the probability of obtaining exactly 8 failures and 2 successes in a single throw of ten dice.

Notice that the number of failures and the number of successes in a single throw of ten dice correspond exactly to the exponents in the various terms. For example, the fourth term tells us the probability that in a single throw of ten dice the result will be exactly 7 failures and 3 successes. Likewise, the last term of the bionmial expansion tells us that the probability that all ten dice will be successes in a single throw of ten dice is $(\frac{1}{6})^{10}$, and this is the same answer that we would obtain by applying the third rule of probability to this situation.

What is the probability of obtaining two or more successes in a single throw of ten dice? The second law of probability tells us that we will find the answer by summing the last nine terms of the binomial expansion.

Here is a way to evaluate each of the terms of this binomial expansion. First, notice that the denominator of each term contains the factor 6^{10}. Therefore, compute $6^{10} = 60,466,176$. The numerators of the terms contain powers of 5 from 1 to 10. Build a table of the first 10 powers of 5. Then, simplify the terms. The result is

$$\left(\frac{5}{6} + \frac{1}{6}\right)^{10} = \frac{1}{60,466,176} \begin{array}{l} (9,765,625 + 19,531,250 + 17,578,125 + \\ 9,375,000 + 3,281,250 + 787,500 + 131,250 \\ + 15,000 + 1,125 + 50 + 1) \end{array}$$

The sum of the last nine terms will tell us the probability of obtaining at least two successes in a single throw of ten dice. This sum is $31,169,301/60,466,176$.

But, remember that because we always have $P + Q = 1$, and, therefore, $P = 1 - Q$, we can find the desired probability of obtaining at least two successes in a single throw of ten dice by adding the first two terms and subtracting the sum from 1. The first two terms tell us the probability of obtaining less than two successes in a throw. The sum of these two terms is $29,296,875/60,466,176$. Then, $1 - 29,296,875/60,466,176 = 31,169,301/60,466,176$. Other probabilities could be computed in the same way.

You are familiar already with the idea of relative frequency. The probabilities that are determined by the terms of the binomial expansion may be thought of as theoretical relative frequencies, that is, relative frequencies in the universe of repetitions of the throw, corresponding to the different

TABLE 4.2

Possible Numbers of Successes in a Single Throw of Ten Dice, and Their Relative Frequencies

No. of Successes x	Relative Frequency f
0	9,765,625/60,466,176
1	19,531,250/60,466,176
2	17,578,125/60,466,176
3	9,375,000/60,466,176
4	3,281,250/60,466,176
5	787,500/60,466,176
6	131,250/60,466,176
7	15,000/60,466,176
8	1,125/60,466,176
9	50/60,466,176
10	1/60,466,176
Total	60,466,176/60,466,176 = 1

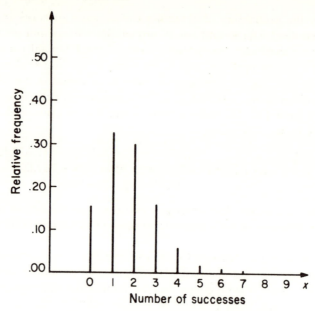

Fig. 4.1.   The Binomial Distribution $(\frac{5}{6} + \frac{1}{6})^{10}$

numbers of successes that are possible in a single throw of ten dice. Table 4.2 shows all the numbers of successes that are possible in a single throw of ten dice and the relative frequencies with which they are likely to occur.

Figure 4.1 is a graph of the distribution in Table 4.2. The relative frequencies are represented by the ordinates in the graph.

The mean or average number of successes per throw in the universe of repetitions of a single throw of ten dice can be found from Table 4.3 by formula 3.7. $N = \Sigma f = 1$ here, because we are dealing with relative frequencies expressed as proportions.

$$\mu_x = \frac{1\%}{1} = \frac{10}{6} = 10\left(\frac{1}{6}\right)$$

In other words, the mean number of successes per throw to be expected in a large number of repetitions of a single throw of ten dice is equal to the number of dice (10) multiplied by the probability that the chosen face will turn up in a single throw or trial of one die ($\frac{1}{6}$).

The standard deviation of the distribution of successes in the universe of repetitions of a single throw of ten dice can be found by formula 3.8.

$$\sigma_x = \frac{1}{1}\sqrt{(1)(25/6) - (10/6)^2} = \sqrt{\frac{50}{36}} = \sqrt{10\left(\frac{1}{6}\right)\left(\frac{5}{6}\right)}$$

Notice that the theoretical standard deviation turns out to be the

TABLE 4.3

**Tabulation for Finding the Mean and the Standard Deviation of
the Number of Successes in a Single Throw of Ten Dice**

No. of Successes x	Relative Frequency f	xf	x²f
0	9,765,625/60,466,176	0	0
1	19,531,250/60,466,176	19,531,250/60,466,176	19,531,250/60,466,176
2	17,578,125/60,466,176	35,156,250/60,466,176	70,312,500/60,466,176
3	9,375,000/60,466,176	28,125,000/60,466,176	84,375,000/60,466,176
4	3,281,250/60,466,176	13,125,000/60,466,176	52,500,000/60,466,176
5	787,500/60,466,176	3,937,500/60,466,176	19,687,500/60,466,176
6	131,250/60,466,176	787,500/60,466,176	4,725,000/60,466,176
7	15,000/60,466,176	105,000/60,466,176	735,000/60,466,176
8	1,125/60,466,176	9,000/60,466,176	72,000/60,466,176
9	50/60,466,176	450/60,466,176	4,050/60,466,176
10	1/60,466,176	10/60,466,176	100/60,466,176
Total	60,466,176/60,466,176	100,776,960/60,466,176	251,942,400/60,466,176
	$\Sigma f = 1$	$\Sigma xf = 10/6$	$\Sigma x^2 f = 25/6$

square root of the product of the number of dice, 10; the probability of success in a single trial or throw of one die, $\frac{1}{6}$; and the probability of failure in a single trial or throw of one die, $\frac{5}{6}$.

If we translate into algebra the process that we have just completed in the illustration involving the binomial distribution $(\frac{5}{6} + \frac{1}{6})^{10}$ we will obtain general formulas by means of which we can write without tedious computations the theoretical mean and the theoretical standard deviation of any binomial distribution.

4.3 THE BINOMIAL DISTRIBUTION $(Q + P)^n$

Let the probability of success in a single trial be P and let the probability of failure in a single trial be Q. Then $P + Q = 1$. Let n be the number of trials. Then, following the same steps that we used in the illustration given in section 4.2.1, the distribution of successes is the binomial $(Q + P)^n$. The expansion of this binomial is

$$(Q + P)^n = Q^n + \frac{n}{1} Q^{n-1}P + \frac{n(n-1)}{1(2)} Q^{n-2}P^2 + \frac{n(n-1)(n-2)}{1(2)(3)} Q^{n-3}P^3 + \cdots$$

$$+ \frac{n(n-1)(n-2)\cdots(n-k+1)}{1(2)(3)\cdots(k)} Q^{n-k}P^k + \cdots + P^n \qquad (4.1)$$

Each term of formula 4.1 tells us the probability of obtaining the number of failures and the number of successes indicated by the exponents of Q and P, respectively, in the term. The general term is an expression for the probability that the result of n trials will be $n - k$ failures and k successes. Notice that the sum of the exponents in each term of formula 4.1 is n.

In order to find the mean and the standard deviation of the number of successes in the binomial distribution represented by equation 4.1, we need to construct a table similar to Table 4.3. Therefore, we construct Table 4.4.

TABLE 4.4

Tabulation for Finding the General Formulas for the Mean and the Standard Deviation of the Binomial Distribution $(Q + P)^n$

No. of Successes x	Relative frequency f	xf	x^2f
0	Q^n	0	0
1	$nQ^{n-1}P$	$nQ^{n-1}P$	$nQ^{n-1}P$
2	$\dfrac{n(n-1)}{1(2)}Q^{n-2}P^2$	$n(n-1)Q^{n-2}P^2$	$2n(n-1)Q^{n-2}P^2$
3	$\dfrac{n(n-1)(n-2)}{1(2)(3)}Q^{n-3}P^3$	$\dfrac{n(n-1)(n-2)}{1(2)}Q^{n-3}P^3$	$\dfrac{3n(n-1)(n-2)}{1(2)}Q^{n-3}P^3$
. . .	. . .	. . .	. . .
k	$\dfrac{n(n-1)...(n-k+1)}{1(2)(3)...(k)}Q^{n-k}P^k$	$\dfrac{n(n-1)...(n-k+1)}{1(2)(3)...(k-1)}Q^{n-k}P^k$	$\dfrac{kn(n-1)...(n-k+1)}{1(2)(3)...(k-1)}Q^{n-k}P^k$
. . .	. . .	. . .	. . .
n	P^n	nP^n	n^2P^n
Total	$\Sigma f = (Q + P)^n$ $= 1$ because $(Q+P)=1$	$\Sigma xf = nP(Q + P)^{n-1}$ $= nP$	$\Sigma x^2f = nP[(Q+P)^{n-1}$ $+ (n-1)P(Q+P)^{n-2}]$ $= nP[1+(n-1)P]$

In Table 4.4, the possible numbers of successes in n trials are written in the first column. In the second column, opposite each possible number of successes, is written the term of the formula 4.1 that tells us the relative frequency or probability of obtaining that particular number of successes in n trials. The third column is formed by multiplying the values of x by the corresponding relative frequencies. The fourth column is formed by multiplying each quantity in the third column by the corresponding number in the first column.

4.3.1 Mean and standard deviation of binomial distribution. The mean number of successes and the standard deviation of the number of successes in the binomial distribution represented by $(Q + P)^n$ can be determined from Table 4.4 by formulas 3.7 and 3.8.

$$\mu_x = \frac{nP}{1} = nP \text{ because } \Sigma xf = nP \text{ and } \Sigma f = 1 \text{ here} \tag{4.2}$$

$$\sigma_x = \frac{1}{1}\sqrt{(1)(nP)\ [1 + (n-1)P] - (nP)^2}$$
$$= \sqrt{nP + n^2P^2 - nP^2 - n^2P^2} = \sqrt{nP - nP^2} = \sqrt{nP(1-P)}$$
$$= \sqrt{nPQ} \tag{4.3}$$

Similarly, it can be proved that the mean number of failures in the binomial distribution represented by $(Q + P)^n$ is $\mu_{n-x} = nQ$.

Also, the standard deviation of the number of failures in the binomial distribution represented by $(Q + P)^n$ is $\sigma_{n-x} = \sqrt{nPQ}$. In other words, the standard deviation of the number of failures in the binomial distribution is identical in value with the standard deviation of the number of successes in the binomial distribution.

In practical statistical situations, we often need to know the mean and the standard deviation of the proportion p of successes rather than the mean and the standard deviation of the number x of successes. All that we need to do is change x to p in the column headings of Table 4.4, replace the numbers of successes in column one by the proportions of successes, namely, $0/n$, $1/n$, $2/n$, ..., k/n, ..., n/n, and adjust the products and the totals in columns three and four accordingly. The relative frequencies in column two remain unchanged and the total of that column remains unchanged.

Now, because we have divided every number in column one by n, every product in column three must be divided by n. Consequently, the total of column three must be divided by n, giving the new total P. Similarly, every product in column four must be divided by n^2. Consequently, the total of column four must be divided by n^2, giving the new total $P[1 + (n-1)P]/n$.

By substituting the new totals into formulas 3.7 and 3.8, we find that in the binomial distribution represented by $(Q + P)^n$ the theoretical means of the proportion p of successes and the proportion q of failures are, respectively,

$$\mu_p = P, \quad \mu_q = Q \tag{4.4}$$

and the theoretical standard deviation of the proportion p of successes and the proportion q of failures is

$$\sigma_p = \sigma_q = \sqrt{\frac{PQ}{n}} \tag{4.5}$$

EXERCISES

1. Consider a true–false type of test. Assume that if you do not know the answer to a question, the probability is ½ that you will give the correct answer if you make a guess. Suppose that on the test there are ten questions for which you guess the answers. Use the binomial distribution $(\frac{1}{2} + \frac{1}{2})^{10}$ to determine the relative frequencies with which a person would be expected to give the different possible numbers of correct answers in such a situation. Your solution to this problem ought to be a table similar to Table 4.2.

2. Determine from the table that you constructed in exercise 1 how much more likely it is that you will make four correct guesses than it is that you will make only two correct guesses. (*Ans.* 210/1024 as compared with 45/1024)

3. What is the probability that you will make exactly three correct guesses in the situation described in exercise 1? (*Ans.* 120/1024) Exactly seven correct guesses? Exactly ten correct guesses? Exactly ten incorrect guesses?

4. In a multiple-choice type of test in which there are three choices of answers for each question, assume that the probability of giving a correct answer by guessing is $\frac{1}{3}$ and the probability of giving an incorrect answer by guessing is $\frac{2}{3}$. Construct a table showing the distribution of relative frequencies of the possible numbers of correct guesses if there are eight questions for which you guess the answers.

5. On the average, in situations such as that described in exercise 1, how many correct answers would you expect a person to make by guessing? (*Ans.* 5) What would be the standard deviation of the distribution of the number of successes for such situations? (*Ans.* $\sqrt{2.5} = 1.58$)

6. Find the mean proportion of successes to be expected in situations such as that described in exercise 1. (*Ans.* $\frac{1}{2}$) Find the standard deviation of the proportion of successes for situations such as that described in exercise 1. (*Ans.* $\sqrt{10}/20 = 0.158$)

7. Find the mean number of successes to be expected in situations such as that described in exercise 4. (*Ans.* 8/3) Find the standard deviation of the distribution of the number of successes in situations such as that described in exercise 4. (*Ans.* 4/3)

8. Find the mean proportion of successes to be expected in situations such as that described in exercise 4. (*Ans.* $\frac{1}{3}$) Find the standard deviation of the distribution of the proportion of successes in situations such as that described in exercise 4. (*Ans.* 1/6)

9. What per cent of the distribution in Table 4.2 corresponds to less than two successes? (*Ans.* 48.5 per cent) To less than two failures? (*Ans.* 0.000084 per cent) To less than three successes? (*Ans.* 77.5 per cent) To less than three failures? (*Ans.* 0.0019 per cent) Is it possible to have exactly 5 per cent of successes (or 5 per cent of failures) in an experiment such as that illustrated in Table 4.2? (*Ans.* No) Why?

10. Using the table that you constructed for the binomial distribution $(\frac{1}{2} + \frac{1}{2})^{10}$ in exercise 1, find the per cent of the distribution corresponding to less than one correct guess. (*Ans.* 0.1 per cent) To less than one incorrect guess. (*Ans.* 0.1 per cent) To less than two correct guesses. (*Ans.* 1.1 per cent) To less than two incorrect guesses. (*Ans.* 1.1 per cent) To less than three correct guesses. (*Ans.* 5.5 per cent) To less than three incorrect guesses. (*Ans.* 5.5 per cent) See if you can find these answers in Table 5.2. Is it possible to have exactly 15 per cent of correct guesses (or exactly 15 per cent of incorrect guesses) in a binomial situation such as that described in exercise 1? (*Ans.* No) Why?

4.3.2 Sampling for attributes. If the proportion of the elementary units in a universe that possess an attribute is π, then the probability P that one elementary unit drawn at random from the universe will possess the attribute is equal to π and, consequently, $Q = 1 - \pi$.

By substituting π for P and $1 - \pi$ for Q in formulas 4.4 and 4.5, we can obtain formulas for the mean and the standard deviation of the distribution of the proportion p of the elementary units that possess the attribute in random samples of size n, and for the mean and standard deviation of the distribution of the proportion q of the elementary units that do not possess the attribute in random samples of size n. The results are

$$\mu_p = \pi, \quad \mu_q = 1 - \pi \tag{4.6}$$

$$\sigma_p = \sigma_q = \sqrt{\frac{\pi(1 - \pi)}{n}} \tag{4.7}$$

If the procedure by which the sample is drawn is the procedure described in section 2.8.1, namely, random sampling with replacement after each drawing of an elementary unit, formulas 4.6 and 4.7 are the correct formulas to use, regardless of the size n of the sample or the size N of the universe from which the sample is drawn. The probabilities or relative frequencies here follow the binomial probability law exactly.

4.3.3 Adjustment for sampling without replacement. In section 2.8.2, a procedure called "sampling without replacement" was described. For random sampling without replacement, an adjustment factor, called the *finite multiplier*, is needed in formula 4.7. If a random sample of size n is drawn without replacement from a universe of size N, then

$$\sigma_p = \sigma_q = \sqrt{\frac{\pi(1-\pi)}{n} \frac{(N - n)}{(N - 1)}} \tag{4.8}$$

The purpose of the finite multiplier, $(N-n)/(N-1)$, is to make allowance for the fact that the number of elementary units in the universe is finite and the fact that the sampling was done without replacement after drawing each elementary unit for the sample.

If N, the size of the universe, is very much greater than n, the size of the sample, the amount of adjustment brought about by the finite multiplier is so insignificant that it may be omitted. In other words, if N is very much greater than n, the fraction represented by the finite multiplier is very nearly equal to 1.

A few simple rules may enable you to avoid considerable useless computation. It can be shown easily that if N is at least 100 times as great as n, then the finite multiplier can be replaced by 1 without affecting the first two decimal places in the values of σ_p and σ_q. Similarly, if N is at least 1000 times as great as n, the finite multiplier can be replaced by 1 without affecting the first three decimal places in the values of σ_p and σ_q. In practical problems, we rarely need to retain more than two decimal places in the values of σ_p and σ_q.

4.3.4 Method for practical application. In most practical statistical situations, all that we know about the universe is the size N of the universe and the proportion p of the elementary units that possess the attribute in a sample of size n drawn from the universe. In such situations, we cannot use formulas 4.7 and 4.8 as they stand. But we may use the observed proportion p in the sample as the best available estimate of the proportion π in the universe.

For estimating the standard deviation of the distribution of p or of q, (a) in random sampling, with replacement after each draw, use

$$s_p = s_q = \sqrt{\frac{p(1-p)}{n}} \qquad (4.9)$$

(b) in random sampling, without replacement after each draw, use

$$s_p = s_q = \sqrt{\frac{p(1-p)}{n}\frac{(N-n)}{(N-1)}} \qquad (4.10)$$

The finite multiplier may be omitted from formula 4.10 under the conditions mentioned in the last paragraph of section 4.3.3.

EXERCISES

1. Use the column totals of table A.13 (p. 383) and the method of section A.6.16 (p. 381) with appropriate values of i and c to find the means and the standard deviations for the two sampling distributions in Tables 3.12 and 3.13. You should obtain exactly the same value for the standard deviation in both distributions. Why? (*Ans.* For Table 3.12, $i = 0.012$, $c = 0.360$, $m = 0.360$, $s = 0.030$; for Table 3.13, $i = 0.012$, $c = 0.000$, $m = 0.000$, $s = 0.030$)

2. Compare the value of the standard deviation obtained in exercise 1 for the distributions in Tables 3.12 and 3.13 with the value of σ_p computed by formula 4.8 when $\pi = 0.36$, $n = 250$, and $N = 5000$. (*Ans.* $\sigma_p = 0.030$)

3. Use the distributions that the members of the class were asked to build up by drawing samples of red beads and white beads in Chapter 1 and make the same kind of comparison as you made in exercise 2.

4.4 GEOMETRIC REPRESENTATION OF PROBABILITIES

The type of geometric representation that is appropriate for the probabilities involved in a particular situation depends upon the type of distribution that is involved. For a discrete distribution, such as a binomial distribution, a correct geometric representation of the probabilities is a series of points on a graph, along with the ordinates of the points. The

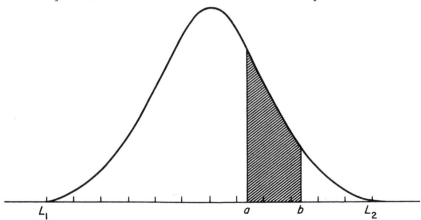

Fig. 4.2. Geometric Representation of Relative Frequency and Probability for a Distribution of a Continuous Variable

abscissas of the points are the theoretically possible values of the discrete variable. Figure 4.1 is a graph of this type.

For a discrete distribution each ordinate represents the relative frequency corresponding to a possible value of the variable, and, consequently, each ordinate represents the probability corresponding to a possible value of the variable. Therefore, the sum of all the ordinates of a binomial distribution represents the sum of the probabilities for all possible values of the variable, and each of these two sums always is equal to 1. The probability corresponding to a specified interval of values of a discrete variable is equal to the sum of all the ordinates in the specified interval.

Figure 4.2 is the graph of the smooth frequency curve for a distribution of a continuous variable. The variable is measured along the horizontal line or axis. The limits of the range of the variable are marked L_1 and L_2. The total relative frequency in the distribution is represented by the total area enclosed by the smooth curve and the horizontal straight line. Consequently, the total area under the curve is equal to 1.

The relative frequency corresponding to any specified interval of values of the continuous variable is represented by the area under the curve in that interval. For example, in Figure 4.2, the relative frequency or probability in the interval from a to b of the variable is represented by the shaded area. Similarly, the relative frequency or probability for all the values of the variable that are greater than b, that is, for all the values of the variable from b to L_2, is represented by the small unshaded area under the curve to the right of the shaded area.

Sometimes, the total area under the smooth curve is made equal to Σf, the total frequency. Then the relative frequency in the interval from a to b

of the values of the variable is equal to the ratio of the shaded area to the total area under the curve. Similarly, in this situation the relative frequency or probability for the interval from b to L_2 is equal to the ratio of the area in the right-hand tail of the curve to the total area under the curve.

When we use a normal distribution as a good approximation to a binomial distribution, we are assuming that the total area under the smooth normal curve is equal to the sum of all the ordinates of the binomial distribution. Furthermore, we are assuming that there is a specific part of the area under the normal curve that is approximately equal in numerical value to a specific ordinate of the binomial distribution. No doubt you will agree readily that the area under the curve on both sides of the specified binomial ordinate, extending half way to the next binomial ordinate on each side, must be a good approximation to the numerical value of the specified binomial ordinate. In this way, the total area under the normal curve can be divided into a series of parts each of which corresponds to an ordinate of the binomial distribution.

In Figure 4.3, the specified value of the discrete variable is a. The binomial probability of drawing the value a is represented by the solid-line ordinate erected at a. Notice that the point z_1 is half an interval to the left of a and the point z_2 is half an interval to the right of a. The shaded area under the smooth curve between the dashed-line ordinates at z_1 and z_2 is the partial area under the normal curve that corresponds to the binomial ordinate at a.

It follows easily now that if we wish to determine the normal probability corresponding to any specified binomial interval, such as the in-

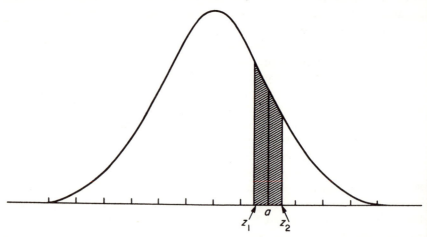

Fig. 4.3.   The Area Under the Normal Probability Curve Corresponding to the Binomial Ordinate and the Binomial Probability for a Specified Value of the Variable

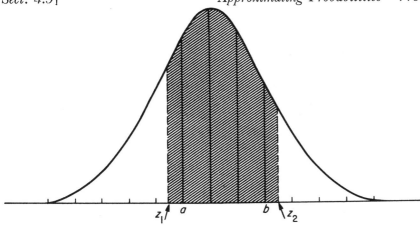

Fig. 4.4.   The Area Under the Normal Probability Curve Corresponding to the Binomial Ordinates and the Binomial Probabilities for All Values of the Variable in a Specified Interval

terval from a to b in Figure 4.4, that includes several possible values of the discrete variable, we must find the area under the normal curve that is made up of all the strips of area corresponding to all the values of the discrete variable in the specified interval. This set of strips will begin a half-step to the left of the first (smallest) value a, that is, at z_1, and it will end a half-step to the right of the last (largest) value b, that is, at z_2. This is the shaded area in Figure 4.4. Most statisticians refer to this method of approximation as a "correction for continuity."

Some examples will be given now to illustrate the ideas that have been discussed in this section.

4.5 APPROXIMATING BINOMIAL PROBABILITIES BY NORMAL PROBABILITIES

In section 4.2, it was stated that all the probabilities involved in the game of "twenty-six" could be computed by using the binomial distribution

$$\left(\frac{5}{6} + \frac{1}{6}\right)^{130} = \left(\frac{5}{6}\right)^{130} + 130\left(\frac{5}{6}\right)^{129}\left(\frac{1}{6}\right) + \frac{130(129)}{1(2)}\left(\frac{5}{6}\right)^{128}\left(\frac{1}{6}\right)^2 + \cdots$$

There are 131 terms in the expansion and each term indicates the probability of obtaining a certain number of successes and failures, where "success" means that the chosen face of a die turns upward. For example, the probability of obtaining exactly 26 successes in 13 rolls of 10 dice is given by the twenty-seventh term, namely,

$$\frac{130(129)(128)\ldots(106)(105)}{(1)(2)(3)\ \ \ldots\ (25)(26)}\left(\frac{5}{6}\right)^{104}\left(\frac{1}{6}\right)^{26}$$

The probability of obtaining from 0 to 11 successes is given by the sum of the first 12 terms. The probability of obtaining from 12 to 25 successes is given by the sum of the thirteenth to the twenty-sixth terms. The probability of obtaining from 26 to 32 successes is given by the sum of the twenty-seventh to the thirty-third terms. The probability of obtaining from 33 to 35 successes is given by the sum of the thirty-fourth to the thirty-sixth terms. The probability of obtaining from 36 to 130 successes is given by the sum of the thirty-seventh to the one hundred thirty-first terms.

Now, let us approximate these binomial probabilities by using the normal distribution. First, we need to know the mean and the standard deviation of the binomial distribution. By formula 4.2, the mean number of successes is $\mu_x = 130(\frac{1}{6}) = 21.7$, approximately. By formula 4.3, the standard deviation of the number of successes is $\sigma_x = \sqrt{130(\frac{1}{6})(\frac{5}{6})} = 4.25$, approximately.

Consider the probability of obtaining from 26 to 32 successes. We need to use the normal distribution for values of the continuous variable from 25.5 to 32.5. In order to be able to use the table of normal probabilities, we must change the values of our variable into standard units. By formula 3.5, we find that

$$z_1 = \frac{25.5 - 21.7}{4.25} = 0.89 \quad \text{and} \quad z_2 = \frac{32.5 - 21.7}{4.25} = 2.54$$

From Table 6.7 (p. 172), we find that

the area to the right of $z_1 = 0.89$ is   0.187
the area to the right of $z_2 = 2.54$ is   0.006
the area between $z_1 = 0.89$ and $z_2 = 2.54$ is   0.181

Therefore, the probability of obtaining from 26 to 32 successes in the game of "twenty-six" is approximately 0.181.

The computations of the normal-probability approximations for all the possible numbers of successes in which we are interested are summarized in Table 4.5. All of the intervals and probabilities are represented graphically in Figure 4.5.

TABLE 4.5

Normal Approximations to the Probabilities in the Game of "Twenty-six"

Number of Successes (Discrete Variable)	Class Limits for Continuous Variable	Normal Abscissas		Probability of z Being Between z_1 and z_2	Amount To Be Won	Player's Expected Payoff
		z_1	z_2			
0–11	Below 11.5	$-\infty$	-2.40	0.008	4	0.032
12–25	11.5–25.5	-2.40	$+0.89$	0.805	0	0.000
26–32	25.5–32.5	0.89	2.54	0.181	4	0.724
33–35	32.5–35.5	2.54	3.25	0.005	8	0.040
36–130	35.5 upward	3.25	$+\infty$	0.001	12	0.012
Total				1.000		0.808

In the fifth column of the table, the amount that the banker pays to the player if he obtains the different numbers of successes is indicated. In the last column, you will find what are called the "player's expected payoffs." Each expected payoff is the product of the probability of obtaining a specific number of successes and the amount to be won with that number of successes.

The sum of the player's expected payoffs is 0.808. This means that if a person plays the game of "twenty-six" a very large number of times with unbiased dice, he can expect to win back only 0.808 of what he deposits or pays for the "privilege" of playing. In other words, in the long run, for every dollar that he pays to the banker the player can expect to win back only about 81 cents. The banker can expect to make a profit in the long run of approximately 19 per cent of all the amounts that he collects from players.

The player's *mathematical expectation* in a game of chance is defined as the product of the probability of winning and the amount that may be won minus the product of the probability of losing and the amount that may be lost. Consequently, in this game of "twenty-six," the player's mathematical expectation is $(1.000)(0.808) - (1)(1) = 0.808 - 1 = -0.192$, the amount that he deposits in order to play the game being 1 unit and the probability that he will lose that unit being 1, that is, certainty.

Here is another example. Suppose that a manufacturer knows from past experience that, on the average, 3 per cent of the items produced by one of his automatic machines are defective. What is the probability that in a batch of 5000 items made by the machine there will be *more than* 175 defective items? The exact probability is obtainable by finding the sum of

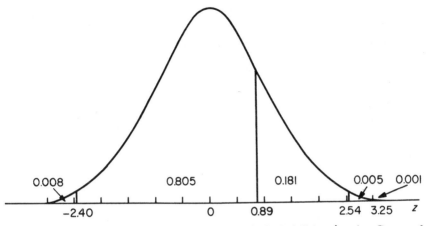

Fig. 4.5. Normal Approximations to Binomial Probabilities in the Game of Twenty-Six

the last 176 terms of the binomial expansion of $(0.03 + 0.97)^{5000}$ and subtracting the sum from 1.

We can find an excellent approximation to the exact probability by using the normal probability table. First, we need to know the mean number of defective items in a batch of 5000 items and the standard deviation of the number of defective items in a batch of 5000 items. By formulas 4.2 and 4.3, we find that $\mu_x = 5000(0.03) = 150$ and $\sigma_x = \sqrt{5000(0.97)(0.03)} = 12.1$. Consequently, by formula 3.5 the abscissa of the normal curve at the beginning of the area corresponding to more than 175 defective items is $z = (175.5 - 150)/12.1 = +2.11$. Now, we need to find the probability of drawing a value of z as great as or greater than 2.11 from a normal distribution. Table 6.7 shows that this probability is 0.017. In other words, in the long run about 17 batches out of every 1000 batches containing 5000 items will be expected to contain *more than* 175 defectives per batch.

Although the normal probabilities are excellent approximations to binomial probabilities when n is very large, the approximation is fairly close even if n is as small as 12, provided P is not far from 0.5. For example, let us examine the situation in which we desire to know the probability of obtaining exactly 7 heads in 12 tosses of an unbiased coin. The binomial distribution in this case is $(\frac{1}{2} + \frac{1}{2})^{12}$ and the probability of obtaining exactly 7 heads is given by the term

$$\frac{12(11)(10)(9)(8)(7)(6)}{7(6)(5)(4)(3)(2)(1)}\left(\frac{1}{2}\right)^5\left(\frac{1}{2}\right)^7 = \frac{792}{4096} = 0.193$$

Now, let us find the normal-probability approximation to this binomial probability. Here $\mu_x = 12(\frac{1}{2}) = 6$ and $\sigma_x = \sqrt{12(\frac{1}{2})(\frac{1}{2})} = \sqrt{3} = 1.732$.

$$z_1 = \frac{6.5 - 6}{1.732} = \frac{0.5}{1.732} = 0.29 \quad \text{and} \quad z_2 = \frac{7.5 - 6}{1.732} = \frac{1.5}{1.732} = 0.87$$

Table 6.7 shows that the area to the right of $z_1 = 0.29$ is 0.386 and that the area to the right of $z_2 = 0.87$ is 0.192. Consequently, the area between $z_1 = 0.29$ and $z_2 = 0.87$ is $0.386 - 0.192 = 0.194$. This is a fairly close approximation to the true binomial probability, 0.193.

Many statisticians have adopted the practical rule that the normal probabilities are sufficiently good approximations of the binomial probabilities whenever both nP and nQ are greater than five.

EXERCISES

1. Compute the probability of obtaining from 12 to 25 successes in the game of "twenty-six." (*Ans.* See Table 4.5.) Construct a graph similar to Figure 4.4 for this exercise, using Table 6.6 to plot the normal curve.

2. Compute the probability of obtaining from 33 to 35 successes in the game of "twenty-six." Construct a graph similar to Figure 4.4 for this exercise.

3. Compute the probability of obtaining from 36 to 130 successes in the game of "twenty-six." Construct a graph for this exercise.

4. Compute the probability of obtaining less than 12 successes in the game of "twenty-six." Construct a graph for this exercise.

5. If it is known that a machine produces defective items at the rate of 3 per cent on the average, what is the probability that in a batch of 5000 items made by the machine there will be *more than* 175 defective items but *less than* 180 defective items? (*Ans.* 0.010)

6. If it is known that, on the average, 3 per cent of the items produced by an automatic machine are defective, what is the probability that in a batch of 5000 items produced by the machine there will be (a) *less than* 125 defective items (*Ans.* 0.017); (b) *more than* 175 defective items (*Ans.* 0.017); (c) from 125 to 175 defective items? (*Ans.* 0.966)

7. If the probability that a person nineteen years of age will live to age seventy is 0.59680, what is the probability that in a group containing 178,359 randomly selected nineteen-year-old people, from 106,000 to 106,500 of them will survive to age seventy? (*Ans.* 0.590) What is the probability that more than 106,500 of the people will survive to age seventy? (*Ans.* 0.394) What is the proability that less than 106,000 of the people will survive to age seventy? (*Ans.* 0.016)

8. A local carnival will give a new $3000 automobile to the holder of the winning ticket in a lottery. If 10,000 tickets were sold at $1.00 each and you bought 10 of the tickets, what is your mathematical expectation? (*Ans.* −$7.00)

9. A person is allowed one free throw of a penny and is to get a dime if the penny turns up head. What is his mathematical expectation? (*Ans.* 5 cents)

10. A person pays 5 cents for one chance to toss a penny and is to get a dime if the penny turns up tail. What is his mathematical expectation? (*Ans.* 0) Is this a fair proposition? (*Ans.* Yes)

4.6 CONFIDENCE LIMITS FOR A PROPORTION

Assume for a moment that we have drawn a large number of random samples, each of which is large enough so that both np and nq are greater than 5, from a universe in which π is the proportion of elementary units that possess a specified attribute. Assume also that we have computed the proportion p and the standard deviation s_p corresponding to each of the samples. We expect to find that about 95 per cent of the intervals $p - 1.96s_p$ to $p + 1.96s_p$ include the value π. See Figure 1.1, a graphical illustration of this idea. Similarly, we expect to find that about 99 per cent of the intervals $p - 2.58s_p$ to $p + 2.58s_p$ include the value π.

In general, approximate confidence limits for the unknown proportion π at any specified probability level P can be determined easily from a proportion p observed in a random sample if both np and nq are greater than 5. Substitute the appropriate value of z from table 6.7 and the value of s_p computed by formula 4.9 or 4.10 into the inequality

$$p - zs_p < \pi < p + zs_p \qquad (4.11)$$

Of course, if the confidence limits are desired for $1-\pi$ instead of for π, simply replace p by q in formula 4.11.

If we introduce the correction for continuity, a formula for determining confidence limits for a proportion π when both np and nq are greater than 5 is

$$p - \frac{1}{2n} - zs_p < \pi < p + \frac{1}{2n} + zs_p \qquad (4.12)$$

We have seen in exercises 9 and 10 on page 114 that, in general, it is not possible to cut off terms at the ends of a binomial distribution which will amount to exactly a specified part—say, 2 per cent or 5 per cent or 10 per cent—of the total distribution. Consequently, the correct interpretation of formula 4.12 is that the probability is *at least P* that the value of π is in the interval computed by formula 4.12; the actual confidence level may be considerably greater than the required confidence level P. The situation here is similar to that shown in Table 5.4.

These confidence intervals are said to be central confidence intervals because they always have p at the midpoint of the interval. Some examples of the practical use of confidence intervals of this type were given in section 1.12.3.

The question probably arises in your mind: What confidence interval should a person use? That is, should one always compute 95 per cent confidence limits for an estimate? 99 per cent confidence limits for an estimate? The correct answer is that it depends upon how much risk one is willing to take of making an incorrect decision. In other words, the answer to the question depends upon how costly it would be if an interval that does not contain the true value of the parameter in the universe were accepted as containing the true value of the parameter.

In the day-to-day affairs of many statisticians, a 95 per cent confidence interval is widely used. Sometimes, when no serious consequences could result from a wrong decision, a 90 per cent confidence interval is used. In a situation such as a medical experiment involving life and death, a central 99 per cent confidence interval might not be strong enough, that is, might involve too much risk. In such a situation, you might consider it necessary to use a noncentral 99 per cent confidence interval or even a 99.99 per cent noncentral confidence interval.

Data for 100 per cent of the elementary units that were scheduled to be in the sample are necessary for the purpose of determining confidence limits to be associated with the estimate computed from the sample. It may be possible, however, to determine outer bounds for the sampling errors if almost all but not all of the elementary units intended to be in the sample actually are included in the sample. This may be done by first assuming that all of the missing elementary units are of one kind, for example, possessing the attribute, and then assuming that all of the missing elementary units are of another kind, for example, not possessing the attribute. In general, of course, the actual sampling error will be smaller than these outer bounds indicate.

EXERCISES

1. The value of the proportion p observed in a random sample containing 500 elementary units from a very large universe is $p = 0.64$. Use formula 4.11 to compute 95 per cent confidence limits for π. (*Ans.* $0.598 < \pi < 0.682$)

2. If 44 per cent of the people in a random sample containing 900 people from a universe of 25,000 people are in favor of a certain proposal, determine 95 per cent confidence limits for π. (*Ans.* $0.408 < \pi < 0.472$) Explain briefly the meaning of this confidence interval.

3. Use formula 4.10 and the proportion p of red beads in your first sample of red and white beads to estimate σ_p. Then use formula 4.12 to determine 95 per cent confidence limits and 99 per cent confidence limits for the proportion π of red beads in the universe from which you drew the sample. Do these confidence limits include the value of π?

4. In the test of the two coffee formulas mentioned in section 1.12.3, it was found that 54 per cent of the 1000 coffee drinkers in the sample preferred coffee made by the new formula. Find 99 per cent confidence limits for π. (*Ans* 0.50 and 0.58)

5. In the survey to test the color of the lining of the cans for coconut mentioned in section 1.12.3, it was found that 65 per cent of the housewives in the sample preferred the coconut in the can with the blue lining. Compute 95 per cent confidence limits for π. (*Ans.* 0.62 and 0.68)

6. A coin is tested by tossing it fairly 200 times; it turns up heads 106 times. Use formula 4.12 to determine 95 per cent confidence limits for the proportion π of times the coin will turn up heads in the long run. (*Ans.* $0.458 < \pi < 0.602$)

7. In a pre-election poll in a very large school, 160 students in a random sample containing 300 students stated that they intended to vote for candidate C. Use formula 4.12 to compute 95 per cent confidence limits for the proportion of all the students in the school who intended, at the time of the survey, to vote for candidate C. (*Ans.* $0.475 < \pi < 0.591$)

8. A florist is experimenting with a crossing of certain kinds of flowers in order to develop a flower with a special new color. In a random sample containing 80 plants from a very large group, he observes that 22 plants have the desired color. Use formula 4.12 to determine 90 per cent confidence limits for the proportion π of all the plants that have the desired color. (*Ans.* $0.13 < \pi < 0.42$)

9. Use section A.7.2 to determine 99 per cent noncentral confidence limits if a random sample of size $n = 50$ produces the proportion $p = 0.06$. (*Ans.* 0.02 and 0.21)

10. Use section A.7.2 to determine 95 per cent noncentral confidence limits if $p = 0.06$, $n = 50$, and $N = 500$. (*Ans.* $0.022 < \pi < 0.155+$) Compare these limits with the limits obtained in section A.7.2.

11. Use section A.7.2 to determine 99 per cent noncentral confidence limits if $n = 50$, $p = 0.06$, and $N = 1000$. (*Ans.* $0.016 < \pi < 0.203$)

12. Begin with the equations $p - \frac{1}{2} - \pi = z\sigma_p$ and $p + \frac{1}{2} - \pi = -z\sigma_p$ and show that approximate noncentral confidence limits to be associated with a pro-

portion p observed in a random sample of size n are the smaller of the two values of π obtained by solving

$$n(n + z^2)\pi^2 - n(2np + z^2 - 1)\,\pi + (n^2p^2 - np + \tfrac{1}{4}) = 0$$

and the larger of the two values of π obtained by solving

$$n(n + z^2)\pi^2 - n(2np + z^2 + 1)\,\pi + (n^2p^2 + np + \tfrac{1}{4}) = 0$$

Use these equations to find 95 per cent noncentral confidence limits for π if it is observed that $p = 0.06$ in a sample of size $n = 50$. (*Ans.* 0.016 and 0.175) Compare these limits with the limits that would have been obtained by using formula 4.12.

4.6.1 Statistical quality control.

The statistical quality control technique is one of the most spectacular practical applications of statistical method. It may be applied to an almost unlimited number of different types of process. Up to the present time its most frequent applications have been in connection with the control of the quality of manufactured products, especially those that are mass produced. But the method can be applied to great advantage in many other types of situation. For example, statistical quality control techniques can be used to control the quality of the clerical work performed by groups of office workers, for instance, a group of statistical clerks editing or coding questionnaires completed in a consumer research survey.

In carrying out the statistical quality control procedure, statistical quality control charts, such as Figure 4.6 and Figure 7.4, are the principal statistical tools.

Each sample of work that is inspected is plotted as a point on the control chart. Whenever a plotted point falls outside the control limits marked on the chart, the production process is assumed to be out of control, and production may be stopped until the cause of the trouble has been discovered and corrected.

In American industry, the most commonly used control limits for the proportion of defective items produced are the 3-sigma limits, namely, $\pi \pm 3\sigma_p$. If the lower limit $\pi - 3\sigma_p$ turns out to be negative, then zero is used as the lower control limit because the proportion of defective items produced never can be less than zero.

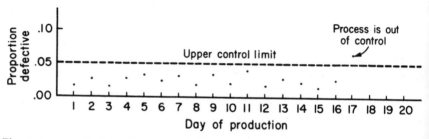

Fig. 4.6.   Statistical Quality Control Chart for the Proportion of Defective Items Produced

4.6.2 Further remarks about confidence limits. We can feel reasonably sure that the value of the parameter in the universe is between the 95 per cent confidence limits because those limits were determined by a procedure that can be depended upon to produce correct decisions about 95 per cent of the times in which it is used. Think of a long series of confidence intervals determined by this procedure. It is not necessary that even as many as two of the samples for the series of confidence intervals be of the same size or from the same universe. You may think of the series as being all the confidence intervals that a statistician might determine in a lifetime of varied research projects.

There usually is no way to predict or know in advance, or even to know afterward, when one of the 5 per cent of incorrect decisions happens. It is very rare indeed for a statistician to have an opportunity to compare his computed confidence interval with the true value of the parameter in the universe. Consequently, we usually operate in the dark with respect to knowing when a confidence interval fails to include the value of the parameter. Although an unbiased coin may be expected to turn up head about 50 per cent of the time in a long series of tosses, we cannot know in advance which particular tosses will result in heads.

Imagine that you have a bag containing a large number of beads, of which exactly 95 per cent are white beads and the remaining 5 per cent are red beads. Mix the beads thoroughly. Now consider yourself to be blindfolded and then to draw one bead at random from the bag. It is bound to be either red or white, but because you are blindfolded you cannot know the color of the bead. Now, make the statement "This is a white bead." Drop the bead back into the bag and mix the beads thoroughly again. Now draw a bead at random from the bag. Again it is bound to be either red or white, but because you are still blindfolded you cannot know the color of the bead. Make the statement "This is a white bead." Repeat this process a large number of times, preferably many thousands, making the statement "This is a white bead" after each drawing.

Now, what can you say about the long series of statements, "This is a white bead," that you made? Certainly, you cannot know on which of the drawings your statement was correct and on which of the drawings your statement was incorrect. But you can feel reasonably sure that about 95 per cent of the time you drew a white bead and that, consequently, about 95 per cent of the time your statement, "This is a white bead," was correct. Likewise, you can feel reasonably sure that you drew a red bead about 5 per cent of the time and that, consequently, your statement, "This is a white bead," was incorrect about 5 per cent of the time. There is no way for you to know exactly which drawings produced these occasional red beads and caused your statement, "This is a white bead," to be incorrect. All you can know is that these incorrect cases probably happened, but you cannot know exactly when. It is in this sense of uncertainty that the word probability is used in connection with statements of confidence limits for parameters.

4.7 TESTS OF HYPOTHESES ABOUT PROPORTIONS

In section 1.12.3, you were given two illustrations of situations in which businessmen were interested in knowing how much reliance to place on an estimate of a proportion obtained from a sample. It was stated there that often what the businessman wants is assurance that the true proportion in the universe is not less than (or not greater than) some specified proportion. Often, a good way to handle such a situation is to formulate it in terms of a hypothesis that can be tested.

For example, 65 per cent of the 1000 housewives in the sample preferred the coconut in the can with the blue lining. Because of chance in random sampling we cannot assure the management that the true proportion in the universe who would prefer the coconut in the can with the blue lining is exactly 65 per cent. Can we assure the management that the true value of π in that universe is greater than 0.62? To answer this question, let us formulate the null hypothesis that $\pi \leq 0.62$ and take as the alternative the statement that $\pi > 0.62$. This requires a one-tail test. Suppose that we desire to use the $\alpha = 0.025$ significance level for the test. To test this hypothesis, we compute

$$z = \frac{0.6495 - 0.62}{\sqrt{\dfrac{(0.62)(0.38)}{1000}}} = \frac{+0.0295}{+0.015} = +1.97$$

Table 6.7 shows that the probability of obtaining a value of z as great as or greater than $+1.97$ by chance in random sampling from a normal universe is $P = 0.024$. Consequently, we can reject the null hypothesis here at the $\alpha = 0.025$ level. That is, we can assure the management that there is only about a 2.4 per cent risk involved in accepting the hypothesis that the percentage of housewives in the universe who would prefer the coconut in the blue-lined can is greater than 62.

In some other situations, the hypothesis is that the proportion π in the universe is some specified proportion. To test this hypothesis, we must make a two-tails test because the proposed hypothesis will not distinguish between sample deviations of p above π and below π. For example, suppose that the null hypothesis is that the proportion of red beads in a very large universe of red beads and white beads is exactly 0.40. Let us use the $\alpha = 0.05$ significance level for this two-tails test. Suppose that we draw a simple random sample of 100 beads from the universe and that the sample contains 36 red beads and 64 white beads. Then, we compute

$$z = \frac{0.365 - 0.40}{\sqrt{\dfrac{(0.40)(0.60)}{100}}} = \frac{-0.035}{+0.049} = -0.71$$

Table 6.7 shows that the probability of obtaining a value of z as great as or greater than $+0.71$ by chance in random sampling from a normal distribution is 0.239. Similarly, the probability of obtaining a value of z as small as or smaller than -0.71 is 0.239. Consequently, the probability in the two tails is $P = 0.239 + 0.239 = 0.478$. This value of P is much larger than $\alpha = 0.05$ and therefore we cannot reject the null hypothesis that $\pi = 0.40$ in this example.

Notice that in the first example the observed value of p was 0.65, which is larger than the hypothetical value $\pi = 0.62$. There the value of p was decreased a half-step to 0.6495 in computing z. Here is the reasoning: 65 per cent of 1000 housewives is 650 housewives. The proper amount of area in the right-hand tail of the normal probability curve in this case begins at the point corresponding to 649.5 housewives, that is, at the point corresponding to $p = 649.5/1000 = 0.6495$.

In the second example, the observed value of p was 0.36, which is smaller than the hypothetical value $\pi = 0.40$. Here the value of p was increased a half-step to 0.365 in computing z. Thirty-six per cent of 100 beads is 36 beads. The proper amount of area in the left-hand tail of the normal probability curve in this case begins at the point corresponding to 36.5 beads, that is, at the point corresponding to $p = 36.5/100 = 0.365$.

In general, the formula for computing z here is

$$z = \frac{p \pm \dfrac{1}{2n} - \pi}{\sigma_p} \qquad \begin{array}{l} \text{using } +\dfrac{1}{2n} \quad \text{if } p < \pi \\[2mm] \text{using } -\dfrac{1}{2n} \quad \text{if } p > \pi \end{array} \qquad (4.13)$$

The procedure to be used in testing a hypothesis should be selected during the planning stage of the investigation: (1) Choose a null hypothesis and an alternative. Notice whether these require a one-tail test or a two-tails test. (2) Choose a significance level α for the test. (3) Determine the critical value(s) of z corresponding to the chosen significance level. (4) Draw a diagram similar to one of those in Figures 2.1 to 2.4. (5) Choose a formula for computing the value of z.

EXERCISES

1. Test the null hypothesis that $\pi \leq 0.60$ if the value of p observed in a random sample containing 500 elementary units is $p = 0.64$. Use a significance level that will permit you to have at least 95 per cent confidence that your conclusion from the test is correct. (*Ans.* Reject null hypothesis and accept alternative.)

2. If in a random sample $n = 256$ and the proportion is $p = 0.36$, test the null hypothesis that the proportion possessing the attribute in the universe is exactly $\pi = 0.40$. Use a significance level such that there is no more than a 5 per cent risk of rejecting the null hypothesis when it is true. (*Ans.* Do not reject null hypothesis.)

3. If 44 per cent of the people in a random sample containing 900 people from a universe of 25,000 people are in favor of a certain proposal, test at the $\alpha = 0.05$ significance level the null hypothesis that the proportion of the people in the universe who favor the proposal is $\pi = 0.50$. (*Ans.* Reject null hypothesis and accept alternative.)

4. In the market research for the two coffee formulas mentioned in section 1.12.3, it was found that 54 per cent of the 1000 coffee drinkers in the sample preferred the coffee made by the new formula. Test the null hypothesis that the true proportion π of coffee drinkers who would prefer the new formula is not greater than 0.525. Use the $\alpha = 0.005$ significance level for this test. (*Ans.* $z = 0.92$; do not reject null hypothesis.)

5. A coin is tossed fairly 200 times and turns up heads 106 times. Test the null hypothesis that the coin will turn up heads exactly 50 per cent of the time in the long run. Use the $\alpha = 0.05$ significance level. (*Ans.* Do not reject null hypothesis.)

6. In a pre-election poll in a very large school, 160 students in a random sample containing 300 students stated that they intended to vote for candidate C. Test the null hypothesis that not more than 50 per cent of the students in the school intended, at the time of the survey, to vote for candidate C. Use the $\alpha = 0.025$ significance level for this test. (*Ans.* Do not reject null hypothesis.)

7. A florist is experimenting with a crossing of certain kinds of flowers in order to develop a flower with a special new color. Genetic principles indicate that one-fourth of the new crop of flowers should have the desired color. In a random sample containing 80 plants from a very large group, the florist observes that 22 plants have the desired color. Should this result be regarded as merely a chance deviation from the expected proportion? Test the null hypothesis that $\pi = 0.25$ for the universe of plants from which the sample was drawn. Use the $\alpha = 0.10$ significance level for this test. (*Ans.* Do not reject null hypothesis.)

8. Suppose that your school contemplated introducing a new service for the students but that the proposed service could not be operated successfully unless more than 60 per cent of all the students in the school agreed to use the service. Assume that after you conducted a sample survey among 400 students chosen at random you found that 64 per cent of the students in the sample agreed to use the service. Test the null hypothesis that not more than 60 per cent of all the students in the school would agree to use the service. Use the $\alpha = 0.01$ significance level for this test. (*Ans.* Do not reject null hypothesis.)

4.8 SAMPLE SIZE FOR ESTIMATING A PROPORTION

One of the questions that usually should be answered during the planning stage of a statistical investigation is: What is the proper size of sample to use in this investigation? The amount of sampling error that we are willing to risk in the estimate to be obtained from the sample is one of the most important factors involved in the determination of the proper size of sample.

Let us suppose that we want to make a sample survey in our community to obtain an estimate of the proportion of senior-high-school boys who intend to look for jobs next summer. Suppose, further, that we want to be reasonably certain that the proportion we obtain from the sample as an estimate of the true proportion π for all the senior-high-school boys in the community does not contain a sampling error greater than 0.03 in absolute value. By "reasonably certain" let us mean that the probability is 0.95. That is, we want the probability to be 0.95 that $|p - \pi| \leq 0.03$.

Now, you may recall that about 95 per cent of the values of p in a large number of large random samples of size n from a universe in which the proportion that possess the attribute is π may be expected to fall within the interval that extends from $\pi - 1.96\sigma_p$ to $\pi + 1.96\sigma_p$. Consequently, we require that $1.96\sigma_p = 0.03$ here. Using formula 4.7 for σ_p, we require that $1.96\sqrt{\pi(1 - \pi)/n} = 0.03$.

If we knew the value of π for the universe, we could substitute that value of π into the last equation given above and solve that equation for n. Usually, of course, we do not know the value of π in practical sampling situations. If we have even a rough idea of its value we can substitute that value for π and solve the equation for n. This may tell us at least whether we need 50 elementary units, 500 elementary units, or 1000 elementary units in the sample. To know that much is better than to have no idea whatsoever about the proper sample size for the survey.

There are at least three kinds of rough guesses of the value of π that may be used for determining sample size. (1) We may be able to use our own past experience or the experience of someone who is available to advise us. For example, someone in vocational guidance work might know that about 80 per cent of the boys in his school tried to obtain jobs for the previous summer. (2) We may be able to guess an upper limit or a lower limit for the value of π. For example, the vocational guidance counselor might tell us that in the last five years the proportion of boys who tried to find summer jobs varied from 70 per cent to 85 per cent. In such a case, we might use 70 per cent as a guess of the lower limit for π. (3) As a last resort, when no other information is available, use 0.50 as the value of π in the equation to solve for n. Let us see how to solve the equation by each of these three methods.

(1) Substitute 0.80 for π and solve the equation for n.

$$1.96\sqrt{\frac{(0.80)(0.20)}{n}} = 0.03$$

$$\sqrt{\frac{(0.80)(0.20)}{n}} = \frac{0.03}{1.96}$$

$$\frac{(0.80)(0.20)}{n} = \left(\frac{0.03}{1.96}\right)^2$$

$$\frac{n}{(0.80)(0.20)} = \left(\frac{1.96}{0.03}\right)^2$$

$$n = \left(\frac{1.96}{0.03}\right)^2 (0.80)(0.20) = 4264(0.80)(0.20) = 682$$

(2) Substitute the lower limit 0.70 for π and solve the equation for n. The result is

$$n = 4264(0.70)(0.30) = 895$$

(3) Substitute 0.50 for π and solve the equation for n. The result is

$$n = 4264(0.50)(0.50) = 1066$$

Notice that the value of n, the sample size, is larger for $\pi = 0.50$ than it is for $\pi = 0.70$ or for $\pi = 0.80$.

In general, an appropriate sample size n to use for estimating a proportion π can be found by solving the equation $z\sigma_p = k$ for n, that is, by computing

$$n = \left(\frac{z}{k}\right)^2 \pi(1 - \pi) \tag{4.14}$$

where the value of k is the specified tolerable amount of error in the estimate p to be obtained from the sample. It is not difficult to prove that the value of n obtained from formula 4.14 is greater for $\pi = 0.50$ than for any other value of π.

The appropriate value of z to be substituted into formula 4.14 may be determined in the following way. If the plan of the investigation indicates that either (1) an estimate of a parameter and a pair of confidence limits are to be computed, or (2) a two-tails test of significance is to be made, or (3) a two-tails test of a hypothesis is to be made, use either one of the two values of z for which $\alpha/2$ is in each tail of the normal curve. (See Table 6.18.) For example, in investigations of types (1), (2), and (3), if $\alpha = 0.05$, you may use $z = 1.960$, and if $\alpha = 0.10$, you may use $z = 1.645$. If the plan of the investigation calls for either (4) a one-tail test of significance or (5) a one-tail test of a hypothesis, use a value of z that puts all of α in one tail of the normal curve. (See Table 6.19.) For example, in investigations of types (4) and (5), if $\alpha = 0.05$, you may use $z = 1.645$, and if $\alpha = 0.10$, you may use $z = 1.282$.

For a specified value of α, the value of n obtained by formula 4.14 always is smaller for a one-tail test than it is for a two-tails test. In this sense, a research plan that requires a one-tail test may be considered to be more efficient than a plan that requires a two-tails test at the same significance level.

If a situation arises in which the value of n obtained by formula 4.14 is greater than 10 per cent of the number of elementary units in the universe from which the sample is to be drawn, then you ought to use formula 4.8 rather than formula 4.7 for σ_p. Let

$$z\sqrt{\frac{\pi(1 - \pi)(N - n_1)}{n_1\,(N - 1)}} = k$$

and substitute the best available information for π or use 0.50 for π if no information about π is available. Solving this equation for n_1, we find

$$n_1 = n\left(\frac{N}{N + n - 1}\right) \tag{4.15}$$

where n is the value obtained from formula 4.14. The value of n_1 obtained from formula 4.15 always is smaller than the value of n obtained from formula 4.14. For example, let us solve case (1) again. This time we shall use equation 4.15. Let us suppose that the number N of senior-high-school boys in the community is 10,000. For $z = 1.96$, $\pi = 0.80$, and $k = 0.03$, formula 4.14 gives $n = 682$ and that is the value of n for formula 4.15. Consequently, our estimate of an appropriate sample size is

$$n_1 = 682\left(\frac{10,000}{10,000 + 682 - 1}\right) = 682\left(\frac{10,000}{10,681}\right) = 639$$

It often is desirable to obtain an estimate for each of two or more parts of a universe as well as for the whole universe. For example, you might need an estimate for the combined group of boys and girls in a school. In a public opinion poll, you might need separate estimates for the North, the South, the East, and the West, as well as for the whole United States. The methods described above can be used to determine an appropriate size of the subsample for each part of the universe so that an estimate with a specified degree of reliability can be obtained for the part. If the reliability of an estimate for the combined universe is more important than the reliability of the estimates for the separate parts of the universe, then you can accept either a larger permissible amount of error or a lower confidence level, or both of these, in the estimates for the parts than you demand in the estimate for the whole universe, in order to keep the sizes of the subsamples and the total sample within practicable bounds. (See also exercise 6 below.)

EXERCISES

1. Determine the size of a random sample that would provide an estimate of the proportion of senior-high-school boys in the community who intend to look for jobs next summer, if the probability is to be 0.99 that the sampling error in the estimate p obtained from the sample is not greater than 0.03 in absolute value. Use formula 4.14 and consider two situations: (a) Assume that you have information that makes it safe to feel that the value of π for the universe is not less than 0.70. (b) Assume that no information about the value of π is available. (*Ans.* (a) 1553; (b) 1849)

2. Repeat exercise 1, except that the probability is to be 0.99 that the sampling error in p is not greater than 0.06 in absolute value. (*Ans.* Divide the value of n obtained in exercise 1 by 4: (a) 388; (b) 462)

3. Repeat exercise 2, except that the probability is to be 0.95 that the sampling error in p is not greater than 0.06 in absolute value. (*Ans.* Divide the values of n obtained in section 4.8 for these two situations by 4: (a) 224; (b) 267)

4. Repeat exercises 1, 2, and 3 using equation 4.15 and assuming that $N = 10,001$. (*Ans.* 1(a) 1344; 1(b) 1561; 2(a) 374; 2(b) 442; 3(a) 219; 3(b) 260)

5. Assume that experienced interviewers would charge \$2.50 per interview of the boys on a list selected at random from the senior-high-school boys in the community. What would the interviewing cost for each of the surveys in exercises 1, 2, 3, and 4?

6. Suppose that there is a proposal of compulsory military drill for all the men in the school. The success or failure of the activity is much more likely to depend upon the attitudes of the men than upon the attitudes of the women or of the combined group, although it would be desirable if each group had a favorable attitude. There are 650 women and 750 men in the school. Design a sample to satisfy the following specifications: (a) The probability is to be 0.90 that the sampling error in the estimate p for the combined group will not be greater than 0.03 in absolute value. (b) The probability is to be 0.95 that the sampling error in the estimate p_1 for the women will not be greater than 0.05 in absolute value. (c) The probability is to be 0.99 that the sampling error in the estimate p_2 for the men will not be greater than 0.06 in absolute value. Assume that on the basis of similar investigations in other schools it is safe to state that the proportion π_1 of the women who favor the proposal is not less than 0.65 but the proportion π_2 of the men who favor the proposal may be anywhere from 0.40 to 0.60. Use formula 4.15. (*Ans.* 228 women and 286 men)

7. If the results of a sample survey such as that in exercise 6 show that the estimated proportion of the women who favor the proposal is 0.68 and the estimated proportion of the men who favor the proposal is 0.55, what is the estimated proportion of the combined universe of men and women that favor the proposal? See section A.2. (*Ans.* 0.61)

8. Rephrase some of exercises 1 to 4 so as to require one-tail tests of significance or one-tail tests of hypotheses. Then compute appropriate sample sizes.

REFERENCES

Sir Ronald A. Fisher, *Statistical Methods and Scientific Inference*. New York: Hafner Publishing Co., 1956.

Eugene L. Grant, *Statistical Quality Control*. New York: McGraw-Hill Book Co., Inc., 1952.

Chapter 5

ADDITIONAL APPLICATIONS OF BINOMIAL DISTRIBUTIONS

5.1 TESTS OF SIGNIFICANCE FOR DIFFERENCES IN RELATED SAMPLES

Sometimes a statistical investigation begins by the selection of two groups of elementary units from the universe by a random process. If the conditions of the investigation permit, the most efficient plan is to have the same number of elementary units in each group. One of the two groups is designated the "control" group and the other the "experimental" group. The treatment, or whatever is involved in the investigation, is administered to the elementary units in the experimental group but it is not administered to the elementary units in the control group. It is assumed that any significant difference between the two groups at the end of the investigation is due to the treatment administered to the elementary units in the experimental group.

In this type of investigation, the two samples are independent of each other. That is, the selection of the elementary units for the experimental group is in no way dependent upon the elementary units that happen to be in the control group. Such groups are called *independent* samples. Tests of significance for independent samples will be discussed in sections 5.2 and 7.6.

Another type of plan for statistical investigations is to use two related samples rather than two independent samples. The relationship between the two samples usually is established by matching and pairing each elementary unit in the control group with an elementary unit in the experimental group. Consequently, in this type of statistical investigation, there must be the same number of elementary units in the control group as there are in the experimental group.

The statistical efficiency of a plan that employs two related samples depends upon the degree to which the investigator is successful in matching the pairs of elementary units. Sometimes it is possible and appropriate to use the same elementary units in the control group and the experimental group. In this type of investigation, each elementary unit in the experimental group serves as its own control and the matching is perfect. This plan often is used in what are called "before-and-after" experiments.

One of the best ways to match the elementary units for two related samples is to use identical twins. By some random procedure, such as tossing a coin, one of the two individuals in each pair of identical twins is assigned to the control group and the other to the experimental group. Of course, only rarely is it possible to find enough available identical twins for a statistical investigation. Consequently, other ways to match and pair the elementary units must be adopted in most statistical investigations that use matched samples.

For example, in consumer surveys, market researchers sometimes select two groups of families that are matched so that the two families in each pair are as nearly as possible alike in a number of characteristics such as ages of the husband and wife, income of the family, number of children in the family, education of the husband and wife, occupation of the husband, type of neighborhood in which the family resides, and other relevant characteristics for the survey, as, for instance, having a television set. In educational experiments, two groups of students often are selected so that the two students in each pair are matched in characteristics such as sex, age, IQ, mark on an examination, aptitudes, and interests.

An appropriate method of analysis for truly quantitative data of this type often is to find the difference between the two measurements for each matched pair and then, by the method to be discussed in section 7.5.5, test the hypothesis that the mean of the universe of such differences is zero.

In the following subsection, a test of significance for differences observed in two matched samples will be given. It is a simple application of the binomial probability law and it illustrates the practical importance of the idea of binomial probability in statistics. Furthermore, the sign test of significance does not require that any assumption whatsoever be made about the shape of the distribution in the universe from which the samples are drawn.

This is a very important characteristic of the sign test, because it makes the test much more widely applicable than many other tests of significance. There are many practical statistical situations in which the investigator does not know what the shape of the distribution in the universe is or suspects that it may be definitely unlike a normal distribution. The sign test of significance is not affected by the shape of the distribution in the universe and it is equally valid no matter what that shape may be.

The sign test has another very important property, namely, that it is not necessary that the data for the sign test be in the form of a truly quantitative classification.

5.1.1 The sign test. The sign test is one of the easiest to apply of all the tests of significance because it requires very little arithmetic computation. Furthermore, it is one of the most widely applicable of all the tests of significance available in statistics; the number and variety of the experiments and investigations in which the sign test may be used are practically unlimited. In other words, the sign test is both extremely simple and extremely flexible. Several illustrations of its usefulness will now be given.

Suppose that some perfectly matched pairs of elementary units have been selected for an experiment. Suppose, further, that the experiment results in two scores or two measurements for each matched pair; one score or one measurement indicates the condition of one member of the matched pair and the other score or measurement indicates the condition of the other member of the matched pair. The data resulting from such an experiment or investigation might be similar in form to that in Table 5.1.

TABLE 5.1

Data for the Sign Test for Two Matched Samples

Matched Pair	Score of One Member under Condition 2	Score of Other Member under Condition 1	Difference for Matched Pair	Sign for Matched Pair
A_1 B_1	$x_1 = 68$	$y_1 = 62$	$x_1 - y_1 = +6$	+
A_2 B_2	$x_2 = 69$	$y_2 = 56$	$x_2 - y_2 = +13$	+
A_3 B_3	$x_3 = 64$	$y_3 = 65$	$x_3 - y_3 = -1$	−
A_4 B_4	$x_4 = 67$	$y_4 = 67$	$x_4 - y_4 = 0$	
A_5 B_5	$x_5 = 70$	$y_5 = 63$	$x_5 - y_5 = +7$	+
A_6 B_6	$x_6 = 66$	$y_6 = 64$	$x_6 - y_6 = +2$	+
A_7 B_7	$x_7 = 61$	$y_7 = 61$	$x_7 - y_7 = 0$	
A_8 B_8	$x_8 = 73$	$y_8 = 68$	$x_8 - y_8 = +5$	+
A_9 B_9	$x_9 = 71$	$y_9 = 67$	$x_9 - y_9 = +4$	+
A_{10} B_{10}	$x_{10} = 64$	$y_{10} = 69$	$x_{10} - y_{10} = -5$	−
A_{11} B_{11}	$x_{11} = 65$	$y_{11} = 60$	$x_{11} - y_{11} = +5$	+
A_{12} B_{12}	$x_{12} = 70$	$y_{12} = 67$	$x_{12} - y_{12} = +3$	+
A_{13} B_{13}	$x_{13} = 71$	$y_{13} = 62$	$x_{13} - y_{13} = +9$	+
A_{14} B_{14}	$x_{14} = 60$	$y_{14} = 59$	$x_{14} - y_{14} = +1$	+
A_{15} B_{15}	$x_{15} = 62$	$y_{15} = 65$	$x_{15} - y_{15} = -3$	−
A_{16} B_{16}	$x_{16} = 62$	$y_{16} = 58$	$x_{16} - y_{16} = +4$	+

The two conditions might be, for example, (1) two different methods of teaching a course in school, or (2) two different methods of on-the-job training in a factory or an office, or (3) two different medical treatments for reducing the fever of sick people, or (4) two different merchandising methods for a manufactured product. For each matched pair, the member assigned to each of the two treatments or conditions is determined by pure chance, for example, by tossing a coin.

The plus and minus signs in the last column of Table 5.1 are the raw material for the sign test of significance. The principle that is the basis for the sign test is simply that if the two conditions are equivalent, plus and minus signs would be equally likely to occur and, if it were not for chance in random sampling, one-half of the signs would be plus and the other half would be minus. In other words, the sign test is based on the binomial probability distribution $(\frac{1}{2} + \frac{1}{2})^n$ where n is the total number of plus and minus signs for the matched pairs in the investigation.

The only assumptions involved in the sign test are that the underlying variable is continuous and that the difference observed for each matched pair is not dependent upon the difference observed for any of the other matched pairs. The assumption that the variable is continuous implies that, theoretically, no matched pair could have tied scores or measurements, that is, equal scores or measurements under the two conditions. However, in practical situations, perfectly precise measurement of a continuous variable is not to be expected and, therefore, it is possible for the two members of a matched pair to obtain the same observed score under the two conditions. For such a matched pair, the difference between scores will be zero. This happened to the fourth and the seventh matched pairs in Table 5.1. In the sign test, all matched pairs that produce zero differences are dropped from the analysis. Consequently, the data for A_4B_4 and A_7B_7 are discarded and the remaining 14 matched pairs are used for the sign test.

In Table 5.1, there are three minus signs and 11 plus signs. Let $n_1 = 3$, the smaller number of signs. Let $n_2 = 11$. Then $n = n_1 + n_2 = 3 + 11 = 14$.

The true binomial probabilities for the application of the sign test here are given by the binomial distribution $(\frac{1}{2} + \frac{1}{2})^{14}$. The probability of obtaining *not more than* three minus signs in a random sample of 14 plus and minus signs if $P = Q = \frac{1}{2}$ may be found by adding the first four terms of the binomial expansion. This sum is

$$\left(\frac{1}{2}\right)^{14} + 14\left(\frac{1}{2}\right)^{13}\left(\frac{1}{2}\right) + \frac{14(13)}{1(2)}\left(\frac{1}{2}\right)^{12}\left(\frac{1}{2}\right)^2 + \frac{14(13)(12)}{1(2)(3)}\left(\frac{1}{2}\right)^{11}\left(\frac{1}{2}\right)^3$$

$$= \frac{1}{16384}(1 + 14 + 91 + 364) = \frac{470}{16384} = 0.029$$

Consequently, the probability for our one-tail test here is 0.029, and this is greater than $\alpha = 0.01$. Therefore, we cannot reject the null hypothesis at the $\alpha = 0.01$ level of significance.

Table 5.2 enables us to find the true binomial probabilities involved in the sign test without arithmetical computation, if n is not greater than 31. For example, to find the probability for our illustration here with $n_1 = 3$ and $n = 14$, all that we need to do is look at Table 5.2 on the line for $n = 14$ and in the column for $n_1 = 3$. There we find the number 0.29, which is the probability that was computed above.

TABLE 5.2

Some Useful Binomial Probabilities When $P = Q = \frac{1}{2}$.

Probability of not more than n_1 occurrences in n trials when $P = Q = \frac{1}{2}$. That is, the sum of the first (or last) $n_1 + 1$ terms of the expansion of $(\frac{1}{2} + \frac{1}{2})^n$.

n \ n_1	0	1	2	3	4	5	6	7	8	9	10	11	12	13	14	15
1	.500															
2	.250	.750														
3	.125	.500														
4	.062	.312	.688													
5	.031	.188	.500													
6	.016	.109	.344	.656												
7	.008	.062	.227	.500												
8	.004	.035	.145	.363	.637											
9	.002	.020	.090	.254	.500											
10	.001	.011	.055	.172	.377	.623										
11		.006	.033	.113	.274	.500										
12		.003	.019	.073	.194	.387	.613									
13		.002	.011	.046	.133	.291	.500									
14		.001	.006	.029	.090	.212	.395	.605								
15			.004	.018	.059	.151	.304	.500								
16			.002	.011	.038	.105	.227	.402	.598							
17			.001	.006	.025	.072	.166	.315	.500							
18			.001	.004	.015	.048	.119	.240	.407	.593						
19				.002	.010	.032	.084	.180	.324	.500						
20				.001	.006	.021	.058	.132	.252	.412	.588					
21				.001	.004	.013	.039	.095	.192	.332	.500					
22					.002	.008	.026	.067	.143	.262	.416	.584				
23					.001	.005	.017	.047	.105	.202	.339	.500				
24					.001	.003	.011	.032	.076	.154	.271	.419	.581			
25						.002	.007	.022	.054	.115	.212	.345	.500			
26						.001	.005	.014	.038	.084	.163	.279	.423	.577		
27						.001	.003	.010	.026	.061	.124	.221	.351	.500		
28							.002	.006	.018	.044	.092	.172	.286	.425	.575	
29							.001	.004	.012	.031	.068	.132	.229	.356	.500	
30							.001	.003	.008	.021	.049	.100	.181	.292	.428	.572
31								.002	.005	.015	.035	.075	.141	.237	.360	.500

The sign test can be used for either a one-tail test or a two-tails test. If our alternative to the null hypothesis in the illustration based on Table 5.1 had been simply that in the universe from which the matched pairs were drawn the number of plus signs is not equal to the number of minus signs, then we would be making a two-tails test and would have to double the probability that was obtained for the one-tail test. In other words, we would add the first four terms and the last four terms of the binomial expansion. The result would be $0.029 + 0.029 = 0.058$. In this two-tails test, we could not reject the null hypothesis at the 5 per cent significance level because $P = 0.058$ is not as small as $\alpha = 0.05$.

If the value of n for the sign test is greater than 31, we cannot use Table 5.2 to find the required probability. But we have seen already in section 4.5 that binomial probabilities for large values of n can be approximated very satisfactorily by using the normal probability table, especially if $P = Q = \frac{1}{2}$, which is the hypothesis for the sign test. Here we have $n = n_1 + n_2$. By formulas 4.2 and 4.3, we get $\mu = nP = n/2$ and $\sigma = \sqrt{nPQ} = (\frac{1}{2})\sqrt{n}$.

Because we are using a continuous normal distribution to approximate a discrete binomial distribution, we need to use the correction for continuity that was explained in section 4.4. That is, we must begin a half-step above or below $x = n_2$, depending on whether $n_2 < n/2$ or $n_2 > n/2$. Consequently, for the sign test, formula 3.5 becomes

$$z = \frac{(n_2 \pm 0.5) - n/2}{(\frac{1}{2})\sqrt{n}} \qquad \begin{array}{l} \text{using } +0.5 \text{ if } n_2 < n/2 \\ \text{using } -0.5 \text{ if } n_2 > n/2 \end{array}$$

which simplifies to

$$z = \frac{n_2 - n_1 \pm 1}{\sqrt{n_1 + n_2}} \qquad \begin{array}{l} \text{using } +1 \text{ if } n_2 < n_1 \\ \text{using } -1 \text{ if } n_2 > n_1 \end{array}$$

Find the probability P associated with this value of z in Table 6.7 (p. 172).

For example, suppose that in an investigation in which you desire to use the sign test you find that $n_1 = 19$ and $n_2 = 30$. Then $n = n_1 + n_2 = 49$ and $z = (30 - 19 - 1)/\sqrt{49} = 10/7 = 1.43$. Table 6.7 shows that the normal probability associated with $z = 1.43$ is $P = 0.076$.

Suppose that the people in a randomly selected sample from a universe are examined and classified according as they are in favor of or opposed to some proposal. This provides "before" data for the people in the sample. Next, suppose that all the people in the sample are exposed to a publicity campaign intended to develop a favorable attitude toward the proposal. Finally, each person in the sample is examined and classified again according as he or she is in favor of or opposed to the proposal. This provides "after" data for the people in the sample. Assign a plus sign to each person who changed in a desirable way and assign a minus sign to each person who changed in an undesirable way. The data resulting from such an investigation might be arranged as in Table 5.3.

TABLE 5.3

Before-and-After Data

Person	Attitude before Treatment	Attitude after Treatment	Sign
A	In favor of	Opposed to	−
B	Opposed to	Opposed to	
C	Opposed to	In favor of	+
D	In favor of	In favor of	
E	Opposed to	In favor of	+
F	In favor of	Opposed to	−
G	Opposed to	In favor of	+
H	Opposed to	In favor of	+

To carry out the sign test of significance for such data, first, count the numbers of plus and minus signs and let n_1 equal the smaller of the two numbers. For Table 5.3, $n_1 = 2$ and $n_2 = 4$. Let $n = n_1 + n_2 = 6$. Disregard the two people whose attitudes toward the proposal were the same before and after the treatment.

The null hypothesis for this investigation is that the publicity campaign dealing with the proposal had (or would have) no effect on the people in the universe from which the sample was drawn. The alternative probably would be that the treatment caused (or would cause) more people in the universe to change from "opposed to the proposal" to "in favor of the proposal" than from "in favor of the proposal" to "opposed to the proposal." In this alternative we have predicted the direction in which the change would tend. Consequently, we must use a one-tail test. Let us make the test at the $\alpha = 0.10$ level of significance.

The exact probability required in this sign test could be computed by adding the first $n_1 + 1 = 2 + 1 = 3$ terms of the binomial expansion of $(\frac{1}{2} + \frac{1}{2})^6$. We can find the probability in Table 5.2 with no computation. On the line for $n = 6$ and in the column under $n_1 = 2$, we find the number .344 which is the required probability. Because $P = 0.344$ is larger than $\alpha = 0.10$, we cannot reject the null hypothesis in this illustration.

It is important to remember that the sign test would have been equally applicable if the values of x and y in columns two and three of Table 5.1 for each matched pair had been their ranks with respect to each other instead of scores. In fact, the sign test is especially useful in investigations in which truly quantitative measurement, although theoretically possible, is impracticable, but in which the investigator is able to rank with respect to each other the two members of each matched pair or is able to rank the status of each person after treatment with respect to his or her prior status.

In general, the null hypothesis for the sign test is that the median of the differences in the universe is zero. At the same time, the sign test provides an estimate of the probability of a positive reaction.

Sometimes the data are such that, or the scales can be transformed so that, the sign test can be used to test the hypothesis that treatment A is better than treatment B by a specified number of percentage points, say, 5 (multiply each value of y by 1.05 and apply the sign test), or that treatment A is better than treatment B by a specified number of units, say, 10 (add 10 to each value of y and apply the sign test).

Here are a few technical remarks about the sign test. It is necessary to have at least six signs in order to reject the null hypothesis at a significance level $\alpha \leq 0.05$, and even with six signs, rejection by the sign test requires that all six signs be of the same kind. If $n_1 = 0$ and $n = 6$, the probability is $1/64 = 0.016$ represented by the first term of the expansion of $(\frac{1}{2} + \frac{1}{2})^6$. If $n_1 = 1$ and $n = 6$, the probability is $7/64 = 0.109$ represented by the sum of the first two terms of that expansion. In other words, if $n_1 = 0$ and $n = 6$, the associated probability is $P = 0.016$, which is less than 0.05. If $n_1 = 1$ and $n = 6$, the associated probability is $P = 0.109$, which is greater than 0.05. Because n_1 is a discrete variable and cannot have any value between 0 and 1, there cannot be an associated probability P which is exactly 0.05; in fact, P cannot have a value between 0.016 and 0.109 if $n = 6$. Therefore, if we wish to make the sign test at a level $\alpha \leq 0.05$ when $n = 6$, we are forced to make the test at the level $\alpha = 0.016$, and this requires that $n_1 = 0$ if the null hypothesis is to be rejected.

EXERCISES

1. In Table 5.3, change the attitude of person B after treatment to "in favor of" and carry out the binomial test. Is the conclusion the same as before? (*Ans.* Yes)

2. In Table 5.1, change x_8 to 68 and change y_8 to 73. Carry out the sign test with the revised data. Is your conclusion the same as that given for the original data? (*Ans.* Yes)

3. In an experiment to determine ways to improve the working relationships in its factories, a large corporation selected 25 foremen at random from the several hundred in its plants. Then, without the foremen or their men knowing it, a record was kept of all the complaints made to management by the workmen about the 25 foremen during a period of six months. After that, the 25 foremen were sent to a management training school maintained by the corporation; there they attended for four weeks a program of lectures and discussion meetings dealing with human relations. For the next six months after the 25 foremen returned to their jobs a record was again kept of all the complaints made about them to management by the workmen. A comparison of the records of each of the 25 foremen for the six months prior to the training period and the six months following the training period shows the following results:

Foreman	No. of Complaints after Training	Foreman	No. of Complaints after Training
A	Less	N	Same
B	Same	O	Less
C	More	P	More
D	Less	Q	Less
E	Less	R	Less
F	Less	S	More
G	Same	T	More
H	Less	U	Less
I	More	V	Less
J	Less	W	Less
K	Less	X	Same
L	Less	Y	Less
M	More		

Assuming that all other conditions of work remained the same during the two six-month work periods, test the null hypothesis that education in human relations has no effect on the ability of foremen to get along with the men who work for them. Use as alternative the statement that foremen who have attended a training course in human relations tend to get along better with their men. Make the test at the $\alpha = 0.05$ level of significance. Is this a one-tail test or a two-tails test? (*Ans.* Table 5.2 shows that $P = 0.039$ when $n_1 = 6$, $n = 21$; reject the null hypothesis in this one-tail test.)

4. In order to study the relative advantages of two different types of dials on the instrument panel in a jet fighter plane, 12 matched pairs of pilot trainees were given the same series of 100 test readings on each of the two types of instrument dials under simulated operating conditions. One member of each matched pair made the 100 test readings on type I dials and the other member of each matched pair made the 100 test readings on type II dials. The score of each individual is the number of errors that he made during the 100 test readings of the instrument dials.

Matched Pair	Score of A on Type I Dials	Score of B on Type II Dials
$A_1 B_1$	7	9
$A_2 B_2$	3	1
$A_3 B_3$	5	5
$A_4 B_4$	10	3
$A_5 B_5$	6	7
$A_6 B_6$	8	4
$A_7 B_7$	1	4
$A_8 B_8$	4	2
$A_9 B_9$	2	1
$A_{10}B_{10}$	9	4
$A_{11}B_{11}$	8	2
$A_{12}B_{12}$	3	3

Apply the sign test to these data to test the null hypothesis that there is no difference in the readability of the two types of dials. Use as alternative the statement that there is a difference in readability between the two types of dials. Is this one-tail test or a two-tails test? Use the $\alpha = 0.05$ level of significance for the test. (*Ans.* Do not reject the null hypothesis in this two-tails test.)

Do you accept the null hypothesis about the readability of the dials or do you think that the experiment should be repeated with a larger sample of pilot trainees?

5. A random sample containing 1000 of the households in a large city was surveyed immediately before the beginning of and immediately after the conclusion of a special advertising campaign and sale for a certain brand "X" of toothpaste. The results of the two surveys were as follows: 230 of the sample households had brand "X" toothpaste on hand at the time of the first survey; of these 230 households, 218 had brand "X" toothpaste on hand at the time of the second survey. Of the 770 households that did not have brand "X" toothpaste on hand at the time of the first survey 42 households had brand "X" toothpaste on hand at the time of the second survey. Test the significance of these findings at the $\alpha = 0.01$ level of significance for a one-tail test. What is the null hypothesis here? What is the alternative that you used here? (*Ans.* $z = 4$ approximately, which is highly significant.)

6. In a large military training school for radiomen, an experiment was carried out to determine whether by installing electric typewriters for all trainees the length of the training period could be shortened without lowering the standard of skill in typewriting. Matched pairs of radioman trainees were selected and one member of each matched pair was assigned to a class in which he studied typewriting of radio code on a manual typewriter for four weeks. The other member of each matched pair was assigned to a class in which he studied typewriting of radio code on an

Matched Pair	Score of Member Who Learned on Manual Typewriter	Score of Member Who Learned on Electric Typewriter
$A_1 B_1$	18	17
$A_2 B_2$	21	23
$A_3 B_3$	15	12
$A_4 B_4$	23	21
$A_5 B_5$	13	14
$A_6 B_6$	16	16
$A_7 B_7$	19	15
$A_8 B_8$	17	22
$A_9 B_9$	14	18
$A_{10} B_{10}$	20	13
$A_{11} B_{11}$	22	19
$A_{12} B_{12}$	15	20
$A_{13} B_{13}$	21	15
$A_{14} B_{14}$	18	19
$A_{15} B_{15}$	19	17
$A_{16} B_{16}$	17	20
$A_{17} B_{17}$	20	16
$A_{18} B_{18}$	19	18
$A_{19} B_{19}$	18	13
$A_{20} B_{20}$	20	12

electric typewriter for three weeks and used a manual typewriter during the fourth week. Then all the trainees were given the same skill test on manual typewriters. Manual typewriters were used by both classes for the final test because radiomen must use them under most field conditions. A student's score on the test was the average number of words typed per minute by him minus the average number of errors made per minute by him. Five strokes, including the spacer, constituted a word. Use the scores of the 20 matched pairs of trainees given her: to test the hypothesis that the kind of typewriter, manual or electric, on which a radioman trainee learns to typewrite radio code has no effect on the degree of typewriting skill acquired by the trainee. Choose a suitable alternative and make the test of significance at the $\alpha = 0.05$ level. (*Ans.* The sign test does not reject the null hypothesis in a two-tails test at the $\alpha = 0.05$ level.)

7. Suppose that the report of an investigation not conducted by you tells you that 40 randomly selected people in a specified universe were used as their own "controls" in a "before-and-after" experiment. A questionnaire was administered to the 40 people to determine each person's status with respect to tolerance for other people. Then a motion picture designed to inform people about undesirable aspects of intolerant behavior and desirable aspects of tolerant behavior was shown to the group. Finally, the questionnaire (or an equivalent form) was administered to the 40 people again and their new status with respect to tolerance for other people was determined. The report states that 25 people had a more tolerant attitude after being exposed to the motion picture than they had before seeing the picture, 12 people had a less tolerant attitude after being exposed to the picture than they had before seeing it, and that the attitude of 3 people appeared to be unchanged.

You can use the sign test of significance here because a person's degree of tolerance for other people may be regarded as a continuous variable. Consequently, the underlying variable for the sign test, namely, a person's before-and-after difference in tolerance, is a continuous variable, even though the results of the experiment show only that each person was more tolerant or less tolerant or unchanged after being exposed to the motion picture. State the null hypothesis and a suitable alternative for a one-tail test and make the test at the $\alpha = 0.05$ level. (*Ans.* $z = 1.97$, $P = 0.024$; reject the null hypothesis and accept the alternative.)

5.2 TEST OF SIGNIFICANCE FOR DIFFERENCES IN INDEPENDENT SAMPLES

A few years ago, several companies began to broadcast new, short commercial announcements about their products over a local radio station in a large city. The companies and the radio station wished to know how effective this kind of advertising is. They hired a market-research organization to estimate the proportion of the people in the area who remembered hearing the commercial for each product.

At the end of the second week of broadcasting, the researchers made a sample survey of the people living in the metropolitan area of the city.

At the end of the fourth week of broadcasting, the researchers made a second sample survey of the people in the area. A different sample of the population was used in each survey. These were independent samples because the selection of people for the second survey was not influenced in any way by the selection of people who happened to be in the first survey. There were approximately 1000 people in each of the two samples. The results of the two sample surveys for two of the products were:

| | Per Cent of People Who Remembered Hearing the Commercial | |
	For a Motion Picture	For a Tooth Brush
At end of 2 weeks	28.8	21.5
At end of 4 weeks	42.8	20.9

How should the management of the companies and the operators of the radio station interpret these results? Assuming that the two samples were simple random probability samples, we can test the significance of the difference between the two percentages for each product. For the motion picture, the difference in the two sample proportions, which shall be denoted by p_1 and p_2 because they are estimates of the parameters π_1 and π_2, is $p_2 - p_1 = 0.428 - 0.288 = 0.140$.

Notice that although the results were presented in the form of percentages at first, they have been put into the form of proportions for this test of significance. The test can be carried out by using either percentages or proportions, but many statisticians prefer to use proportions in this test. Percentages probably are preferable for the presentation of findings in a report to management.

In the planning stage of a research project such as this, the null hypothesis, the alternative hypothesis, the procedure for testing the null hypothesis, and the significance level for the test would be chosen. Because the company and the radio station would desire to find that $\pi_2 > \pi_1$, the researchers would choose as the null hypothesis the statement that $\pi_2 \leq \pi_1$ and they would choose as the alternative the statement that $\pi_2 > \pi_1$. This pair of hypotheses requires a one-tail test. Let us suppose that the researchers decided that they did not want to take more than a 1 per cent risk of drawing the wrong conclusion, that is, $\alpha = 0.01$.

If the probability is as small as or smaller than 0.01 that a difference as great as the observed difference between the two proportions would be the result of sampling errors alone, then we can conclude that there was a statistically significant change in the population of the city during the final two weeks of the broadcasting for the motion picture.

In order to test the significance of the difference $p_2 - p_1 = 0.140$ for the motion picture, we must change the difference 0.140 into standard units so that we can apply the table of normal probabilities. We can use formula 5.2 to do this. The result is

$$z = \frac{0.428 - 0.288}{\sqrt{(.358)(.642)\left(\dfrac{1}{1000} + \dfrac{1}{1000}\right)}} = \frac{+0.140}{\sqrt{.000460}} = \frac{+0.140}{+0.0214} = +6.54$$

Table 6.7 shows that the normal probability of obtaining a value of z as large as or larger than $+6.54$ is much smaller than 0.01. Consequently, we reject the null hypothesis and accept the alternative hypothesis that there was a statistically significant increase during the final two weeks of broadcasting in the proportion of the population who remembered having heard the commercial announcements for the motion picture.

If you carry out a similar test for the tooth brush commercial, the value of z turns out to be -0.33. Consequently, the small apparent decrease in the proportion of the population who remembered having heard the commercial announcements for the tooth brush easily can be attributed to chance variations in drawing two random samples from a universe in which no change had taken place insofar as the tooth brush commercial was concerned.

Here is the theoretical basis for this test of significance. Imagine that you have drawn a large number of pairs of independent random samples of sizes n_1 and n_2 from two large universes for which you do not know the values of the two proportions π_1 and π_2 that possess a particular attribute. It is not necessary that n_1 be equal to n_2. Imagine also that you have computed the proportions p_1 and p_2 that possess the attribute in each pair of samples and that you have computed the difference $p_2 - p_1$ for each pair of samples. These differences could be tabulated in the form of a frequency distribution and graphed in the form of a histogram.

We need to know the properties of this frequency distribution of the differences $p_2 - p_1$. Here they are: (1) Although we are dealing here with proportions and binomial probabilities, the frequency distribution and the histogram for the differences $p_2 - p_1$ can be approximated satisfactorily by a normal distribution and a normal curve. (2) The mean of this frequency distribution of differences is equal to the difference between the two proportions in the two universes; that is, as shown in section A.9, $\mu_{p_2-p_1} = \pi_2 - \pi_1$. (3) The standard deviation of the frequency distribution of differences is

$$\sigma_{p_2-p_1} = \sqrt{\frac{\pi_1(1-\pi_1)}{n_1} + \frac{\pi_2(1-\pi_2)}{n_2}}$$

The difference $p_2 - p_1$ between a specific pair of sample proportions can be changed into standard units by using formula 3.5 which gives us in this case

$$z = \frac{(p_2 - p_1) - (\pi_2 - \pi_1)}{\sqrt{\dfrac{\pi_1(1-\pi_1)}{n_1} + \dfrac{\pi_2(1-\pi_2)}{n_2}}} \qquad (5.1)$$

The hypothesis most frequently tested here is the null hypothesis that $\pi_1 = \pi_2$. Now, while we are testing a null hypothesis, we assume that the null hypothesis is true, we base the computations on that assumption, and in this way we determine the probability that a value of z as large as or larger in absolute value than the computed value of z would occur by pure chance in random sampling if the null hypothesis were true. Consequently, when we are testing the null hypothesis that $\pi_1 = \pi_2$ or, what is the same thing, that $\pi_2 - \pi_1 = 0$, we may represent these two equal but unknown numbers π_1 and π_2 by π.

To carry out the test of this null hypothesis, we consider p_1 and p_2 to be two independent estimates of π. Under the assumed conditions, a more reliable estimate of π is the weighted average of p_1 and p_2, namely,

$$p = \frac{n_1 p_1 + n_2 p_2}{n_1 + n_2}$$

This value of p is, of course, exactly the same value of the proportion p that would be obtained by combining the two samples into one sample. The test value of z is obtained by substituting the improved estimate p for the unknown π. Then, formula 5.1 becomes

$$z = \frac{p_2 - p_1}{\sqrt{p(1-p)\left(\dfrac{1}{n_1} + \dfrac{1}{n_2}\right)}} \qquad (5.2)$$

If the size of either of the two universes from which the two samples are drawn is relatively small compared with the size of the sample, and if the sample is drawn without replacement, the finite multiplier should be incorporated in the formula for $\sigma_{p_2-p_1}$ and in the denominators of formulas 5.1 and 5.2.

Table 6.7 will give us a satisfactory approximation to the true binomial probability of obtaining, by chance in random sampling, a value of z as large as or larger in absolute value than the value of z computed by formulas 5.1 or 5.2, provided that all the quantities $n_1 p_1$, $n_2 p_2$, $n_1 q_1$, and $n_2 q_2$ are greater than five. If you ever find it necessary to test the significance of the difference between two sample proportions when some of these quantities are less than or equal to five, you should consult a more advanced book on statistical methods.

If the two samples are not independent of each other, formula A.4 must be used in developing new formulas corresponding to formulas 5.1 and 5.2, or you can use the method given for related samples in section 5.1.1.

We must often distinguish between statistical significance and practical significance. Most businessmen are practical men and to them practical significance often means financial significance. Radio and television commercials are expensive. Let us suppose that the advertising for the motion picture mentioned earlier would not be profitable unless there was a gain in familiarity by more than 5 per cent of the universe during the promotion period. Do the results of the two surveys show convincing evidence of this kind of practical significance?

For this test, the null hypothesis is $\pi_2 - \pi_1 \leq 0.05$ and the alternative is $\pi_2 - \pi_1 > 0.05$. We can obtain an approximation by using formula 5.1 with $\pi_2 - \pi_1 = 0.05$. The result is

$$z = \frac{(0.428 - 0.288) - (0.05)}{\sqrt{\dfrac{(0.428)(0.572)}{1000} + \dfrac{(0.288)(0.712)}{1000}}} = \frac{+0.090}{+0.0212} = +4.25$$

Tables 6.7 and 6.19 show that the probability associated with a value of z as great as or greater than $z = +4.25$ is much less than 0.001. Consequently, there is strong evidence that the gain for the motion picture was more than 5 per cent in this universe. That is, the gain for the motion picture was practically significant as well as statistically significant.

In 1937, about 450,000 Americans were stricken with pneumonia and about 120,000 of them died because of that disease. In one hospital, some doctors divided the patients who were victims of the most deadly type of pneumonia into two groups or samples by a random procedure. To one group they gave the newly discovered sulfanilamide treatment. To the other group they gave the same treatment that had been standard in the past. Eighty per cent of the first group got well. Eighty per cent of the other group died.

When this statistical evidence became known to all the doctors, they used the sulfanilamide tablets for all types of pneumonia. Recoveries were so rapid and easy that one doctor exclaimed, "People just won't die of pneumonia any more!" No doubt, some of you or some members of your families have had pneumonia in recent years and have been saved by the knowledge gained in this experiment. Thousands of other statistical experiments have resulted in better things for us to eat and to wear and to give us entertainment. There is no reason under the sun why you cannot use the same kind of statistical thinking to bring benefits to yourself and to your fellow men.

EXERCISES

1. Other products in the test of frequently repeated short commercial radio broadcasts mentioned in section 5.2 showed the following results:

	Per Cent of People Who Remembered Hearing the Commercials	
Product	*Survey No. 1*	*Survey No. 2*
A hair tonic	27.8	26.3
A pie mix	30.8	42.3
An airline	27.8	42.8
A liniment	15.2	22.4
A cigarette	36.0	37.3
An antifreeze	18.3	21.5

Which, if any, of the companies were rewarded with statistically significant increases in familiarity with their commercials? Use the $\alpha = 0.05$ level of significance and a one-tail test. (*Ans.* All except the hair tonic and the cigarette.) Is the apparent loss in familiarity suffered by the hair tonic too great to be attributed to sampling error alone? (*Ans.* No). Which, if any, of the companies were rewarded with financially significant increases, if a gain of more than 3 per cent of the universe was required to make the advertising for a product profitable? (*Ans.* Pie mix, airline, liniment.)

2. The United States Government experimental farm at Beltsville, Maryland, performed an experiment to demonstrate to chicken breeders the importance of vitamin B_{12} in the diet of breeding hens. Two random samples of eggs, each containing 100 fertile eggs, were selected. The eggs in sample number one were laid by a group of hens whose diet was deficient in vitamin B_{12}. The eggs in sample number two were laid by hens whose diet was supplemented by fish meal to supply vitamin B_{12}. The two samples of eggs were set in incubators. Of the 100 eggs in the first sample, 34 failed to hatch and 19 produced chicks that died in the first week after hatching. Of the 100 eggs in the second sample, 15 failed to hatch and 4 produced chicks that died in the first week after hatching.

What is the proportion p_1 of the eggs in the first sample that produced healthy chicks? (*Ans.* $p_1 = 0.47$.) What is the proportion p_2 for the second sample? (*Ans.* $p_2 = 0.81$.) How many standard units are there in the difference between p_1 and p_2? (*Ans.* $z = 5.01$.) Is the evidence strong enough to convince you that chicken breeders should reject at the $\alpha = 0.001$ significance level the hypothesis that vitamin B_{12} has no effect on the proportion of healthy chicks produced by hens? (*Ans.* Yes.)

Suppose that it is considered necessary for π_2 to be more than 10 percentage points greater than π_1 in order to cover the extra cost of the vitamin B_{12}. Test the null hypothesis that $\pi_2 - \pi_1 \leq 0.10$ against the alternative that $\pi_2 - \pi_1 > 0.10$. How would you interpret the result of this test to a chicken breeder? (*Ans.* $z = +3.76$; reject the null hypothesis at the $\alpha = 0.001$ level and accept the alternative.)

3. In a sample containing 6872 boys in Maryland, it was found that "going to the movies" was the main leisure-time activity of 9.4 per cent of the boys. A sample containing 6635 girls in Maryland showed that "going to the movies" was the main leisure-time activity of 12.0 per cent of the girls. Assuming that these were two independent random samples of the boys and the girls in Maryland at the time of the

survey, test the significance of the observed difference in the proportions of boys and girls for whom "going to the movies" was the main leisure-time activity. Choose an appropriate null hypothesis, alternative, and significance level.

4. I. P. Garrod reported in the journal *Tubercle* in September 1952 that in a series of 47 patients with pulmonary actinomycosis the mortality rate was 22 per cent among the 33 patients to whom penicillin was administered and 79 per cent among the 14 patients who were not given an antibiotic. Choose an appropriate null hypothesis, alternative, and significance level and then test the significance of the observed results. What assumptions are required in this test of significance? (*Ans. z* = +3.70.)

5.3 CONFIDENCE LIMITS FOR A MEDIAN

In section 3.3.3, we found that the median pay rate of the 12 truck drivers whose pay rates were given in Table 3.2 was $1.80. Here, the sample size was $n = 12$. Looking at Table 5.4 on the line for $n = 12$ and in the column under "confidence greater than or equal to 95 per cent," we see that $k = 3$, $n - k + 1 = 10$, and the exact degree of confidence is 98.07 per cent. This means that the pay rate of the third truck driver in the sample and the pay rate of the tenth truck driver in the sample are 98.07 per cent confidence limits for the median pay rate of the universe of truck drivers from which the sample of 12 truck drivers was drawn. The pay rate of the third driver in the list is $1.55 and the pay rate of the tenth driver in the list is $2.05.

Similarly, looking at Table 5.4, on the line for $n = 12$ and in the column under "confidence greater than or equal to 99 per cent," we see that $k = 2$, $n - k + 1 = 11$, and the exact degree of confidence is 99.68 per cent. This means that we may have 99.68 per cent confidence that the median pay rate of the universe of truck drivers is between the pay rate of the second driver in the list in Table 3.2 and the pay rate of the eleventh driver in the list, that is, between $1.50 and $2.15.

This method for finding confidence limits to be attached to the median of a random sample is based on the definition of the median, namely, that the median of a distribution is the value of the variable that splits the distribution into two parts each of which contains 50 per cent of the elementary units in the distribution. Consequently, the probability P that an elementary unit drawn at random from a universe will have a value of the variable greater than the median of the universe is $\frac{1}{2}$ and the probability $Q = 1 - P$ that an elementary unit drawn at random from the universe will have a value of the variable that is not greater than the median of the universe is $\frac{1}{2}$. Therefore, in a random sample of size n, the probabilities of obtaining specified numbers of values of the variable greater (or less) than the median of the universe are given by the terms of the expansion of the binomial $(\frac{1}{2} + \frac{1}{2})^n$.

TABLE 5.4

Confidence Limits for the Median*

$x_k <$ median $< x_{n-k+1}$

n	k	Confidence $\geq$ 95 Per Cent $n-k+1$	Confidence	k	Confidence $\geq$ 99 Per Cent $n-k+1$	Confidence
6	1	6	98.44%			
7	1	7	99.22			
8	1	8	99.61	1	8	99.61%
9	2	8	98.05	1	9	99.80
10	2	9	98.93	1	10	99.90
11	2	10	99.41	1	11	99.95
12	3	10	98.07	2	11	99.68
13	3	11	98.88	2	12	99.83
14	3	12	99.35	2	13	99.91
15	4	12	98.24	3	13	99.63
16	4	13	98.94	3	14	99.79
17	5	13	97.55	3	15	99.88
18	5	14	98.46	4	15	99.62
19	5	15	99.04	4	16	99.78
20	6	15	97.93	4	17	99.87
21	6	16	98.67	5	17	99.64
22	6	17	99.15	5	18	99.78
23	7	17	98.27	5	19	99.87
24	7	18	98.87	6	19	99.67
25	8	18	97.84	6	20	99.80
26	8	19	98.55	7	20	99.53
27	8	20	99.04	7	21	99.70
28	9	20	98.22	7	22	99.81
29	9	21	98.79	8	22	99.59
30	10	21	97.86	8	23	99.74
31	10	22	98.53	8	24	99.87
32	10	23	99.00	9	24	99.65
33	11	23	98.25	9	25	99.77
34	11	24	98.78	10	25	99.55
35	12	24	97.95	10	26	99.70
36	12	25	98.56	10	27	99.80
37	13	25	97.65	11	27	99.62
38	13	26	98.32	11	28	99.75
39	13	27	98.81	12	28	99.53
40	14	27	98.08	12	29	99.68
41	14	28	98.62	12	30	99.78
42	15	28	97.82	13	30	99.60
43	15	29	98.42	13	31	99.73
44	16	29	97.56	14	31	99.52
45	16	30	98.22	14	32	99.67

TABLE 5.4 (Continued)

Confidence Limits for the Median*

$$x_k < \text{median} < x_{n-k+1}$$

	Confidence $\geq$ 95 Per Cent			Confidence $\geq$ 99 Per Cent		
n	k	$n-k+1$	Confidence	k	$n-k+1$	Confidence
46	16	31	98.71	14	33	99.77
47	17	31	98.00	15	33	99.60
48	17	32	98.53	15	34	99.72
49	18	32	97.78	16	34	99.53
50	18	33	98.36	16	35	99.67
51	19	33	97.56	16	36	99.77
52	19	34	98.18	17	36	99.61
53	19	35	98.65	17	37	99.73
54	20	35	97.99	18	37	99.55
55	20	36	98.50	18	38	99.68
56	21	36	97.80	18	39	99.77
57	21	37	98.34	19	39	99.62
58	22	37	97.60	19	40	99.73
59	22	38	98.18	20	40	99.57
60	22	39	98.63	20	41	99.69
61	23	39	98.02	21	41	99.51
62	23	40	98.50	21	42	99.64
63	24	40	97.85	21	43	99.74
64	24	41	98.36	22	43	99.59
65	25	41	97.68	22	44	99.70
66	25	42	98.22	23	44	99.54
67	26	42	97.51	23	45	99.66
68	26	43	98.08	23	46	99.75
69	26	44	98.53	24	46	99.62
70	27	44	97.93	24	47	99.72
71	27	45	98.40	25	47	99.57
72	28	45	97.78	25	48	99.68
73	28	46	98.28	26	48	99.52
74	29	46	97.63	26	49	99.65
75	29	47	98.15	26	50	99.74

*K. R. Nair, "Table of Confidence Interval for the Median in Samples from any Continuous Population," *Sankhya: The Indian Journal of Statistics*, Vol. 4, Part 4, March, 1940, pp. 551–558. Reproduced here with the permission of the author and the Statistical Publishing Society, Calcutta, India. If the n values of the variable in a sample are arranged in order of magnitude, so that $x_1 < x_2 < x_3 \ldots < x_k \ldots < x_{n-k+1} \ldots < x_{n-2} < x_{n-1} < x_n$, then x_k and x_{n-k+1} are the desired confidence limits for the median of the variable in the universe from which the sample was drawn.

Because the binomial distribution is not a continuous distribution, it is not possible to find confidence limits such that the degree of confidence is exactly 0.95 or exactly 0.99 at each value of n. Consequently, the exact value of the degree of confidence is shown in Table 5.4 for each value of n and the corresponding pair of values k and $n - k + 1$.

For example, if $n = 35$, the probability is 0.9795 that you will be right if you state that the median of the universe is between the twelfth and the twenty-fourth values of the variable in the sample, the values of the variable in the sample having been arranged in the order of their magnitude. Similarly, if $n = 25$, the probability is 0.9980 that you will be right if you state that the median of the universe is between the sixth and the twentieth values of the variable in the sample, the values of the variable in the sample having been arranged in the order of their magnitude.

The only assumptions involved in this method for determining confidence limits for a median are that the underlying variable is a continuous variable and that the sample is a simple random sample from the universe.

If the value of n, the size of the sample, is greater than 75, Table 5.4 cannot be used to determine confidence limits for a median. However, we have seen already that if n is large, the normal distribution can be used as a good approximation to the binomial distribution, especially if $P = \frac{1}{2}$. In order to use the normal distribution to approximate the binomial distribution $(\frac{1}{2} + \frac{1}{2})^n$ we need to know the values of μ and σ for the binomial distribution so that we can use those values as the values of μ and σ for the normal distribution. Assuming that the finite multiplier may be omitted and substituting $P = 1/2$ and $Q = 1/2$ into formulas 4.2 and 4.3, we find that $\mu = n/2$ and $\sigma = \sqrt{n}/2$.

Then, compute the value of $z\sigma = z\sqrt{n}/2$ where z is the value of the normal abscissa corresponding to the desired confidence level. For example, for 95 per cent confidence limits, let $z = 1.96$, and for 99 per cent confidence limits, let $z = 2.58$. Let g be the integer nearest to but not less than the value of $z\sigma$. Approximate confidence limits for the median of the universe are determined by the two elementary units that are located by counting g elementary units downward from the median of the sample and g elementary units upward from the median of the sample.

Suppose that a restaurant serves from 10,000 to 12,000 customers per week. In order to estimate the amount of the median check for a certain week, let us suppose that all the checks paid during that week are thoroughly shuffled and mixed in a box and that a random sample of 200 checks is drawn from the universe consisting of all the checks for the week.

Suppose that a rapid inspection of the 200 checks in the sample indicates that the amount of some checks is as small as 10 cents and the amount of others is as large as $2.75, but that most of the checks seem to be in the neighborhood of $1.00. Sort the 200 checks into three piles so that the first pile contains all the checks for less than 75 cents, the second pile contains all the checks for from 75 cents to $1.25, and the third pile contains all the checks for more than $1.25. Then count the number of checks in each pile.

Because $n = 200$ is an even number, the median of the sample is the amount midway between the amount of the 100th check and the amount of the 101st check. If the number of checks in the first pile is less than 100 and if the number of checks in the third pile is less than 100, the median check for the sample is in the second pile, and it is not necessary to arrange the checks in the first and third piles in order of magnitude to find the median of the sample. Arrange the checks in the second pile in order of magnitude and then locate the 100th check and the 101st check and determine the median for the sample.

To find 95 per cent confidence limits for the median of the universe, compute $\mu = n/2 = 200/2 = 100$, $\sigma = \sqrt{n}/2 = \sqrt{200}/2 = 14.14/2 = 7.07$, and $1.96\sigma = 1.96(7.07) = 13.96$. The integer nearest to 13.96 but not less than 13.96 is 14. Count 14 checks downward from the median, that is, find the 87th check and write down the amount of the 87th check. Count 14 checks upward from the median of the sample, that is, find the 114th check and write down the amount of the 114th check. The amounts on the 87th check and the 114th check may be used as approximate 95 per cent confidence limits for the median amount of the universe of checks for the week.

5.4 REMARKS

A binomial distribution is completely determined if we know the values of P and n. In other words, for a specified pair of values of P and n, there is one and only one binomial distribution.

The binomial expansion sometimes is called the formula for repeated trials. In order to obtain valid probabilities or relative frequencies by using a binomial expansion, it is always necessary that the results in the successive trials be independent of each other. It is also always necessary that the value of P remain constant throughout the series of n successive trials. Furthermore, it is necessary that n, the number of trials, be specified before the set of trials begins. The sampling ought to be done with replacement after each trial. However, if the size of the universe is so much greater than the size of the sample that the effect of the finite multiplier may be disregarded as negligible, then the terms of the binomial expansion may be expected to yield satisfactory approximations of the true probabilities even though the sampling is done without replacement.

The terms of the binomial expansion were shown to be the probabilities for chance events of the type discussed in Chapters 4 and 5 by the Swiss mathematician Jacques Bernouilli (1654–1706) in *Ars Conjectandi*, a treatise on probability published in 1713. The binomial distribution sometimes is called the Bernouilli distribution, or Bernouilli's formula. The repetitions involved theoretically in an investigation of this kind are often described as Bernouilli trials.

EXERCISES

1. Consider the nine students in Table 3.3 to be a random sample from a lar‍
universe of students and find 95 per cent or better confidence limits and 99 per ce‍
or better confidence limits for the median score of the universe of students on t‍
mathematics test. Indicate the exact degree of confidence to be attached to ea‍
pair of confidence limits. (*Ans.* 39 and 60, $P = 0.9805$; 37 and 60, $P = 0.998($
What assumption about the variable in the universe is required here?

2. Consider the 65 students in Table 3.7 to be a random sample from a lar‍
universe of students and find 95 per cent or better confidence limits and 99 per ce‍
or better confidence limits for the median score of the universe of students on t‍
mathematics proficiency test. Indicate the exact degree of confidence to be attach‍
to each of these pairs of confidence limits. (*Ans.* 51 and 57, $P = 0.9768$; 51 and 5‍
$P = 0.9970$.) What assumption about the universe is required in this method f‍
finding confidence limits for a median? Is it necessary to have truly quantitati‍
classification here? (*Ans.* No.)

3. Determine 99 per cent confidence limits for the median of the universe ‍
meal checks for the week in the last illustration given in section 5.3. (*Ans.* The amour‍
on the 82nd and the 119th checks.)

TABLE 5.5

Reading Preferences of a Sample of the Women in a College

Type of Book	Rank Assigned by a Woman								Total N of Wom‍
	1	*2*	*3*	*4*	*5*	*6*	*7*	*8*	
Humor	10	17	16	9	19	7	2	—	80
History	12	8	18	9	11	8	7	7	80
Biography	19	12	6	14	13	8	8	—	80
Romance	13	11	10	11	8	6	11	10	80
Travel	11	7	10	16	9	10	9	8	80
Mystery	12	9	11	5	5	8	9	21	80
Poetry	3	13	7	9	5	14	16	13	80
Textbooks	—	3	2	7	10	19	18	21	80
Total No. of Women	80	80	80	80	80	80	80	80	

4. Determine 95 per cent confidence limits for the median age at time of fir‍
marriage for the universe from which the sample of 242 men in Table 3.8 was draw‍
(*Ans.* 26.55 years and 28.13 years.)

5. Determine approximate 99 per cent confidence limits for the median age ‍
time of first marriage of the universe of men from which the sample of men in Tab‍
3.8 was drawn. (*Ans.* 26.34 years and 28.40 years.)

6. Determine approximate 95 per cent confidence limits for the median week‍
earnings in the universe of stenographers from which the sample in Table 3.15 w‍
drawn. (*Ans.* $79.55 and $80.59.)

7. Suppose that you are employed in the office of a large department store th‍
has about 25,000 charge accounts, and that your employer wishes to have as quick‍

as possible a reliable estimate of the median amount due on all of the charge accounts. Describe in detail a sampling plan and estimation plan by which you would carry out this assignment.

8. Determine approximate 95 per cent confidence limits for the median rank for each type of book among all the women in the college from which the sample in Table 5.5 was drawn.

REFERENCES

Julius R. Blum and Nicholas A. Fattu, "Nonparametric Methods," *Review of Educational Research* 24 (1954), pp. 467–487.

H. N. David, *Games, Gods and Gambling*. New York: Hafner Publishing Co., 1962.

Wilfrid J. Dixon and Frank J. Massey, Jr., *Introduction to Statistical Analysis*. New York: McGraw-Hill Book Co., Inc., 1957.

Darrell Huff and Irving Geis, *How to Take a Chance*. New York: W. W. Norton and Co., Inc., 1959.

Horace C. Levinson, *Chance, Luck and Statistics*. New York: Dover Publications, Inc., 1963.

Frederick Mosteller, Robert E. K. Rourke, and George B. Thomas, Jr., *Probability and Statistics*. Reading, Mass.: Addison-Wesley Publishing Co., Inc., 1961.

Helen M. Walker and Joseph Lev, *Statistical Inference*. New York: Henry Holt and Co., 1953.

Chapter 6

NORMAL DISTRIBUTIONS AND NORMAL PROBABILITIES

6.1 INTRODUCTION

It seems that nearly everybody wants to be normal. However, most of those who are worried about being "normal" do not understand fully what the word means.

Helen, who is seventeen years old, steps on the scale in the corner drug store and finds that she weighs 5 pounds more than the chart on the scale indicates that she should weigh. As soon as she arrives at home she tells her mother that she does not want any dessert for dinner and that she is going to drink a lot of grapefruit juice in order to reduce her weight. She feels that she is not normal.

The statistical methods contained in Chapters 8 and 9 can be used to show that it is quite impossible to predict by means of an age-weight chart or a height-weight chart exactly what a particular individual should weigh. The weights given in these charts are norms. To be 5 pounds above the weight indicated on the chart may be very desirable for one person of a certain age, body build, national origin, and stage of growth. To be 5 pounds below the weight indicated on the chart may be equally desirable for proper proportionality and health in another person of the same age but different body build, national orign, and stage of growth. Each individual has his or her own growth-rate pattern. There is no such thing as *the* normal man or *the* normal woman.

In Rostand's famous romantic drama, Cyrano de Bergerac is an expert poet, artist, and duelist. In fact, he composed poems while he dueled always finishing his opponent and his poem at exactly the same instant. But he had a very large nose, and it worried him, because he felt that he was not normal. It worried him so much that in spite of his valiance as a fighter, he did not have the courage to try to win the heart of Roxane whom he loved so desperately. One so beautiful as she could never love anyone so homely as he with his big nose, he thought. To make the story short, Cyrano discovered on his death bed that Roxane long ago had fallen in love with him through the poems that he had written to her and had delivered by another man.

Don't be a Cyrano! That is one lesson to be learned from this chapter. Take stock of your individual differences; a little straight thinking, planning, and effort will likely enable you to turn those differences to the benefit of yourself and the world in which you live.

The great emphasis that has been placed on statistical norms and averages in recent years has frequently led to misunderstanding and needless

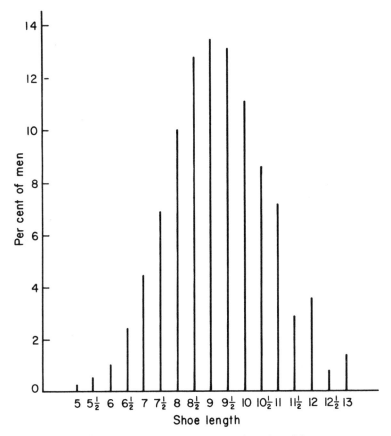

Fig. 6.1.   Shoe Sizes of 10,000 American Men

apprehension among young people. A norm often is thought of as a mark of perfection that everyone should strive to attain. That is a false notion. The norm for intelligence among Americans is IQ 100. But, surely no one would feel badly about being five or ten points above the norm for intelligence. On the other hand, there is no need to be discouraged if one is five or ten, or even more, points below the norm for intelligence. It is a common thing to observe that some persons with very high IQ never amount to much. People with IQ's below average who have a healthy philosophy of life and a willingness to make an effort often behave more intelligently and become more worthy citizens than some of those who are born with high IQ's.

6.2 INDIVIDUAL DIFFERENCES AND NATURAL VARIABILITY

Several tables and graphs will be presented in this chapter to illustrate the shape of distributions of individual differences in physical character- istics in large groups of people. Tables 6.1 and 6.2 and Figure 6.1 show the way American men differ in their shoe sizes. Tables 6.20, 6.21, and 6.24 and Figure 6.22 show the way in which men differ in the neck size and sleeve length of their shirts. Tables 6.22, 6.23, and 6.25 and Figure 6.23 show the way women differ in the length and width of their shoes. All of these distributions are bell-shaped and most of them are approximately symmetrical.

TABLE 6.1

Shoe Sizes of American Men Who Wear Florsheim Shoes*

Shoe Size (Length)	Per Cent of Men
5	0.20
5½	0.41
6	1.18
6½	2.30
7	4.28
7½	6.90
8	9.96
8½	13.01
9	13.51
9½	13.19
10	10.98
10½	8.48
11	7.08
11½	2.75
12	3.61
12½	0.73
13	1.43
Total	100.00

Source: Courtesy of the Flor- sheim Shoe Co., Chicago, Ill. Based on a sample containing 10,000 pairs of shoes.

TABLE 6.2

Shoe Sizes of American Men Who Wear Florsheim Shoes*

Shoe Size (Width)	Per Cent of Men
AAA	2.28
AA	6.05
A	13.20
B	21.64
C	30.20
D	23.58
E	3.05
Total	100.00

Source: Courtesy of the Flor- sheim Shoe Co., Chicago, Ill. Based on a sample containing 10,000 pairs of shoes.

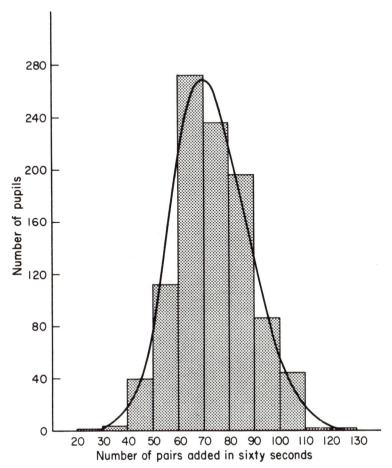

Fig. 6.2. Ability of High-School Pupils in Adding Pairs of One-Place Numbers

Table 6.9 and Figure 6.16 show the distribution of differences in IQ to be expected in a randomly selected group of boys and girls. Table 6.3 and Figure 6.2 show how boys and girls vary in such a simple mental characteristic as the ability to add pairs of one-digit numbers.

There also are personality differences among individuals. For example, there are differences in cooperativeness, courage, loyalty, sociability, and excitability. Table 6.4 illustrates the way we should expect people to be distributed on a test of the dominance–submissiveness trait in personality. Only a small percentage of people would be at the extremely dominant end of the scale, and only a small percentage would be at the extremely submissive end. Most individuals would be only slightly above average or

TABLE 6.3			TABLE 6.4	

Ability of High School Pupils in Adding Pairs of One-Place Numbers*

Distribution of Dominance-Submissiveness Personality Trait

No. of Pairs Added in 60 Seconds	No. of Pupils
20–29	2
30–39	4
40–49	41
50–59	113
60–69	272
70–79	235
80–89	196
90–99	86
100–109	43
110–119	2
120–129	2
Total	996

*Source: S. A. Courtis, "Report on the Courtis Tests in Arithmetic," City of New York, 1911–1912, p. 52.

Score on Test		Per Cent of People
0–4	Dominant	0.3
5–9		0.9
10–14		2.8
15–19		6.6
20–24		12.1
25–29		17.4
30–34	Average	19.8
35–39		17.4
40–44		12.1
45–49		6.6
50–54		2.8
55–59		0.9
60–64	Submissive	0.3
	Total	100.0

slightly below average. In other words, most people are reasonably well-balanced with respect to dominance and submissiveness.

The idea of standard units that we have used often already in this book gives us a better way to study the individual differences in a distribution. For the distribution in Table 6.4, the mean score is 32 and the standard deviation of the scores is 10. Frequently we have said that differences or deviations are not considered significant unless their absolute values are greater than or equal to 1.96 standard units, that is, unless the corresponding normal probability is less than or equal to 0.05. In this example, 1.96 standard units equals $(1.96)(10) = 19.6 = 20$ score units, approximately.

Subtracting 20 from $\mu = 32$ and adding 20 to $\mu = 32$ we obtain 12 and 52, respectively. Consequently, if we use the 5 per cent level of significance, we can say that only those individuals who score 12 or less or who score 52 or more on the test are significantly far from average in the dominance–submissiveness personality trait.

There also are individual differences in performance. Even in tests of proficiency at routine tasks, we should expect to find distributions in which some individuals are about twice as proficient as other individuals in the group. For example, in a certain factory, there were 33 employees whose job was to burn, twist, and solder the ends of insulated wires. A test was made to see how these workers varied in proficiency in this task. While the least proficient worker was completing 65 units of work, the best worker finished 130 units of work. The other members of the group varied all the way from 65 to 130 and the average number of units completed was approximately 100.

This example suggests that selecting people at random for even relatively simple, routine tasks is a poor way to assign people to jobs. Few people are happy at jobs they cannot do well. Individuals need tasks for which they have the appropriate ability, aptitude, and motivation. One secret of successful vocational choice and also of successful personnel administration lies in recognizing and utilizing individual differences.

We conclude, then, that differences are the rule. The most natural thing in the world is not constancy or sameness but variability or difference. Not only is variability more characteristic than constancy in the things that we experience, but variability is highly desirable in general. Instead of urging individuals to try to equal norms or averages, they should be encouraged to deviate from norms—in the right direction.

What a drab world it would be if all of us were exactly alike in every way! Differences in personality give color to the world; they make life interesting. The world needs builders, inventors, and poets, and we should actively strive to develop such differences among ourselves. One might want a herd of cattle to be alike, but one does not want this in people—unless one is the dictator of a totalitarian state. Only such a one could proclaim:

> We shall breed a new race out of an elite, trained to hardness, cruelty, violence; supermen, leading masses. On them we shall found a new Reich that will last for a thousand years.
> The supermen will be ingenious, treacherous, masterful. The masses will be uniform, with arms that rise and fall rhythmically, voices that cry hoarsely, rhythmically, "Sieg Heil"![1]

On the other hand, a democratic state is never exclusive. It respects and invites differences in abilities and attitudes, hoping that such variations will be for the common good by raising the level of living of all the people.

Distributions, not types without variation, are the rule. The normal thing for a person is to be above average in some characteristics, about average in some, and below average in some characteristics, as compared with other people of the same age and sex. To determine the true value of a person, then, one must study the parts that person plays in all the distributions of which he is a member. And everyone is a member of many distributions.

6.3 NORMAL DISTRIBUTIONS

In sections 6.1 and 6.2, the word "normal" has been used in the colloquial

[1] These words were spoken by the Nazi leader in Dorothy Thompson's story "On a Paris Railway Station" for her column "On the Record" in the *New York Herald Tribune* on May 13, 1940. They are reprinted here with the permission of Miss Thompson and of The *New York Herald Tribune*.

sense. In the rest of this book, in such terms as "normal distribution" and "normal probability," the word "normal" is used in a strictly technical sense. Let us proceed now to learn what the technical statistical meaning of the word "normal" is. In the first place, do not adopt the false idea that, because considerable variability is to be expected in most natural phenomena and in most human affairs, the variations in all natural phenomena and in all human affairs form perfect normal distributions. That is far from true; some of those distributions are almost perfectly normal distributions, some are only crude approximations of normal distributions, and some are very unlike normal distributions. Here are a few of the most important properties of normal distributions. Some other characteristics of normal distributions will be given in sections 6.8 and 6.9.

6.3.1 The variable is continuous. A normal distribution is theoretically a distribution of a continuous variable. This requires that (1) the elementary units in the universe are classified in a truly quantitative way, and (2) all values in the range of the variable are theoretically possible values of elementary units drawn from the universe. In practice, of course, we have seen that a normal distribution often can be used to approximate satisfactorily binomial distributions and other distributions of discrete variables.

6.3.2 All normal distributions are symmetrical. In some of the illustrations in this book, smooth frequency curves have been drawn through the tops of the rectangles in the histograms so that the smooth curve encloses with the base line as nearly as possible the same area as the histogram. Figure 6.2 shows the histogram and the smooth frequency curve for the distribution given in Table 6.3.

Notice that the smooth frequency curve is approximately symmetrical or balanced around a vertical line near the middle of the range of the variable. In other words, if you folded the graph in Figure 6.2 along a certain vertical line near the middle of the range of the variable, the left-hand half of the curve would fit almost exactly over the right-hand half of the curve. We say that such distributions are symmetrical.

Symmetry or balance is a very important property of normal distributions and of normal frequency curves. (See Figure 6.11.) If a frequency distribution, the histogram that represents it, and the smooth frequency curve that is the best fit to the distribution are not symmetrical or nearly symmetrical, the distribution is not considered to be a normal distribution. Distributions that are not symmetrical are said to be *skewed*.

However, you should not expect the observed distributions that are obtained by random sampling to be perfectly symmetrical. Sampling errors in the data are likely to produce some irregularities in the distributions that we obtain in practical investigations. Also, errors in reporting sometimes account for the peculiarly large frequencies in some classes and the relatively small frequencies in other classes in a table.

6.3.3 All normal distributions are bell-shaped. If you draw the smooth frequency curves for several normal distributions, you will find that all the curves are bell-shaped, but you will find also that some of the normal curves are flatter than others. That is because the distributions have different standard deviations. The larger the standard deviation the flatter the curve.

6.4 NON-NORMAL DISTRIBUTIONS

Figures 6.3 and 6.4 illustrate the most common types of distributions. You will be given an illustration of the type of histogram shown in Figure

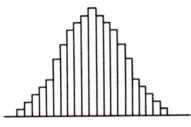

Fig. 6.3. Symmetrical Bell-Shaped Distribution

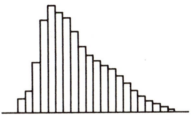

Fig. 6.4. Skewed Bell-Shaped Distribution

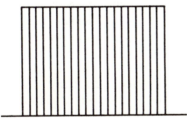

Fig. 6.5. Rectangular Distribution

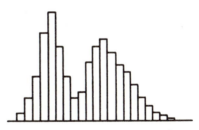

Fig. 6.6. Bimodal Distribution

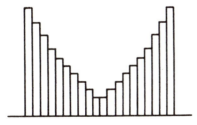

Fig. 6.7. U-Shaped Distribution

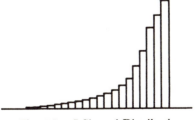

Fig. 6.8. J-Shaped Distribution

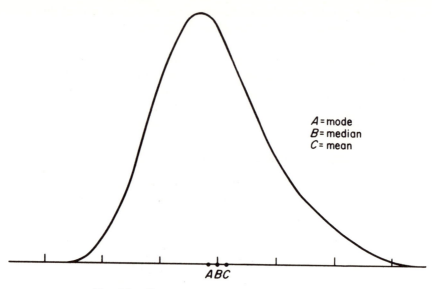

Fig. 6.9.   Frequency Curve with Skewness +0.3

6.5 in connection with the study of distributions of random digits in section 7.2. Distributions of the types shown in Figures 6.6 and 6.7 occur occasionally. For example, some bimodal distributions have appeared in reports on biological research.

Distributions of the form shown in Figure 6.8 occur frequently in certain special fields of statistical work. For example, they are common in certain types of statistical quality control in manufacturing, such as studies of the number of defects per manufactured unit. And if you obtained a frequency distribution showing the number of employees in a large factory who were involved in 0, 1, 2, 3, etc., accidents during a specified period, say, a month, it is quite likely that the distribution would be J-shaped. J-shaped distributions have been encountered also in studies of the frequency of purchase of products by consumers and the period of time elapsed since the last purchase of products—candy bars, for example.

6.4.1 Skewness of distributions. The mode of a distribution is the value of the variable at the point of the range directly below the highest point of the smooth frequency curve that fits the distribution best. If the distribution is perfectly balanced, or symmetrical, the mean and the median are identical with the mode; then all three parameters are represented by the same point on the base line of the graph.

If the distribution is skewed, then the mean, the median, and the mode are not quite identical, and the three points representing them do not coin-

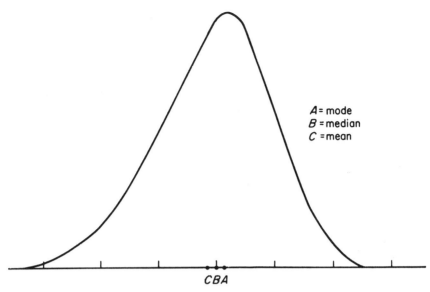

Fig. 6.10.   Frequency Curve with Skewness −0.3

cide on the base line of the graph. Figures 6.9 and 6.10 show the positions
of the mean, the median, and the mode for two skewed frequency curves.
How far apart the three points are depends upon the amount of skewness
in the curve. If the distribution is only slightly skewed, the three points
will fall quite close to each other.

Karl Pearson (1857-1936), a famous English statistician, many years
ago examined a large number of skewed distributions of continuous variables and found that in many of these distributions the median was located
about two-thirds of the way from the mode to the mean. Consequently,
he suggested the formula: mode = 3(median) − 2(mean). This formula
is used sometimes to compute a rough approximation to the mode of a
distribution. However, because of its shortcomings, as indicated in sections
3.3.5 and A.6.8, the mode of a sample is not very useful in practical statistics.

6.4.2 How to estimate skewness. If the mean and the median of a
sample are not identical in value, the distance between the mean and the
median of the sample might be used as an indicator of the amount of skewness in the distribution. But the degree of skewness of a distribution is the
same no matter what you choose as the unit of measurement and no matter
how wide you make the rectangles in the histogram. In other words, the
degree of skewness of a distribution is the same no matter whether you draw
the histogram on a small scale or on a large scale. Consequently, it is best
to think of the skewness of a distribution as not depending on the unit of

measurement in the data. This suggests that it would be a good idea to represent the skewness of distributions in standard units.

The mean, the median, and the standard deviation are always expressed in the same unit of measurement as the original data. Therefore, the difference between the mean and the median will be in the same unit of measurement as the original data. If we divide the difference between the mean and the median by the standard deviation, the result will be in standard units.

The expression (mean − median)/(standard deviation) could be used for indicating the degree of skewness of any distribution. But Karl Pearson's original definition of the skewness was (mean − mode)/(standard deviation). If we substitute the right-hand side of Pearson's formula for the mode into Pearson's definition of the skewness, we obtain

$$\text{skewness} = \frac{3(\text{mean} - \text{median})}{\text{standard deviation}} \tag{6.1}$$

We shall use formula 6.1 for estimating the skewness of distributions.

If the value of the mean is greater than the value of the median, the skewness of the distribution is a positive number. Whenever the value of the median is greater than the value of the mean, the skewness of the distribution is a negative number. Figure 6.9 illustrates positive skewness. Figure 6.10 illustrates negative skewness.

If a distribution has positive skewness, the curve tapers off more gradually toward the right than it does toward the left; it drops down to zero more abruptly on the left-hand side of the mean. Positive skewness indicates that the values of the variable farthest from the mean in the distribution tend to be values that are greater than the mean; that is, the most unusual or most extreme elementary units have values of the variable that are at the high end of the range of the variable. If a distribution has negative skewness, the most unusual or most extreme values of the variable tend to be at the low end of the range of the variable.

In most distributions, the value of the skewness is between −1 and +1. However, it is possible for a distribution to be so extremely skewed that the value of the skewness is greater than +1 or less than −1. If the skewness is between −0.2 and +0.2, the histogram and the smooth frequency curve probably will appear to your eye to be practically symmetrical. Table 6.5 shows the value of the skewness for some of the distributions in this book. If you study these values of the skewness and the shapes of the corresponding histograms, it will help you to grasp the meaning of various amounts of skewness.

Do not make the mistake of assuming that all distributions that have zero skewness are necessarily normal distributions. There are many types of distribution that have zero skewness but are not normal distributions.

TABLE 6.5

The Skewness of Some Distributions in This Book

Distribution	Skewness
Table 3.8, Fig. 3.1	+0.7
Table 3.10, Fig. 3.3, Fig. 3.4	0.0
Table 3.11, Fig. 3.8	0.0
Table 3.12, Fig. 3.6	0.0
Table 3.13, Fig. 3.7	0.0
Table 3.15	+0.4
Table 3.16	+0.5
Table 6.3, Fig. 6.2	+0.2
Table 6.4	0.0
Fig. 6.9	+0.3
Fig. 6.10	−0.3
Table 6.9, Fig. 6.16	0.0
Table 6.20, Fig. 6.22	+1.1
Table 6.21	+0.4
Table 6.22, Fig. 6.23	+0.1
Table 7.1, Fig. 7.1	0.0

Zero skewness is only an indication of symmetry, and a distribution can be perfectly symmetrical without being a normal distribution. For example, the rectangular distribution in Table 7.1 and Figure 7.1 is perfectly symmetrical and has zero skewness. Furthermore, there are bell-shaped frequency distributions that have zero skewness, that is, perfect symmetry, but are not normal distributions. The "Student" distribution of t described in section 7.3 is a perfectly symmetrical, bell-shaped distribution, but it is considerably different from the normal distribution if n' is very small. It is possible, also, to have bimodal distributions and U-shaped distributions for which the skewness is zero.

EXERCISE

1. In section A.6 (p. 00), it is suggested that you compute the median, the mean, and the standard deviation for many of the distributions in Tables 3.11, 3.12, 3.13, 3.15, 3.16, 3.17, 6.1, 6.2, 6.3, 6.14, 6.15, 6.16, 6.20, 6.21, and 6.22. Compute the skewness for each of those distributions for which you already know the values of the median, the mean, and the standard deviation. Construct a table similar to Table 6.5 for the values of the skewness that you compute. Examine the histograms corresponding to the values of the skewness that you have computed so that you will know in the future what a particular amount of skewness signifies in a distribution.

6.5 HOW TO CONSTRUCT A NORMAL DISTRIBUTION

Suppose that someone asked you to estimate the number or the percentage of the boys and girls in your community who have intelligence quotients between 95 and 105, or below 85. It is very easy to answer such

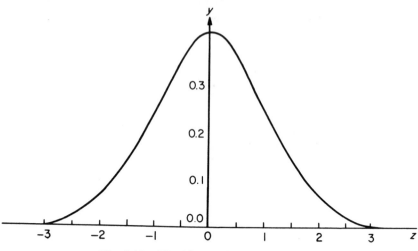

Fig. 6.11.   The Normal Probability Curve

questions if you know how to construct a normal distribution, that is, if you understand the normal probability law and know how to use the table of normal probabilities, Table 6.7.

6.5.1 Graph of the normal curve. Figure 6.11 is a graph of a normal distribution. You can plot such a graph by using Table 6.6, which is a table of the abscissas and ordinates for any normal distribution that has its abscissas, that is, the values of the variable, expressed in standard units and for which the total area under the curve is 1. Each pair of values of z and y determines a point on the curve in Figure 6.11. For example, $z = 0.0$ and $y = 0.399$ determines the point where the curve crosses the y-axis, that is, the highest point on the curve. And $z = 1.0$ with $y = 0.242$ determines the point on the curve directly above the point on the horizontal line where $z = 1$. Because the normal curve is perfectly symmetrical, the points on the left-hand half of the curve must be exactly symmetrical with the points on the right-hand half of the curve. Consequently, for $z = -1.0$ the value of y must be the same as it is for $z = +1.0$, namely, $y = 0.242$. Therefore, in order to save space, only the values of z and y for the right-hand half of the curve are printed in Table 6.6.

6.5.2 Normal probabilities or areas. After you have plotted the normal curve carefully, you can make a rough estimate of the percentage of the total area that lies between any two specified ordinates. To calculate very accurately the area between two specified ordinates is a task for ad-

TABLE 6.6

Abscissas and Ordinates of the Normal Probability Curve

Abscissa z	Ordinate y	Abscissa z	Ordinate y	Abscissa z	Ordinate y	Abscissa z	Ordinate y
0.0	.399	1.0	.242	2.0	.054	3.0	.004
0.1	.397	1.1	.218	2.1	.044	3.1	.003
0.2	.391	1.2	.194	2.2	.035	3.2	.002
0.3	.381	1.3	.171	2.3	.028	3.3	.002
0.4	.368	1.4	.150	2.4	.022	3.4	.001
0.5	.352	1.5	.130	2.5	.018	3.5	.001
0.6	.333	1.6	.111	2.6	.014	3.6	.001
0.7	.312	1.7	.094	2.7	.010	3.7	.000
0.8	.290	1.8	.079	2.8	.008	3.8	.000
0.9	.266	1.9	.066	2.9	.006	3.9	.000

vanced mathematics courses such as the calculus. However, practical statisticians need so urgently to know the areas between pairs of ordinates of the normal curve that mathematicians have worked out all the values that we need and they have published tables of the areas for everyone to use. Table 6.7 is one table of normal probabilities or areas. There are other available tables showing the areas to as many as fifteen decimal places.

Why do statisticians need to know these areas under the normal curve so urgently? We need to know the areas between ordinates of the normal curve in order to be able to determine the probabilities and relative frequencies corresponding to specified intervals in the range of the variable. In other words, we need to know these areas because, as we learned in previous chapters of this book, areas under a relative-frequency curve represent probabilities, and probabilities are equivalent to relative frequencies. Therefore, if a statistician knows the area between a pair of ordinates of the normal probability curve, he can write down the relative frequency for the class represented by the interval of values of the variable between the ordinates.

6.5.3 How we use Table 6.7. Table 6.7 is a table of the areas on the right-hand side of a large number of ordinates in the right-hand half of the normal probability curve. Because the normal probability curve is perfectly symmetrical, the symmetrical areas under the left-hand half of the curve are simply a repetition of these areas. Consequently, in order to save space, only the areas in the right-hand half of the normal curve are printed in Table 6.7. The abscissas are the values of z. The areas or probabilities are the values of P_z.

In Table 6.7, the value of P_z corresponding to any particular value of z tells us the proportion of the area under the normal curve on the right-hand

TABLE 6.7

Areas under the Normal Probability Curve

z	P_z	z	P_z	z	P_z	z	P_z	z	P_z	z	P_z	z	P_z
.00	.500	.50	.309	1.00	.159	1.50	.067	2.00	.023	2.50	.006	3.00	.001
.01	.496	.51	.305	1.01	.156	1.51	.066	2.01	.022	2.51	.006		
.02	.492	.52	.302	1.02	.154	1.52	.064	2.02	.022	2.52	.006		
.03	.488	.53	.298	1.03	.152	1.53	.063	2.03	.021	2.53	.006		
.04	.484	.54	.295	1.04	.149	1.54	.062	2.04	.021	2.54	.006		
.05	.480	.55	.291	1.05	.147	1.55	.061	2.05	.020	2.55	.005		
.06	.476	.56	.288	1.06	.145	1.56	.059	2.06	.020	2.56	.005		
.07	.472	.57	.284	1.07	.142	1.57	.058	2.07	.019	2.57	.005		
.08	.468	.58	.281	1.08	.140	1.58	.057	2.08	.019	2.58	.005		
.09	.464	.59	.278	1.09	.138	1.59	.056	2.09	.018	2.59	.005		
.10	.460	.60	.274	1.10	.136	1.60	.055	2.10	.018	2.60	.005		
.11	.456	.61	.271	1.11	.134	1.61	.054	2.11	.017	2.61	.005		
.12	.452	.62	.268	1.12	.131	1.62	.053	2.12	.017	2.62	.004		
.13	.448	.63	.264	1.13	.129	1.63	.052	2.13	.017				
.14	.444	.64	.261	1.14	.127	1.64	.051	2.14	.016				
.15	.440	.65	.258	1.15	.125	1.65	.049	2.15	.016				
.16	.436	.66	.255	1.16	.123	1.66	.048	2.16	.015				
.17	.433	.67	.251	1.17	.121	1.67	.047	2.17	.015				
.18	.429	.68	.248	1.18	.119	1.68	.046	2.18	.015				
.19	.425	.69	.245	1.19	.117	1.69	.046	2.19	.014	2.69	.004		
.20	.421	.70	.242	1.20	.115	1.70	.045	2.20	.014	2.70	.003		
.21	.417	.71	.239	1.21	.113	1.71	.044	2.21	.014				
.22	.413	.72	.236	1.22	.111	1.72	.043	2.22	.013				
.23	.409	.73	.233	1.23	.109	1.73	.042	2.23	.013				
.24	.405	.74	.230	1.24	.107	1.74	.041	2.24	.013				
.25	.401	.75	.227	1.25	.106	1.75	.040	2.25	.012				
.26	.397	.76	.224	1.26	.104	1.76	.039	2.26	.012				
.27	.394	.77	.221	1.27	.102	1.77	.038	2.27	.012				
.28	.390	.78	.218	1.28	.100	1.78	.038	2.28	.011				
.29	.386	.79	.215	1.29	.099	1.79	.037	2.29	.011			3.29	.001
.30	.382	.80	.212	1.30	.097	1.80	.036	2.30	.011	2.80	.003	3.30	.000
.31	.378	.81	.209	1.31	.095	1.81	.035	2.31	.010	2.81	.002		
.32	.374	.82	.206	1.32	.093	1.82	.034	2.32	.010				
.33	.371	.83	.203	1.33	.092	1.83	.034	2.33	.010				
.34	.367	.84	.200	1.34	.090	1.84	.033	2.34	.010				
.35	.363	.85	.198	1.35	.089	1.85	.032	2.35	.009				
.36	.359	.86	.195	1.36	.087	1.86	.031	2.36	.009				
.37	.356	.87	.192	1.37	.085	1.87	.031	2.37	.009				
.38	.352	.88	.189	1.38	.084	1.88	.030	2.38	.009				
.39	.348	.89	.187	1.39	.082	1.89	.029	2.39	.008				
.40	.345	.90	.184	1.40	.081	1.90	.029	2.40	.008				
.41	.341	.91	.181	1.41	.079	1.91	.028	2.41	.008				
.42	.337	.92	.179	1.42	.078	1.92	.027	2.42	.008				
.43	.334	.93	.176	1.43	.076	1.93	.027	2.43	.008				
.44	.330	.94	.174	1.44	.075	1.94	.026	2.44	.007				
.45	.326	.95	.171	1.45	.074	1.95	.026	2.45	.007				
.46	.323	.96	.169	1.46	.072	1.96	.025	2.46	.007	2.96	.002		
.47	.319	.97	.166	1.47	.071	1.97	.024	2.47	.007	2.97	.001		
.48	.316	.98	.164	1.48	.069	1.98	.024	2.48	.007				
.49	.312	.99	.161	1.49	.068	1.99	.023	2.49	.006				

P_z = the probability or area to the right of the value of z in one tail of the normal distribution.

If you need the probability or area outside of the interval from $-z$ to $+z$, multiply P_z by 2.

side of the ordinate erected at z. For example, the first value of z in the table is $z = 0$. The corresponding value of P_z is 0.500. This means that 0.500, that is, one-half of the area of the normal curve is on the right-hand side of the ordinate erected at $z = 0$. It also means that one-half of the relative frequencies in a normal distribution are to the right of $z = 0$.

Similarly, when $z = 0.10$ the value of P_z is 0.460. This means that 0.460 of the area of the normal curve is on the right-hand side of the ordinate erected at $z = 0.10$. It also means that the total of the relative frequencies on the right-hand side of $z = 0.10$ is 0.460. Also, it means that the probability of drawing a value of z as great as or greater than 0.10 in a single random drawing from a normal distribution is 0.460.

6.5.4 An illustration. Let us suppose that we have been asked to estimate the distribution of IQ among the children less than twelve years old in a certain community.

In attempting to solve any problem, it is important to specify the essential conditions of the situation. The basic information for the task that we are undertaking here consists of the following facts and assumptions: (1) The distribution with which we are concerned consists of all the children less than twelve years old in the community. (2) The number of children in the distribution is 937; that is, $N = 937$. (3) Intelligence is a continuous variable, although in practice a person's IQ usually is expressed as a whole number. (4) The mean of the intelligence quotients of the children in the distribution is 100; that is, $\mu = 100$. (5) The variability of intelligence among the children in the distribution is represented by a standard deviation of 16 IQ units; that is, $\sigma = 16$. (6) The distribution of IQ among the children is a normal distribution.

Statements 1, 2, and 3 are statements of facts. Statements 4, 5, and 6 are assumptions. The first statement defines the distribution precisely. If we did not know that $N = 937$, we could derive relative frequencies but not actual frequencies. The third statement will be helpful in determining appropriate class limits for the distribution. Assumptions 4, 5, and 6 enable us to apply the normal probability law to our problem.

The solution of this problem is summarized in Table 6.9. Here are the details of the procedure for constructing Table 6.9. The first step in the planning of the table is the determination of the limits of the IQ classes into which we wish to distribute the children. Ten IQ units is a convenient class interval for our purposes.

Because we are assuming that the mean IQ in the group of children is 100, it will be convenient and reasonable to arrange the classes so that the middle value of the variable in the middle class will be 100. In other words, picture in your mind a histogram with IQ = 100 at the midpoint of the base of the middle and highest rectangle. Figure 6.16 shows the histogram that was constructed after Table 6.9 had been completed.

The middle class will contain all the boys and girls in the community who have IQ's between 95 and 105, that is, including 95 and all values up to but not including 105. The limits of the other classes can now be determined.

How many classes should be included in the table? The answer to this question is obtained by using common sense and the idea of a normal distribution. In the first place, we know that we do not need to have any classes with negative limits, for no one can have an IQ below zero. Furthermore, because the normal distribution is perfectly symmetrical, we would not expect to need any classes with limits more than 100 units above the mean, that is, greater than 200.

However, a better answer to the question about the number of classes to be included in the table may be obtained from a normal probability table such as Table 6.7. We must include any class for which the corresponding probability indicates that at least one child in the group of 937 children belongs in that class. Table 6.7 shows that for $z = 3.30$ the value of P_z is 0.000. In other words, to three decimal places, the probability of there being a person more than 3.30 standard units above the mean is zero. Consequently, if we determine the IQ that is 3.30 standard units above the mean, then we know that in a group of 937 children no one would be expected to have an IQ that high or higher. To find this upper end of the range of the variable, multiply the standard deviation by 3.30 and add the product to the mean, that is, to 100. The result is: $(3.30)(16) = 52.8 = 53$; $100 + 53 = 153$; also, $100 - 53 = 47$.

Therefore, we ought to expect to find children in the community with IQ's all along the range from about 47 to about 153. And we ought not to expect to find any children in the community with IQ's lower than 47 or higher than 153.

However, because it is desirable that all the class intervals have the same length, namely, 10 IQ units, the highest class will be taken as 145–155 rather than 145–153. Similarly, the lowest class will be taken as 45–55 rather than 47–55. The values of all the class limits are shown in the first column of Table 6.9.

If there were approximately 100,000 children in the group instead of about 1000, we would need to include both higher and lower classes than those in the first column of Table 6.9. If we used a table of areas of the normal curve to seven decimal places, we would find that in a group of 100,000 elementary units there are likely to be a few values of the variable more than four standard units above the mean and a few values of the variable more than four standard units below the mean.

In dealing with normal distributions, Table 6.8 is helpful as a guide in determining the range to be expected for the variable in a distribution. For example, Table 6.8 indicates that for a normal distribution containing between 1000 and 10,000 elementary units, we would expect to find values of the variable anywhere up to four standard units above the mean and below the mean. And we would not expect to find any elementary units with values of the variable more than four standard units above or below the mean.

TABLE 6.8

Expected Limits of the Range in a Normal Distribution

Number of Elementary Units in the Distribution	*Limits of the Range of the Variable in a Normal Distribution*
Not more than 100	$\mu \pm 2.5\sigma$
101–1,000	$\mu \pm 3\sigma$
1,001–10,000	$\mu \pm 4\sigma$
10,001–100,000	$\mu \pm 4.5\sigma$
100,001–1,000,000	$\mu \pm 5\sigma$

In order that the tables of the normal probability law may be applicable to every normal distribution, no matter what the mean and the standard deviation may be, the values of the abscissas are given in standard units in Tables 6.6 and 6.7. Consequently, if we wish to use Table 6.7 in our problem, we must transform our class limits into standard units. You are familiar already with formula 3.5 for changing the values of a variable into standard units. In our problem here, we know that $\mu = 100$ and $\sigma = 16$. The lowest class limit in Table 6.9 is $x = 45$. In standard units, this becomes $z = (45 - 100)/16 = -55/16 = -3.4375$.

If you study Table 6.7 carefully, you will notice that when the value of z is not far from zero a small change in the value of z corresponds to a fairly large change in the value of P_z, the probability or area. Consequently, we shall retain two decimal places in the values of z in the second column of Table 6.9. However, the computations should be carried to three or four decimal places and then rounded off. Thus, we record -3.44 opposite the first class to correspond to the class limit $x = 45$.

For the upper limit of the first class, we can find the value in standard units by computing $z = (55 - 100)/16 = -45/16 = -2.8125$ and we record this value to two decimal places as -2.81 in Table 6.9. Similarly, we could compute the remainder of the values of z for the class limits.

Because addition is easier than division, we can use a short cut here. The class interval is 10 units in length. That is, $i = 10$. Now $i/\sigma = 10/16 = 0.6250$, and if we add 0.6250 to -3.4375 we obtain -2.8125, which is the value of z at the upper limit of the first class. Similarly, the value of z at the class limit $x = 65$ may be obtained by adding 0.6250 to -2.8125, obtaining -2.1875 which is rounded off to -2.19. The other class limits may be changed into standard units by successive additions of 0.6250 to the value of z at the preceding limit.

TABLE 6.9

Normal Distribution of the Intelligence Quotients of 937 Children

IQ Class	Class Limits in Standard Units z	Probability or Relative Frequency P	Computed Frequency 937P	Expected No. of Children f
45–55	−3.44 to −2.81	0.002	1.874	2
55–65	−2.81 to −2.19	0.012	11.244	11
65–75	−2.19 to −1.56	0.045	42.165	42
75–85	−1.56 to −0.94	0.115	107.755	108
85–95	−0.94 to −0.31	0.204	191.148	191
95–105	$\begin{cases} -0.31 \text{ to } 0.00 \\ 0.00 \text{ to } 0.31 \end{cases}$	$\left. \begin{matrix} 0.122 \\ 0.122 \end{matrix} \right\} 0.244$	228.628	229
105–115	0.31 to 0.94	0.204	191.148	191
115–125	0.94 to 1.56	0.115	107.755	108
125–135	1.56 to 2.19	0.045	42.165	42
135–145	2.19 to 2.81	0.012	11.244	11
145–155	2.81 to 3.44	0.002	1.874	2
Total		1.000	937.000	937

Now we are ready to use Table 6.7 to determine the probability or relative frequency for each class in Table 6.9. Begin with the highest value of z in Table 6.9, $z = 3.44$. Table 6.7 shows that the value of P_z is 0.000 for all values of z from $z = 3.30$ upward. Consequently, the relative frequency above $z = 3.44$ is zero to three decimal places. Table 6.7 shows that the area above $z = 2.81$ is 0.002. Figure 6.12 is a graphical representation of this small area. Therefore, we record 0.002 in the third column of Table 6.9 opposite the 145–155 class.

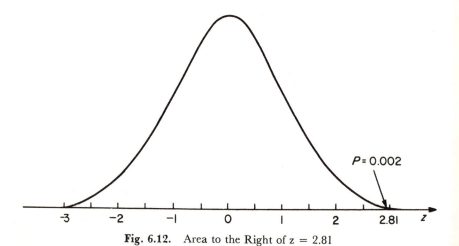

Fig. 6.12.   Area to the Right of $z = 2.81$

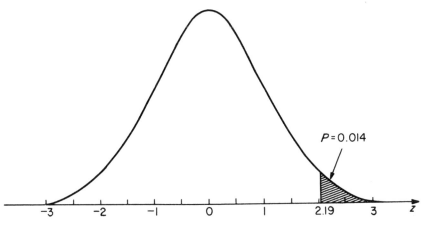

$P = 0.014$

Fig. 6.13.   Area to the Right of $z = 2.19$

Table 6.7 shows that the area to the right of $z = 2.19$ is 0.014. (See Figure 6.13.) But we have used 0.002 of this area in the 145–155 class. Therefore, we have $0.014 - 0.002 = 0.012$ as the relative frequency for the 135–145 class. The area to the right of $z = 1.56$ is 0.059. But 0.014 is to the right of $z = 2.19$. Therefore, the area between $z = 1.56$ and $z = 2.19$ is $0.059 - 0.014 = 0.045$. (See Figure 6.14.) This is the relative frequency for the 125–135 class. The relative frequency for the 115–125 class is found by subtracting the value of P_z at $z = 1.56$ from the value of P_z at $z = 0.94$. This gives us $0.174 - 0.059 = 0.115$. Similarly, the relative frequency for the 105–115 class is $0.378 - 0.174 = 0.204$.

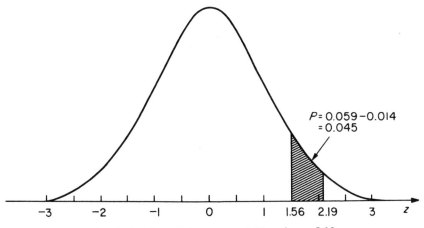

$P = 0.059 - 0.014$
$= 0.045$

Fig. 6.14.   Area Between $z = 1.56$ and $z = 2.19$

We come now to the middle class. The lower limit of z in this class is negative and the upper limit of z in this class is positive. Consequently, it will simplify the work here if we split this interval into two parts, namely, the part from $z = -0.31$ to $z = 0.00$ and the part from $z = 0.00$ to $z = +0.31$. First, let us find the area for the part of the interval from $z = 0.00$ to $z = +0.31$. The value of P_z at $z = 0.00$ in Table 6.7 is 0.500 and the value of P_z at $z = 0.31$ is 0.378. Consequently, the area between $z = 0.00$ and $z = +0.31$ is $0.500 - 0.378 = 0.122$. (See Figure 6.15.) The relative frequency 0.122 is recorded in column three of Table 6.9 opposite the limits 0.00 and $+0.31$ of z.

It is not necessary to do any more computation in order to fill in the relative frequencies in the upper half of column three. All that we need to do is remember that the normal distribution is symmetrical and in the upper half of column three we can repeat the numbers that we have computed already. Therefore, we write 0.122 as the relative frequency for the part of the middle class from $z = -0.31$ to $z = 0.00$. Combining the two halves of the middle class, we have 0.244 as the relative frequency for the 95–105 class. (See Figure 6.15.)

We record 0.204 as the relative frequency for the 85–95 class. Similarly, we record 0.115 as the relative frequency for the 75–85 class, and so on, until we record 0.002 as the relative frequency for the 45–55 class.

The total of the probability column is 1.000. This is a partial check that no mistake has been made so far. In column four, the relative frequencies are changed into computed frequencies by multiplying each of the relative frequencies by 937, that is, by the value of N for the distribution. In column five, the computed frequencies are rounded off to the nearest whole number of children.

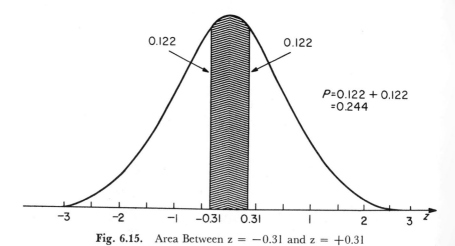

Fig. 6.15.   Area Between $z = -0.31$ and $z = +0.31$

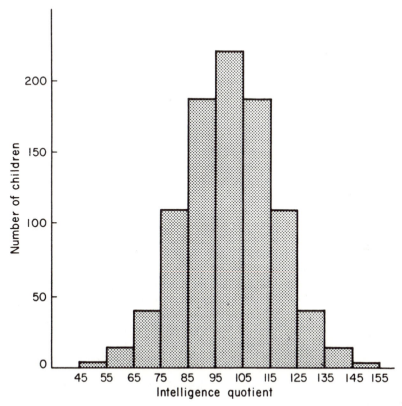

Fig. 6.16. Normal Distribution of IQ Among 937 Children

The solution of the problem is complete now, except that it may help us to visualize and remember the distribution if we construct a histogram by using the data in the first and the fifth columns of Table 6.9. Figure 6.16 is the histogram.

EXERCISES

1. Use Table 6.6 to determine the ordinates of a normal curve for the following abscissas: $z = 0.7$, $z = 1.3$, $z = 1.8$, $z = -0.7$, $z = -1.3$, and $z = -1.8$. Plot the points of the normal curve that are determined by these abscissas and ordinates.

2. Plot enough additional points on the normal curve in exercise 1 so that you can draw a smooth normal curve through the set of points. You should obtain a graph similar to that in Figure 6.11.

3. Use Table 6.7 to find the probability that in a normal distribution a value of the variable chosen at random will be (a) at least 0.70 standard units greater than the mean of the distribution; (b) at least 0.70 standard units less than the mean of the distribution. Draw a graph similar to that in Figure 6.20 to show the area corresponding to the probabilities in this problem.

4. Use Table 6.7 to find the probability that in a normal distribution a value of the variable chosen at random will be (a) between 1.34 standard units greater than the mean and 1.83 standard units greater than the mean; (b) between 1.34 standard units less than the mean and 1.83 standard units less than the mean; (c) between 1.34 standard units less than the mean and 1.83 standard units greater than the mean. Draw graphs to show the areas representing the probabilities in these three situations.

5. In order to assist the clothing industry to design and manufacture garments that fit well for American boys and girls, the Bureau of Home Economics of the United States Department of Agriculture measured large samples of boys and girls of all ages from four to seventeen years. For the sample of 2771 boys seventeen years of age, the Bureau found that the mean height of the boys was 173.75 cms. and that the standard deviation of their heights was 6.49 cms. Use Table 6.7 to estimate the number of boys in the sample who were more than 180.24 cms. tall. (*Ans.* 441.) More than 186.73 cms. tall. (*Ans.* 64.) Less than 167.26 cms. tall. (*Ans.* 441.) Less than 160.77 cms. tall. (*Ans.* 64.)

6: The Bureau of Home Economics found that the mean height of the 2115 girls seventeen years of age in their sample was 161.13 cms. and that the standard deviation of the distribution of the heights of these girls was 5.94 cms. Use Table 6.7 to estimate the number of girls in the sample who were more than 167.07 cms. tall (*Ans.* 336.) More than 173.01 cms. tall. (*Ans.* 49.) Less than 155.19 cms. tall. (*Ans.* 336.) Less than 149.25 cms. tall. (*Ans.* 49.)

7. Draw graphs to represent the numbers of boys in the four groups mentioned in exercise 5.

8. Draw graphs to represent the numbers of girls in the four groups mentioned in exercise 6.

9. Construct a normal distribution of the heights of the 2771 boys in the sample described in exercise 5. Construct a histogram for this distribution.

10. Construct a normal distribution of the heights of the 2115 girls in the sample described in exercise 6. Construct a histogram for this distribution.

11. Use Table 6.6 to assist you in drawing a smooth normal curve over the histogram that you constructed in exercise 9.

12. Use Table 6.6 to assist you in drawing a smooth normal curve over the histogram that you constructed in exercise 10.

13. It has been stated frequently in educational circles that if a very large class of students takes a well-constructed test, the scores of the students should form approximately a normal distribution. Show how to construct a normal distribution of marks for grading a school test. Assume that the passing marks are A, B, C, and D, and that all other marks are F. (*Ans.* A, 7%; B, 24%; C, 38%; D, 24%; F, 7%.)

6.6 TESTS OF GOODNESS OF FIT

The binomial test described in section 5.1.1 is a test of the goodness of fit of a specified binomial distribution to the distribution observed in a sample. The rejection of the null hypothesis in connection with the binomial test is equivalent to stating that the specified binomial distribution is not a satisfactory fit to the distribution observed in the sample.

The binomial test of goodness of fit is applicable, of course, only if there are exactly two categories in the classification of the elementary units. Sometimes when there are more than two categories in the classification, we are willing and able to combine some of the categories and in this way reduce the number of categories to two.

There are many other kinds of practical situations in which a person needs to know whether or not a distribution that he has found in a sample may be considered to have been drawn from some specific type of universe. The Kolmogorov-Smirnov test is a simple and powerful test of the goodness of fit of any theoretical or hypothetical distribution to an observed frequency distribution. This test may be interpreted as answering the question of whether or not it is permissible to think that the distribution in the universe from which the sample was drawn is of some specific type, such as, for example, a normal distribution.

The Kolmogorov-Smirnov test is made by comparing the cumulated relative frequencies for the sample with the cumulated relative frequencies for the theoretical or hypothetical distribution in the universe. In other words, in order to apply the Kolmogorov-Smirnov test of goodness of fit to a sample we need to know the relative frequencies of the theoretical or hypothetical distribution in the universe for the same categories or classes that are used in the frequency distribution for the sample. The test is more sensitive, and therefore more useful, when there are a large number of categories or classes in the frequency distribution.

The artificial data in Table 6.10 will be used as the first illustration of procedure in the Kolmogorov-Smirnov test of goodness of fit. The null hypothesis to be tested here is that the universe from which the random sample was drawn is a rectangular universe, that is, that the elementary units in the universe are uniformly distributed among the five classes,

TABLE 6.10

**Tabulation for the Kolmogorov-Smirnov
Test of Goodness of Fit**

Class Limits	Frequency in Sample	Rel. Freq. in Sample	Cum. Rel. Freq. in Sample S	Rel. Freq. in Universe	Cum. Rel. Freq. in Universe U	U − S
9.5–12.5	3	0.12	0.12	0.20	0.20	0.08
12.5–15.5	1	0.04	0.16	0.20	0.40	0.24
15.5–18.5	7	0.28	0.44	0.20	0.60	0.16
18.5–21.5	5	0.20	0.64	0.20	0.80	0.16
21.5–24.5	9	0.36	1.00	0.20	1.00	0.00
Total	25	1.00		1.00		

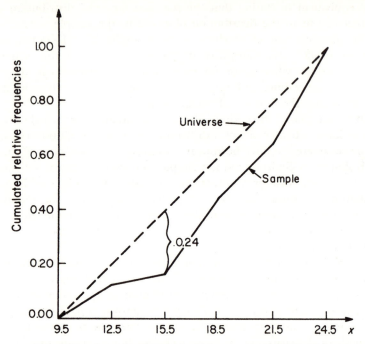

Fig. 6.17.   Cumulated Relative Frequencies of Table 6.10

with 20 per cent of the elementary units of the universe in each class. The alternative hypothesis is that the universe from which the random sample was drawn is not a rectangular universe. This pair of hypotheses requires a two-tails test.

If the two cumulated relative frequency distributions differ by too much in either direction at any particular value of the variable, we should reject the null hypothesis and accept the alternative. In Table 6.10 and Figure 6.17, none of the values of $U - S$ is negative. Table 6.13 and Figure 6.18 show an example in which some of the values of $U - S$ are positive and others are negative. Because we are not concerned about the direction of the difference between the two cumulated relative frequency distributions at a particular value of the variable, we can disregard any minus signs that happen to be attached to values of $U - S$, focus our attention on the absolute values of the differences $U - S$, and pick out the difference that has the largest absolute value. In Table 6.10, the largest absolute value of $U - S$ is 0.24.

Table 6.11 has been constructed to facilitate this test of the null hypothesis. We compare our observed value 0.24 with the critical values on the line for $n = 25$. We see that the probability to be associated with $n = 25$

TABLE 6.11

Critical Values of the Maximum Difference $|U - S|$ in the Kolmogorov-Smirnov Test of Goodness of Fit*

Sample Size n	Significance Level				
	.20	.10	.05	.02	.01
	Reject null hypothesis if $|U - S|$ is greater than or equal to				
1	.900	.950	.975	.990	.995
2	.684	.776	.842	.900	.929
3	.565	.636	.708	.785	.829
4	.493	.565	.624	.689	.734
5	.447	.509	.563	.627	.669
6	.410	.468	.519	.577	.617
7	.381	.436	.483	.538	.576
8	.358	.410	.454	.507	.542
9	.339	.387	.430	.480	.513
10	.323	.369	.409	.457	.489
11	.308	.352	.391	.437	.468
12	.296	.338	.375	.419	.449
13	.285	.325	.361	.404	.432
14	.275	.314	.349	.390	.418
15	.266	.304	.338	.377	.404
16	.258	.295	.327	.366	.392
17	.250	.286	.318	.355	.381
18	.244	.279	.309	.346	.371
19	.237	.271	.301	.337	.361
20	.232	.265	.294	.329	.352
21	.226	.259	.287	.321	.344
22	.221	.253	.281	.314	.337
23	.216	.247	.275	.307	.330
24	.212	.242	.269	.301	.323
25	.208	.238	.264	.295	.317
26	.204	.233	.259	.290	.311
27	.200	.229	.254	.284	.305
28	.197	.225	.250	.279	.300
29	.193	.221	.246	.275	.295
30	.190	.218	.242	.270	.290
35	.18	.20	.22	.25	.27
40	.17	.19	.21	.23	.25
45	.16	.18	.20	.22	.24
50	.15	.17	.19	.21	.23
60	.14	.16	.17	.19	.21
70	.13	.14	.16	.18	.19
80	.12	.13	.15	.17	.18
90	.11	.13	.14	.16	.17
100	.11	.12	.13	.15	.16
$n > 100$	$\dfrac{1.07}{\sqrt{n}}$	$\dfrac{1.22}{\sqrt{n}}$	$\dfrac{1.36}{\sqrt{n}}$	$\dfrac{1.52}{\sqrt{n}}$	$\dfrac{1.63}{\sqrt{n}}.$

*Adapted from Leslie H. Miller, "Tables of Percentage Points of Kolmogorov Statistics," *Journal of the American Statistical Association* 51 (1956), pp. 111–121. Reproduced here by permission of the author and the editor of the *Journal*.

and $|U - S| = 0.24$ is 0.10. Consequently, the null hypothesis that is being tested here cannot be rejected at any significance level less than 0.10.

This means that if we repeated a very large number of times the process of drawing a random sample of 25 elementary units from the rectangular universe indicated in Table 6.10, we should expect to find that 10 per cent of the samples would have cumulated relative frequency distributions for which the maximum absolute value among the differences $U - S$ would be as great as or greater than 0.24.

The Kolmogorov-Smirnov test theoretically is appropriate for the test of goodness of fit whenever the underlying variable in the investigation is continuous. But it may also be used in situations in which the underlying variable is discrete. If the test is applied to a distribution of a discrete variable in a sample and the decision indicated by Table 6.11 is to reject the hypothesis about the distribution in the universe, the decision to reject is a safe one—at least, as safe as it would be if the underlying variable were continuous; this means that the actual significance level will be at least as small as that indicated in Table 6.11.

In some applications of the Kolmogorov-Smirnov test, it may be necessary to use the sample data to estimate the parameters of the hypothetical universe. For example, in a test of normality we might, because of lack of knowledge about the values of the mean and the standard deviation in the universe, use the values of the mean m and the standard deviation s computed from the sample as the values of μ and σ for the hypothetical

TABLE 6.12

Mileage at First Failure of Bus Motor*

Distance Driven (Thousands of Miles)	Observed Number of Failures	Expected Number of Failures†
Less than 20	6	3.6
20–40	11	8.2
40–60	16	18.5
60–80	25	31.3
80–100	34	39.9
100–120	46	38.3
120–140	33	27.6
140–160	16	15.0
160–180	2	6.1
180 and more	2	2.5
Total	191	191.0

*Source: D. J. Davis, "An Analysis of Some Failure Data," *Journal of the American Statistical Association* 47, (1952), p. 145. Reproduced here with the permission of D. J. Davis and the editor of the *Journal*.

†Based on the normal probability law.

normal distribution in the universe. Obviously, this ought to produce the best-fitting normal distribution for the sample. Consequently, if, in such a situation, the decision indicated by Table 6.11 is to reject the hypothetical distribution for the universe, then the actual significance level at which the rejection is made will be at least as small as that indicated by Table 6.11.

The first two columns of Table 6.12 show the number of miles that 191 new buses operated by a large city bus company were driven before the first motor failure of each bus. Failure was either abrupt, in which some part broke and the motor would not run, or, by definition, when the maximum power produced, as measured by a dynamometer, fell below a fixed percentage of the standard rated value. Failures of motor accessories that could be replaced easily were not included in these data.

Davis computed the expected frequencies in the third column of Table 6.12 to show that a normal distribution is a reasonably good fit to the observed data. This fact must not be interpreted as indicating that *all* distributions of mechanical or industrial failures are normal distributions. Davis gave some other illustrations of failure distributions that the normal distribution did not fit satisfactorily.

Now let us use the Kolmogorov-Smirnov test to see whether or not the normal distribution computed by Davis is a good fit to the observed distribution of bus motor failures. In order to apply the test, we need to compute the cumulated relative frequencies for the observed sample and the cumulated relative frequencies for the proposed normal distribution. These cumulated relative frequencies are shown in Table 6.13 and Figure 6.18. Table 6.13 also shows the values of $U - S$.

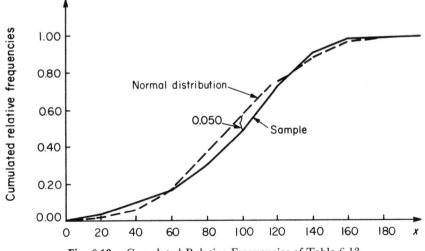

Fig. 6.18. Cumulated Relative Frequencies of Table 6.13

TABLE 6.13

Kolmogorov-Smirnov Test of Goodness of Fit for Table 6.12

Distance Driven (Thousands of Miles)	Relative Frequency in Sample	Cumulated Rel. Freq. in Sample S	Rel. Freq. in Normal Distribution	Cum. Rel. Freq. in Normal Dist. U	U − S
Less than 20	0.031	0.031	0.019	0.019	−0.012
20–40	0.058	0.089	0.043	0.062	−0.027
40–60	0.084	0.173	0.097	0.159	−0.014
60–80	0.131	0.304	0.164	0.323	0.019
80–100	0.178	0.482	0.209	0.532	0.050
100–120	0.241	0.723	0.200	0.732	0.009
120–140	0.173	0.896	0.144	0.876	−0.020
140–160	0.084	0.980	0.079	0.955	−0.025
160–180	0.010	0.990	0.032	0.987	−0.003
180 and more	0.010	1.000	0.013	1.000	0.000
Total	1.000		1.000		

The critical values for the Kolmogorov-Smirnov test in this example are determined by the expressions in the last row of Table 6.11. These critical values range from $1.07/\sqrt{191} = 1.07/13.83 = 0.077$ for $\alpha = 0.20$ to $1.63/\sqrt{191} = 0.118$ for $\alpha = 0.01$. The largest absolute value of $U - S$ in Table 6.13 is 0.050, which is less than any of the critical values for $n = 191$. Consequently, we do not reject the hypothesis that the observed distribution in the sample was drawn from a universe that has the normal distribution given by Davis. In fact, because the largest absolute value of $U - S$ is considerably less than the critical value 0.077 for $\alpha = 0.20$, we can accept the normal distribution given by Davis as a very good fit for the observed sample.

The Kolmogorov-Smirnov test of goodness of fit is not intended for application to truly nonquantitative classifications. For a truly nonquantitative distribution in which there are only two categories or which can be reduced to two categories satisfactorily, a binomial test of goodness of fit may be appropriate. For a truly nonquantitative distribution that cannot be reduced to two categories satisfactorily, a test known as "the chi square test of goodness of fit" may be appropriate. This test is described in most of the advanced books on statistical methods.

EXERCISES

1. Use the Kolmogorov-Smirnov test to test the hypothesis that the distribution in Table 6.14 is a random sample from a normal universe of shirts with 33-inch sleeves

in which the mean neck size is 15.19 inches and the standard deviation of the neck sizes is 0.70 inches.

TABLE 6.14

**Neck Size of Men's Shirts That Have
33-Inch Sleeve Length***

Neck Size (Inches)	No. of Shirts (in Units of 100 Dozen)
13½	½
14	2
14½	9
15	11
15½	10
16	5
16½	2
17	1
Total	40½

Source: Columns 1 and 4 of Table 6.24 (p. 00).

2. Use the Kolmogorov-Smirnov test to test the goodness of fit of a normal distribution to the distribution of sleeve lengths of the shirts in Table 6.15.

TABLE 6.15

**Sleeve Length of Men's Shirts That
Have 15-Inch Neck Size***

Sleeve Length (Inches)	No. of Shirts (in Units of 100 Dozen)
31	1
32	6
33	11
34	5
35	2
Total	25

Source: Table 6.24.

3. Use the Kolmogorov-Smirnov test to test the goodness of fit of a normal distribution to the distribution of vehicle speeds given in Table 6.16.

TABLE 6.16

**Speed of South-Bound Passenger Vehicles Passing
a Highway Reference Point in a Test Period***

Speed in Miles Per Hour	No. of Vehicles Observed
10–15	1
15–20	3
20–25	4
25–30	16
30–35	25
35–40	56
40–45	79
45–50	28
50–55	24
55–60	8
60–65	3
65–70	1
70–75	1
Total	249

*Source: National Safety Council, Committee on Speed Regulation, "Methods for Making Speed Observations," (tentative draft), p. 5, August, 1940. Reprinted by permission of the Council.

4. Determine the normal distribution that best fits the distribution of men's regular suit sizes in Table 6.26. In other words, use the normal probability law to smooth out the irregularities in the distribution of the sizes of men's regular suits. For example, in the table there are proportionately fewer size 39 and size 41 suits than there probably are men in the universe of men who wear "regular" suits. These irregularities in the distribution in Table 6.26 can probably be accounted for, in part at least, by the fact that many small-store operators consider it unwise to carry sizes 39 and 41 in stock. They know that they can make a size 40 suit fit a size 39 man easily by some simple alterations. They probably consider it better business to carry more color choices in a smaller number of sizes. Test the goodness of fit of the normal distribution to the observed distribution.

5. Choose a random starting point in Table A.15 and then record the next 100 digits that occur along some chosen route. Use the Kolmogorov-Smirnov test to test the hypothesis that the large universe of random digits of which Table A.15 is one page is a rectangular distribution.

6. The null hypothesis for the test of goodness of fit that was made in connection with Table 6.10 could have been: If an elementary unit is chosen at random from the universe, the value of the variable for the chosen elementary unit is equally likely to be in each of the five classes.   True ☐ False ☐

6.7 A TEST OF RANDOMNESS

Sometimes it is desirable to be able to examine data after they have been collected for evidence sufficient to force us to reject the hypothesis that the data are the kind that might reasonably be expected to result from a random process. For example, if we use 100 numbers from a table of random digits to determine the elementary units to be included in a sample, is there any way to test those 100 numbers to see if they may safely be regarded as a random sequence of numbers?

Also, there are situations in which it is useful and desirable to know whether certain events naturally occur in random fashion or whether we should look for some systematic law or formula to account for the sequence in which they occur. For example, do accidents in a factory happen to employees in random fashion or are certain employees more likely than others to be involved in accidents?

There are several available tests of the randomness of a sequence of events. One of these tests is called a "runs" test. A *run* is defined as a sequence of identical symbols that are preceded by and followed by different symbols or by no symbols at all. For example, the following sequence of plus and minus signs contains eight runs. The eight runs have been bracketed and numbered.

The symbols that are used may be any pair of appropriate opposites, such as the plus signs and minus signs used above. If you were investigating the randomness of the tossing of a coin, you probably would use H and T as the symbols for heads and tails. If you were investigating the randomness of a sequence of random digits, you might examine the sequence to see how many odd digits and how many even digits occurred in the sequence, and in that case you might use O and E as the symbols for odd and even.

If there are too many runs or too few runs in a sequence of events, then we reject the hypothesis that the events occur in random order. For example, if in a sequence of twenty successive tosses of a coin we observed either of the following sequences, we most likely would reject the idea that the heads and tails occurred in a random sequence. We probably would suspect that either the coin was not fair or that the tosser was not using a fair method of tossing the coin. In other words, we do not expect random events to turn out this way:

```
T T T T T T T T T T H H H H H H H H H H   ( 2 runs)
T H T H T H T H T H T H T H T H T H T H   (20 runs)
```

In both of these sequences, the percentage of heads is 50 and the percentage of tails is 50. But that does not convince us that the sequences are random.

Where should we draw the line in deciding whether or not a sequence of events contains too many runs or too few runs to be considered random?

Twenty runs in a sequence of twenty tosses of a coin obviously are too many. And two runs in a sequence of twenty tosses of a coin obviously are too few for a random process. No doubt, by now you have become sufficiently well acquainted with the statistical method of thinking to say that we should reject the hypothesis of randomness in a sequence if the number of runs in the sequence is so large that it would occur less than some specified percentage of the time, e.g., 5 per cent, or $2\frac{1}{2}$ per cent, or 1 per cent, and that we should reject the hypothesis of randomness in a sequence if the number of runs in the sequence is so small that it would occur less than some specified percentage of the time. That is exactly how the runs test works.

Let n_1 = the number of symbols of the less frequent kind in the sequence, n_2 = the number of symbols of the more frequent kind in the sequence, $n = n_1 + n_2$, and x = the number of runs in the sequence. If neither n_1 nor n_2 is greater than 20, we use Table 6.17 to test the randonmess of the sequence.

In the first sequence of 20 heads and tails given above, there are 10 heads (H), 10 tails (T), and only 2 runs. The upper section of Table 6.17 shows that 6 is the lower critical value for the number of runs if $n_1 = 10$ and $n_2 = 10$. Because the observed number of runs, namely, 2, is not greater than 6, we reject the hypothesis that the tossing process is producing a random sequence of heads and tails.

In the second sequence of 20 heads and tails, there are 10 heads (H), 10 tails (T), and 20 runs. The lower section of Table 6.17 shows that the upper critical value for the number of runs is 16 if $n_1 = 10$ and $n_2 = 10$. Because the observed number of runs in the second sequence of heads and tails, namely, 20, is not less than 16, we reject the hypothesis that the tossing process is producing a random sequence of heads and tails.

Consider the sequence of 16 plus signs and minus signs given on page 00. There, $n_1 = 6$ (the number of plus signs), $n_2 = 10$ (the number of minus signs), $n = 6 + 10 = 16$, and $x = 8$ (the number of runs).

The hypothesis to be tested is that the 16 plus and minus signs in the sequence are in random order. Suppose that we choose as the alternative the statement that the 16 plus signs and minus signs in the sequence are not in random order. This requires a two-tails test because the alternative does not indicate the direction of any lack of randomness that may be present. Let $\alpha = 0.05$ be the significance level for this two-tails test.

Table 6.17 shows that the critical values of x, the number of runs in a sequence, when $n_1 = 6$ and $n_2 = 10$ are 4 and 13. Because the value of x, namely, 8, for the sequence being tested is between 4 and 13, we do not reject the hypothesis that the sequence of signs is in random order.

If the alternative that is chosen includes the prediction that the number of runs is (or will be) too small, then a one-tail test is required and the

TABLE 6.17

Critical Values of x in the Runs Test at the $\alpha = 0.05$ Significance Level for a Two-Tails Test and at the $\alpha = 0.025$ Significance Level for a One-Tail Test*

Reject the null hypothesis if x is equal to or less than

Value of n_2	Value of n_1																		
	2	3	4	5	6	7	8	9	10	11	12	13	14	15	16	17	18	19	20
5		2	2																
6		2	2	3	3														
7		2	2	3	3	3													
8		2	3	3	3	4	4												
9		2	3	3	4	4	5	5											
10		2	3	3	4	5	5	5	6										
11		2	3	4	4	5	5	6	6	7									
12	2	2	3	4	4	5	6	6	7	7	7								
13	2	2	3	4	5	5	6	6	7	7	8	8							
14	2	2	3	4	5	5	6	7	7	8	8	9	9						
15	2	3	3	4	5	6	6	7	7	8	8	9	9	10					
16	2	3	4	4	5	6	6	7	8	8	9	9	10	10	11				
17	2	3	4	4	5	6	7	7	8	9	9	10	10	11	11	11			
18	2	3	4	5	5	6	7	8	8	9	9	10	10	11	11	12	12		
19	2	3	4	5	6	6	7	8	8	9	10	10	11	11	12	12	13	13	
20	2	3	4	5	6	6	7	8	9	9	10	10	11	12	12	13	13	13	14

Reject the null hypothesis if x is equal to or greater than

Value of n_2	Value of n_1																		
	2	3	4	5	6	7	8	9	10	11	12	13	14	15	16	17	18	19	20
2	5																		
3	6	7																	
4	6	8	9																
5	6	8	9	10															
6	6	8	9	10	11														
7	6	8	10	11	12	13													
8	6	8	10	11	12	13	14												
9	6	8	10	12	13	14	14	15											
10	6	8	10	12	13	14	15	16	16										
11	6	8	10	12	13	14	15	16	17	17									
12	6	8	10	12	13	14	16	16	17	18	19								
13	6	8	10	12	14	15	16	17	18	19	19	20							
14	6	8	10	12	14	15	16	17	18	19	20	20	21						
15	6	8	10	12	14	15	16	18	18	19	20	21	22	22					
16	6	8	10	12	14	16	17	18	19	20	21	21	22	23	23				
17	6	8	10	12	14	16	17	18	19	20	21	22	23	23	24	25			
18	6	8	10	12	14	16	17	18	19	20	21	22	23	24	25	25	26		
19	6	8	10	12	14	16	17	18	20	21	22	23	23	24	25	26	26	27	
20	6	8	10	12	14	16	17	18	20	21	22	23	24	25	25	26	27	27	28

*Adapted from Frieda S. Swed and C. Eisenhart, "Tables for Testing Randomness of Grouping in a Sequence of Alternatives," *Annals of Mathematical Statistics* 14 (1943), Table II, pages 83–86, with the permission of the authors and the editor of the *Annals*.

critical value of x is found in the upper section of Table 6.17. If the observed number of runs is equal to or less than this critical value, the hypothesis of randomness is rejected at the 0.025 significance level and the alternative is accepted.

If the alternative that is chosen includes the prediction that the number of runs is (or will be) too large, then a one-tail test is required and the critical value of x, the number of runs, is found in the lower section of Table 6.17. If the observed number of runs in the sequence is equal to or greater than this critical value, the hypothesis of randomness for the process is rejected at the 0.025 significance level and the alternative is accepted.

If either n_1 or n_2 is greater than 20, or if both n_1 and n_2 are greater than 20, we cannot use Table 6.17 for the test of randomness. If neither n_1 nor n_2 is less than 10, we can obtain a good approximation to the true probability that is involved in the test of randomness by using the normal probability table, Table 6.7. For this purpose, we compute

$$\mu = \frac{2n_1n_2}{n} + 1, \quad \sigma = \sqrt{\frac{2n_1n_2(2n_1n_2 - n)}{n^2(n-1)}}$$

and

$$z = \frac{x \pm 0.5 - \mu}{\sigma} \qquad \begin{array}{l} \text{using } +0.5 \text{ if } x < \mu \\ \text{using } -0.5 \text{ if } x > \mu \end{array}$$

For example, suppose that the number of boys in a large group is the same as the number of girls in the group and that the members of the group have been requested to sign their names to a list if they are interested in attending a dance for the group. Suppose that, beginning at a randomly selected name in the list, the next 50 individuals are boys and girls in the following sequence:

G G B G B B G G G G B B G B G B B G G G G G B B B B B B G G B B B
G G G B G B G B B G G G G G B B G G G

Test the hypothesis that the boys and the girls are listed in random order against the alternative that the sequence is not random because there are too few runs. This requires a one-tail test. Use the significance level $\alpha = 0.05$.

In this sequence, $n_1 = 22$ (boys), $n_2 = 28$ (girls), $n = 22 + 28 = 50$, and the number of runs is $x = 23$. Consequently, $\mu = 25.64$, $\sigma = 3.45$, and $z = -0.62$. Table 6.7 shows that the probability in the tail associated with $z = -0.62$ is $P = 0.268$. Because this value of P is greater than the chosen $\alpha = 0.05$, we do not reject the hypothesis that the boys and the girls are listed in random order. If the alternative had not included a prediction of the direction of departure from randomness, a two-tails test would have been used and the value of P obtained from Table 6.7 would need to be multiplied by 2.

If neither n_1 nor n_2 is less than 10, satisfactory approximations for the critical values of x, the number of runs, can be computed by substituting for z the values of the abscissa in Table 6.7 associated with the chosen value of α and then solving the equation for x. For example, if the area in each of the two tails of the normal curve is to be 0.025, then $z = \pm 1.96$. If $n_1 = 20$ and $n_2 = 20$, then $\mu = 21$ and $\sigma = 3.12$. The solutions of the two equations $-1.96 = (x + 0.5 - 21)/3.12$ and $+1.96 = (x - 0.5 - 21)/3.12$ are $x = 14.4$ and $x = 27.6$, respectively. We may take 14 and 28 as the critical values for any runs test in which $n_1 = 20$ and $n_2 = 20$ and the probability in each tail is 0.025. You can compare these critical values with the critical values given in Table 6.17 for $n_1 = 20$ and $n_2 = 20$.

EXERCISES

1. Suppose that a department store has sold an equally large number of brand X and brand Y television sets. The manager of the store begins to suspect that there are beginning to be considerably more calls for service on brand X sets than there are on brand Y sets. A record is kept for a week showing the type of set on which each service call was received and the time at which the call was received. The resulting sequence of 23 service calls is

X Y X X X Y X X Y Y X X X X X X Y X X X X X Y

Use the runs test to test the randomness of this sequence of service calls and interpret for the manager of the store the conclusion indicated by the runs test.

2. Beginning at a randomly selected moment, record the sex of the individuals entering a theater to attend the showing of a motion picture. Stop when you have recorded a sequence of, say, 36 persons. Test the hypothesis that the patrons of the theater on that occasion are a randomly ordered sequence of males and females.

3. In the sequence of 50 boys and girls given earlier, there were 22 boys and 28 girls. Compute approximate critical values for x, the number of runs, in any sequence in which $n_1 = 22$ and $n_2 = 28$ and the probability in each tail is to be 0.025.

4. Choose at random a digit in Table A.15 and a sequence of, say, 100 digits, beginning with the chosen digit. Test the hypothesis that the table is a random set of odd and even digits. Compute critical values of x corresponding to the observed values of n_1 and n_2 and the value of α that you chose for the test.

6.8 CENTRAL LIMIT THEOREM

The normal distribution would have a central place in the study of statistical methods even if none of the natural or social or intellectual distributions that we find all around us were approximately normal.

The normal probability law is the very foundation on which rest such important statistical processes as the determination of confidence limits and the testing of hypotheses about universes. These statistical processes are possible mainly because most of the sampling distributions that are the bases for confidence limits and tests of hypotheses are either normal distributions or can be approximated satisfactorily by normal distributions.

This is true for both discrete distributions and continuous distributions. For example, in determining confidence limits for proportions and in testing hypotheses about proportions, we usually are able to use the continuous normal distribution and the normal probabilities instead of the discontinuous binomial distribution and the binomial probabilities. In determining confidence limits for means and in testing hypotheses about means (to be discussed in Chapter 7) we usually apply the normal probability law regardless of whether the variables involved are discrete or continuous. We will find some other important uses for the normal probability law in Chapters 8 and 9.

The basic principle here may be summarized as follows: If we draw a large number of large random samples of size n from a universe that has the mean μ and the standard deviation σ, the distribution of the means of the samples can be approximated closely by a normal distribution for which the mean and the standard deviation are μ and $\sigma/\sqrt{n}$, respectively. This statement usually is called the *central limit theorem*.

It is a very important theorem in statistics, because it permits us to draw a random sample from almost any universe and apply the table of normal probabilities to determine confidence limits for the mean of the universe and to test hypotheses about the mean of the universe, provided the sample size is large, that is, not less than, say, 30 to 32. In fact, some examples will be cited in section 7.2.1 indicating that the distribution of means of samples may be at least roughly approximated by a normal distribution even if the distribution in the universe is rectangular or a right-triangular distribution and the size of the samples is as small as $n = 4$. The common practice of using samples of size $n = 5$ in statistical quality control inspection of products manufactured in large numbers is based on the central limit theorem.

6.9 OTHER PROPERTIES OF NORMAL DISTRIBUTIONS

The range of the values of z in a normal distribution is infinitely long. It runs infinitely far out to the left through negative numbers and infinitely far out to the right through positive numbers. Tables 6.18 and 6.19 indicate that there are small relative frequencies for values of z far out along the z-axis in both directions. In other words, no matter how far out along the z-axis in either direction you go, the theoretical normal curve does not touch the z-axis, but it comes closer and closer. Such a curve is said to approach the z-axis *asymptotically*, or the z-axis is said to be an *asymptote* of the curve. An asymptote is a straight line that is tangent to a curve, the point of tangency (that is, the point at which the straight line touches the curve) being at infinity.

Consequently, we never can have a complete, perfectly normal distribution in an observed sample, because the range of the variable in an observed sample always is finite. Furthermore, the variable in a theoretical normal distribution always is continuous. But the values of the variable in an observed distribution always are discrete; we can measure values of a continuous variable only with a limited degree of accuracy and we can record the values only to a limited number of significant digits. Therefore, when we state that the distribution in a sample or in a finite universe is a normal distribution, what we mean, at best, is that the relative frequencies in the intervals along the finite range of the variable agree closely with the relative frequencies in the same intervals in that part of the range of the variable in a theoretical or ideal normal distribution.

The normal distribution is, therefore, a theoretical mathematical concept or idea rather than a completely and perfectly observable fact or event, but it is of great importance to practical statisticians. Because it is a theoretical mathematical concept expressible in algebraic and geometric terms, mathematicians have been able to deduce from it results that have far-reaching consequences for scientists and technicians. So much of the theory of statistics is based on the idea of a normal distribution that if this idea and all of the developments in modern statistics that depend in whole or in part for their validity on the idea of a normal distribution were taken away, much of the structure of modern statistical inference about universes from samples would collapse.

The mathematical theorems based on the idea of a normal distribution are as much the foundation of modern practical statistics as the theorems of geometry are the foundation of modern practical engineering. Just as it would be impossible to have an elaborate system of practical engineering principles if there were only observed data and no mathematical theorems such as those in geometry, so also it would be impossible to have an elaborate system of practical statistical principles if there were only observed data and no mathematical theorems such as those based on the idea of a normal distribution. If it were not for our system of statistical principles, we might never know whether or not an inference drawn from a sample and applied to the universe has any more reliability than a sheer guess.

The discovery of the mathematical formula for the normal probability law or normal frequency distribution is one of the most important discoveries in all the history of science. Abraham de Moivre (1667-1754), a French mathematician who moved to England when he was twenty-one years of age, made the basic discovery leading to the equation of the normal curve; the result was published in 1733. However, the curve is called the Gaussian curve by some statisticians because of the important part played by the German mathematician, Karl Friedrich Gauss (1777-1855), in its development. Pierre Simon Laplace (1749-1827), a French mathematician, was another important contributor to the development of the normal probability law.

If all the N values of a continuous variable x in a universe may be considered to form a normal distribution, the equation of the smooth normal frequency curve that fits the distribution is

$$y = \frac{N}{\sigma\sqrt{2\pi}}\, e^{-(x-\mu)^2/2\sigma^2} \tag{6.2}$$

where y is the ordinate corresponding to the abscissa x

> N is the total frequency or total number of elementary units in the distribution and also is the total area under the normal curve
> μ is the mean of the values of x in the distribution
> σ is the standard deviation of the values of x in the distribution
> $e = 2.71828\ldots$, that is, the base of the Napierian system of logarithms
> $\pi = 3.14159\ldots$, that is, the constant involved in the circumference of a circle

For example, the equation of the normal curve that fits the distribution in Table 6.9 and Figure 6.16 is

$$y = \frac{937}{16\sqrt{2(3.1416)}}\, e^{-(x-100)^2/2(16)^2} = 23.36 e^{-(x-100)^2/512}$$

Similarly, the equation of the normal curve that best fits a random sample and smooths out the irregularities in the sample distribution is obtained by substituting the values of n, m, and s in place of N, μ, and σ in equation 6.2.

The points of inflexion for the normal curve determined by equation 6.2, that is, the two points at which the curve changes from concave downward to concave upward, are the points on the curve for which $x = \mu - \sigma$ and $x = \mu + \sigma$. That is an interesting property of the standard deviation of a normal distribution. Furthermore, the two tangents to the normal curve at the points of inflexion cross the x-axis at the places where $x = \mu - 2\sigma$ and $x = \mu + 2\sigma$.

As is true for other equations in algebra, equation 6.2 can be transformed in many ways. Consequently, it often is written in other forms in statistics books. One of the most common and important forms of the equation of the normal curve is the form that is applicable to the relative frequencies for values of the variable that are expressed in standard units. The equation of the normal curve in that case is

$$y = \frac{1}{\sqrt{2\pi}}\, e^{-z^2/2} \quad \text{where} \quad z = \frac{x-\mu}{\sigma} \tag{6.3}$$

and the total relative frequency or the area under the curve is 1. In the normal distribution corresponding to equation 6.3, the mean of the values of the variable z is zero and the standard deviation of the values of the variable z is 1.

Equation 6.3 is the equation from which Tables 6.6 and 6.7 were constructed. If you understand logarithms, it will be easy for you to compute some of the ordinates that are shown in Table 6.6.

The mean and the standard deviation owe much of their importance in statistics to their fundamental connection with the normal probability law. You have seen already in section 6.5 that if we know the mean and the standard deviation of a normal distribution, we can reproduce the whole distribution in the form of relative frequencies for any set of classes that we choose. In other words, the mean and the standard deviation completely determine the shape of a normal distribution. Consequently, if the universe is a normal distribution, the mean and the standard deviation of a random sample contain all the relevant information that the sample can disclose with respect to the distribution of the universe. Therefore, the mean and the standard deviation make a powerful team of parameters whenever we consider the distribution to be a normal distribution.

The normal frequency distribution or normal probability law provides us with a useful model for a great many types of informal but practical applications of the statistical method of thinking about the problems and situations of daily life. For most of these practical purposes, we do not need to do computations with pencil and paper. All that we need to do is fix in memory a picture of a normal curve, such as Figure 6.19, and two or three important probabilities, such as that approximately 32 per cent of the elementary units in a normal distribution are expected to have values of the variable outside the interval from one standard unit below the mean

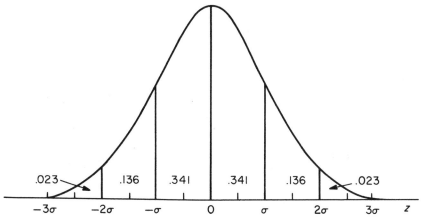

Fig. 6.19. The Normal Probability Distribution

to one standard unit above the mean, and approximately 5 per cent are expected to have values of the variable outside of the interval from two standard units below the mean to two standard units above the mean. Form the habit of making rough estimates in standard units of the approximate positions of the persons or things that make up distributions.

Tables 6.18 and 6.19 contain more than enough normal probabilities for the mental work of understanding most practical statistical situations.

TABLE 6.18

Probability or Area in the Two Tails of the Normal Curve Outside the Interval $-z$ to $+z$

z	$2P_z$
0.674	0.50 or 1 chance in 2
1.645	0.10 or 1 chance in 10
1.960	0.05 or 5 chances in 100
2.326	0.02 or 2 chances in 100
2.576	0.01 or 1 chance in 100
3.2905	0.001 or 1 chance in 1000
3.89059	0.0001 or 1 chance in 10,000
4.41717	0.00001 or 1 chance in 100,000
4.89164	0.000001 or 1 chance in 1,000,000
5.32672	0.0000001 or 1 chance in 10,000,000
5.73073	0.00000001 or 1 chance in 100,000,000
6.10941	0.000000001 or 1 chance in 1,000,000,000

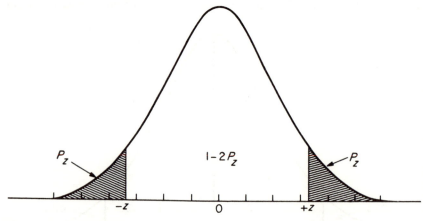

Fig. 6.20. Probability or Area in the Two Tails of the Normal Curve Outside of the Interval $-z$ to $+z$

TABLE 6.19

**Probability or Area in the One Tail of the
Normal Curve to the Right of z**

z	P_z
0	.50 or 50 per cent
0.674	.25 or 25 per cent
1	.16 or 16 per cent
1.282	.10 or 10 per cent
1.645	.05 or 5 per cent
1.960	.025 or 2½ per cent
2	.023 or 2.3 per cent
2.054	.02 or 2 per cent
2.326	.01 or 1 per cent
2.576	.005 or ½ of 1 per cent
3	.00135
3.090	.001 or $\frac{1}{10}$ of 1 per cent
3.5	.00023
4	.00003
4.5	.000003

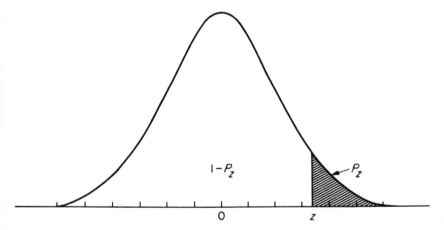

Fig. 6.21. Probability or Area in the One Tail of the Normal Curve to the Right of z

EXERCISES

1. What are the equations of the normal curves that fit the distributions in exercises 9 and 10 on p. 00?

2. Using logarithms and equation 6.3, compute the ordinates of the normal curve, as given in Table 6.6 for $z = 1$, $z = 2$, $z = 3$, $z = 1.96$, and $z = 2.58$. What is the ordinate at $z = 0$?

3. Using logarithms, compute the ordinates of the normal curve whose equation is

$$y = 23.36e^{-(x-100)^2/512}$$

for $x = 100$, $x = 110$, $x = 120$, $x = 130$, $x = 140$, $x = 150$. Plot the curve.

6.10 FURTHER ILLUSTRATIONS OF FREQUENCY DISTRIBUTIONS

Table 6.20 shows a distribution of the neck sizes in a sample containing 120,000 white shirts made by one of the big manufacturers of men's shirts. Figure 6.22 is a graph for the distribution. This is a relative frequency distribution, thus, the numbers in the "per cent of sales" column are relative frequencies. The actual frequencies in the sample would add up to 10,000 dozen shirts, that is, 120,000 shirts. For some purposes, relative frequencies or percentages are more useful than the actual frequencies.

TABLE 6.20

Sizes of Arrow White Shirts
Sold to American Men*

Neck Size	Per Cent of Sales
13½	1
14	6
14½	19
15	25
15½	24.5
16	13
16½	6.5
17	3
17½	1
18	1
Total	100.0

*Source: Courtesy of Cluett, Peabody and Co., Troy, N. Y. See Table 6.24.

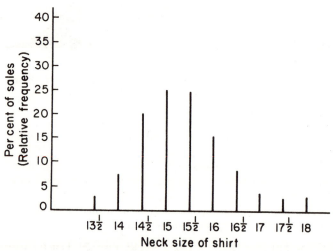

Fig. 6.22. Sizes of 120,000 White Shirts Sold to American Men

For example, if you were the operator of a small haberdashery store, you might not feel justified in keeping all possible sizes of shirts in stock. Sizes 15 and 15½ are the most wanted sizes in white shirts, according to Table 6.20. If your customers were similar to the men who bought the shirts described in the table, and you did not carry any size 13½ shirts, you would be unable to fit 1 per cent of your customers. That is, the probability of not being able to fit a customer would be about 0.01.

Similarly, if you decided not to carry any shirts larger than 16, then 11.5 per cent of the men who came into your store to buy white shirts would be disappointed. In other words, the probability of not being able to fit a customer would be about 0.115.

Table 6.21 is another relative frequency distribution of a variable. It shows how sleeve lengths were distributed among the 10,000 dozen shirts. The most popular sleeve length is 33 inches, because Table 6.21 shows that 40.5 per cent of this large sample of shirts were in the 33-inch sleeve length group.

If a store operator whose customers were similar to the men who bought the shirts described in Table 6.21 did not carry shirts with 31-inch sleeve length, the store would be unable to fit about 3.5 per cent of the men who came in to buy white shirts. In other words, the probability of not being able to fit a customer would be about 0.035 if the store did not carry shirts with 31-inch sleeve length.

TABLE 6.21
Sizes of Arrow White Shirts
Sold to American Men*

Sleeve Length	Per Cent of Sales
31	3.5
32	22.5
33	40.5
34	23.0
35	10.5
Total	100.0

Source: Courtesy of Cluett, Peabody and Co., Troy, N. Y. See Table 6.24.

Tables 6.22 and 6.23 show similar information about the distribution of the shoe sizes of a sample containing 10,000 women. In these two tables, the frequencies are actual rather than relative. However, the actual frequencies could be changed into relative frequencies very easily in this case. All that is necessary is to divide each actual frequency by the total frequency and multiply the quotient by 100. In Tables 6.22 and 6.23, the total frequency is 10,000 and, consequently, the relative frequencies could be obtained by moving the decimal point two digits to the left in the actual frequencies.

TABLE 6.22

Shoe Sizes of American Women Who Wear Florsheim Shoes*

Shoe Size (Length)	No. of Women f
4	31
4½	100
5	258
5½	494
6	820
6½	1,154
7	1,354
7½	1,449
8	1,297
8½	1,054
9	849
9½	491
10	454
10½	92
11	103
Total	10,000

Source: Courtesy of the Florsheim Shoe Co., Chicago, Ill. See Table 6.25.

TABLE 6.23

Shoe Sizes of American Women Who Wear Florsheim Shoes*

Shoe Size (Width)	No. of Women f
AAAAA	67
AAAA	782
AAA	1,425
AA	1,854
A	2,170
B	2,088
C	1,213
D	309
E	92
Total	10,000

Source: Courtesy of the Florsheim Shoe Co., Chicago, Ill. See Table 6.25.

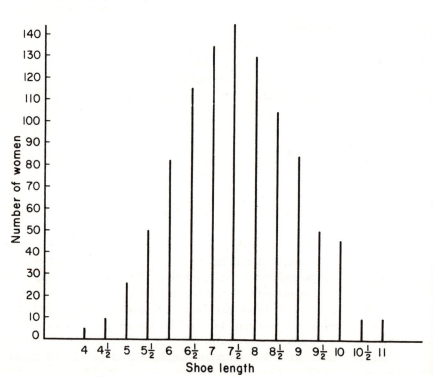

Fig. 6.23. Shoe Sizes of 10,000 American Women

Figure 6.23 is a graph for the distribution in Table 6.22. The most common length of shoe for the women in this sample is 7½. If a store whose customers are similar to these women did not carry any size 4 shoes, about 31 out of every 10,000 customers, that is, about 0.31 per cent of the customers, would not be able to obtain the size they want. Less than one out of every 300 customers would be disappointed if size 4 shoes were not carried in stock.

Similarly, Table 6.23 shows that less than 1 per cent of these women wear size AAAAA. Less than 1 per cent of these women wear size E. Consequently, if the store operator did not carry size AAAAA shoes or size E shoes, the loss in sales would be $0.67 + 0.92 = 1.59$ per cent. The most popular width of shoe for these women is size A.

6.10.1 Two-way, joint, or bivariate frequency distributions. Table 6.20 is a one-way or univariate frequency table showing the distribution of neck sizes in a very large sample of men's white shirts. Suppose that you are the shirt buyer for a department store. Could you make up a reasonable purchase order to a manufacturer for 100 dozen white shirts? You could not do this correctly by using Table 6.20 alone, because you should also tell the manufacturer what the sleeve lengths must be. Can you obtain the necessary information for making up a reasonable purchase order from Tables 6.20 and 6.21? No, you still cannot prepare a reasonable purchase order even if you use both tables.

Table 6.24 is called a two-way frequency distribution, a joint frequency distribution, or a bivariate frequency distribution, that is, a two-variate frequency distribution. It shows not only the distribution of neck sizes and the distribution of sleeve lengths, but also the way in which each neck size is distributed among the sleeve lengths and the way in which each sleeve length is distributed among the neck sizes.

By using Table 6.24 you can make up a good purchase order for 100 dozen shirts, or for any other number of shirts that a store needs. Of course, you probably would make some adjustments in the order to take care of oversupply or shortage in your present stock or other special conditions, such as not wishing to stock some of the extreme sizes.

It is important that you understand that the two-way frequency table, Table 6.24, contains much more information than Tables 6.20 and 6.21.

Table 6.25 is another two-way frequency distribution or two-way frequency table. It shows the way in which the shoe lengths of the women in the sample are distributed among the shoe widths, and it shows also the way in which the shoe widths of these women are distributed among the shoe lengths.

You could not make up a good manufacturing schedule for a shoe manufacturer nor could you make up a good order for the stock of a new store for women's shoes by using Tables 6.22 and 6.23. But you could

TABLE 6.24

Sizes of Arrow White Shirts Sold to American Men*

Neck Size (*Inches*)	Sleeve Length (*Inches*)					Total (*in Units of 100 Dozen*)
	31	32	33	34	35	
13½	—	½	½	—	—	1
14	1	3	2	—	—	6
14½	1	6	9	3	—	19
15	1	6	11	5	2	25
15½	½	4	10	7	3	24½
16	—	2	5	4	2	13
16½	—	1	2	2	1½	6½
17	—	—	1	1	1	3
17½	—	—	—	½	½	1
18	—	—	—	½	½	1
Total (in units of 100 dozen)	3½	22½	40½	23	10½	100

Source: Courtesy of Cluett, Peabody and Co., Inc., Troy, N. Y., makers of Arrow Shirts. Based on a sample containing 10,000 dozen white shirts.

TABLE 6.25

Shoe Sizes of 10,000 American Women Who Wear Florsheim Shoes*

Shoe Length	Shoe Width									Total
	AAAAA	AAAA	AAA	AA	A	B	C	D	E	
4						18	11	2		31
4½				3	16	54	20	7		100
5			3	22	55	105	50	16	7	258
5½			32	69	105	154	95	29	10	494
6		4	70	129	203	224	139	37	14	820
6½		41	130	214	274	267	170	43	15	1,154
7		89	182	267	301	278	175	48	14	1,354
7½	7	119	223	289	317	269	168	46	11	1,449
8	8	122	222	261	289	223	129	36	7	1,297
8½	9	127	199	215	215	168	98	17	6	1,054
9	10	122	153	175	169	136	66	14	4	849
9½	11	75	96	86	97	80	40	5	1	491
10	11	63	81	86	91	74	40	7	1	454
10½	4	9	17	18	18	18	6	1	1	92
11	7	11	17	20	20	20	6	1	1	103
Total	67	782	1,425	1,854	2,170	2,088	1,213	309	92	10,000

Source: Courtesy of The Florsheim Shoe Co., Chicago, Illinois.

make up such a schedule for a manufacturer or such a purchase order for a new store by using Table 6.25. There is much more useful information in Table 6.25 alone than there is in Tables 6.22 and 6.23 together.

Table 6.25 shows that only 2 women out of 10,000 women in the sample wear size 4D. The owner of a small store probably would be a poor manager if he carried size 4D shoes in stock unless he had some regular, dependable customers who wore that size.

Tables 6.24 and 6.25 ought to impress you as illustrations of the many aspects of life in which frequency tables are useful ways of presenting information in compact form. The practical significance of such tables should be obvious. For example, a person who is opening a ladies' shoe store or a men's shirt store would be starting the business with an invitation to bankruptcy if he or she ordered the same number of shoes or shirts in all possible sizes. Perhaps it is lack of knowledge of simple statistical distributions such as these that causes many failures in business.

6.10.2 How to construct a joint frequency table. If you know how to construct a one-way frequency table, it should be easy to learn how to construct a two-way, or joint, frequency table.

First, draw on a sheet of paper a framework of rectangular cells similar to the ruled lines in Tables 6.24 and 6.25. The number of cells in the horizontal and in the vertical direction depends upon the particular problem under consideration. You can determine the number of classes, or the number of different values of the variable if it is discrete, by examining the raw data, by considering the purpose for which the table is to be used, and by considering the size of the sample.

For example, before construction of Table 6.25, it was observed that the variables are discrete, that the raw data included 15 different shoe lengths and 9 different shoe widths, and that the sample was so large that there were substantial numbers of shoes in each shoe length and in each shoe width. Therefore, it was decided not to group any of the shoe lengths or any of the shoe widths. Consequently, the form for Table 6.25 had to contain 15×9 cells in addition to the marginal spaces for totals and descriptive headings.

Next, the descriptive headings are written in their proper places. Then the raw data are read and a tally mark is placed in the cell corresponding to the vertical and horizontal classifications of each elementary unit. After all the tally marks have been entered in the cells of the tabular form, the number of tally marks in each cell is counted and recorded in the cell. Then, the cell frequencies in each column are added and the results are written in the spaces reserved for the column totals. Also, the cell frequencies in each row are added and the results are written in the spaces reserved for the row totals.

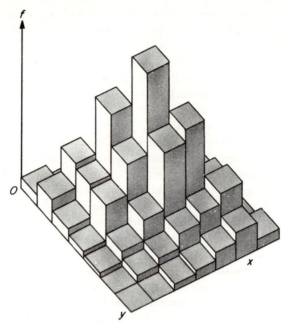

Fig. 6.24. Histogram for a Bivariate Frequency Distribution

6.10.3 Graphs of bivariate frequency distributions. The most common type of graph for a grouped joint or bivariate frequency distribution is a three-dimensional histogram. One way to begin the construction of such a histogram is to procure some long pieces of wood or plastic whose cross section is either square or rectangular. Then lay out on a flat surface a diagram containing as many small areas in each direction as there are cells in the frequency table. Each small area should be the same size and shape as the cross section of the wood or plastic strips. Cut strips of the wood or plastic to lengths that are proportional to the frequencies in the cells of the table. Stand each strip on end in the area corresponding to the cell for which the strip's length is proportional. Glue the strips in place. Figure 6.24 is a picture of such a histogram.

For an ungrouped joint frequency distribution of two discrete variables, such as Tables 6.24 and 6.25, you could cut straight pieces of stiff wire in lengths proportional to the cell frequencies and stand the pieces of wire at the centers of the corresponding areas, in this way producing a three-dimensional set of ordinates.

If a smooth curved surface is fitted through the tops of the strips in the histogram so the smooth curved surface covers the same volume as the histogram encloses, the surface is called a bivariate or joint frequency surface. Figure 6.25 shows two illustrations of what are called normal bivariate frequency surfaces.

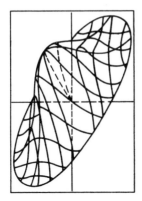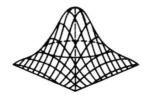

Fig. 6.25. Normal Bivariate Frequency Surfaces (Left-hand illustration reprinted by permission from *Introduction to Mathematical Statistics*, 2d ed., by Paul G. Hoel, New York, Wiley.)

EXERCISES

1. Assuming that the distribution of sleeve length and neck size of shirts for customers of a store is the same as the distribution in Table 6.24, what per cent of the customers would not be able to obtain the sizes of shirts that they need if the store did not stock any of the following sizes: 13½ neck with 32 sleeve; 13½ neck with 33 sleeve; 15½ neck with 31 sleeve; 17½ neck with 34 sleeve; 17½ neck with 35 sleeve; 18 neck with 34 sleeve; 18 neck with 35 sleeve?

2. Assuming that the distribution of shoe width and shoe length for the customers of a store is the same as the distribution in Table 6.25, what per cent of the women who come into the store to buy shoes would not be able to obtain the sizes of shoes that they need if the store did not stock any of the following sizes: 4D, 10½D, 10½E, 11D, 11E?

3. What percentage of the men's suits in Table 6.26 were made in the "long young stouts" model? What percentage of the "regular" men's suits were size 40? Size 41? Size 42?

TABLE 6.26
Sizes of Society Brand Suits Manufactured for American Men*

Model	Size													Total
	34	35	36	37	38	39	40	41	42	44	46	48	50	
Regulars	10	80	300	650	1000	800	1000	200	900	350	100	20		5410
Shorts	4	40	100	200	250	275	300	90	200	30				1489
Longs			15	100	250	275	350	100	300	150				1540
Young Stouts				4	20	70	30	90	60	20	4			298
Short Young Stouts				2	10	20	10	20	15	10	4			91
Long Young Stouts							4	10	10	10	4			38
Total	14	120	415	950	1506	1380	1740	430	1514	615	140	38	4	8866

**Source:* Courtesy of Alfred Decker and Cohn, Inc., Chicago, Illinois, makers of Society Brand Clothes. This is an actual manufacturing schedule for stock, based on the past experience of the company.

4. Which of the following may be considered to be types of individual differences?
  (a) intelligence differences   ☐
  (b) performance differences   ☐
  (c) personality differences   ☐
  (d) physical differences   ☐
  (e) all of the above   ☐

5. Individual differences of any one of the kinds mentioned above are differences in degree only.   True ☐ False ☐

6. Individual differences should *not* be thought of in terms of
  (a) a relative frequency distribution over a range   ☐
  (b) a bell-shaped frequency distribution   ☐
  (c) a distribution clustering around an average   ☐
  (d) either–or categories   ☐

7. If a large number of randomly selected people have been given the same quantity and quality of training for a particular job, it is safe to assume that all will maintain almost exactly the same production records on the job.   True ☐ False ☐

8. The mean of any distribution of standard scores is _____ and the standard deviation of any distribution of standard scores is _____.

9. A standard score of zero in a normal distribution of test scores corresponds to a percentile rank of (a) 25 ☐ (b) 50 ☐ (c) 75 ☐ (d) 100 ☐

10. A person whose standard score is +2.00 in any large normal distribution of ability or achievement is superior in that characteristic to approximately 99.9 per cent of the people in the distribution.   True ☐ False ☐

11. In any curve representing a frequency distribution, regardless of whether or not the distribution is skewed, the point on the horizontal axis for which half of the area is to the right and the other half is to the left is known as the (a) midpoint of the range ☐ (b) median ☐ (c) mode ☐ (d) mean ☐

Score on a mechanical aptitude test

12. The curve shown here represents (a) a bell-shaped distribution ☐ (b) a symmetrical distribution ☐ (c) a distribution of a continuous variable ☐ (d) all of the above ☐

13. The range of the scores on the test of mechanical aptitude is from (a) 0 to 100 ☐ (b) 20 to 80 ☐ (c) 25 to 75 ☐ (d) 22.5 to 77.5☐

14. The mean, the median, and the modal degree of mechanical aptitude for these men are represented by the score of 50. True ☐ False ☐

REFERENCES

Leo A. Goodman, "Kolmogorov-Smirnov Tests for Psychological Research," *Psychological Bulletin* 51 (1954), pp. 160–168.

B. W. Lindgren, *Statistical Theory*. New York: The Macmillan Comany, 1962.

A. N. Lowan (Director), *Tables of Probability Functions*, Vols. 1 and 2, Federal Works Agency, Work Projects Administration for the City of New York. Washington, D. C.: National Bureau of Standards, 1941. (Areas of the normal curve to fifteen decimal places for values of the argument to four significant digits.)

J. Neyman, *First Course in Probability and Statistics*. New York: Henry Holt and Company, 1950.

G. Udny Yule and M. G. Kendall, *An Introduction to the Theory of Statistics*. New York: Hafner Publishing Co., 1950.

Chapter 7

ADDITIONAL APPLICATIONS OF NORMAL AND OTHER PROBABILITIES

7.1 SAMPLES OF RANDOM DIGITS

Another sampling experiment now will give you some concrete experience with the distribution of the sampling errors in the mean of a variable. You can build up a large number of samples by drawing samples of random digits from Table A.15. If there are 25 students in the class and if each student draws 4 samples, you will have 100 samples altogether. It would be better still to have several hundred samples.

Do this experiment in two parts. First, draw 4 samples with 30 random digits in each sample. Second, draw 4 samples with 120 random digits in each sample. Draw the samples from Table A.15 by some random process. For example, to draw a sample of 30 digits, you might find a random starting point in Table A.15 and use that digit and the next 29 digits in some predetermined route. To draw a sample of 120 digits, you might draw random numbers to determine three columns of Table A.15 and then use the 100 digits in the first two columns and the first 20 digits in the third column.

It is not necessary to copy the digits in each sample. You can make the necessary records in either of two simple ways. The first is to add the digits in each sample and record the total. Also, add the squares of the digits in each sample and record the total. Then you can use formulas 3.1 and 3.4 to compute the mean and the standard deviation of each of your samples.

The second way is to build the framework for a table with a large number of columns, say 9, if you are planning to draw 8 samples. At the head of the first column write "Digit" and "x." In the body of the first column write "0, 1, 2, 3, 4, 5, 6, 7, 8, 9." At the head of the second column write "Sample No. 1" and "*f*." In the body of the second column record the number of times that each digit occurs in the first sample. Use the third column to record the number of times that each digit occurs in the second sample, and so on. Then you can use formulas 3.7 and 3.8 to compute the mean and the standard deviation of each of your samples.

The second method probably is the better, especially for samples containing 120 digits, and if you do not have a calculating machine to use for determining the sums of squares for each sample. Besides, the second method will give you some good experience with the construction and practical use of simple frequency distributions.

In a perfect table of random digits, 10 per cent of the digits are zeros, 10 per cent of the digits are ones, 10 per cent of the digits are twos, and so on through nines. In other words, the range of the variable x is from 0 to 9, and the relative frequency of each digit is 0.10. This is shown in Table 7.1 and Figure 7.1. A frequency distribution in which all the frequencies (or relative frequencies) are the same is a *uniform distribution* and often is called a *rectangular distribution*. A glance at Figure 7.1 probably will indicate to you why such distributions are called rectangular distributions.

TABLE 7.1

Relative Frequency Distribution of the Digits in a Perfect Table of Random Digits

Variable (*Digit*) x	*Relative Frequency* (*Per Cent*) f	xf	x^2f
0	10	0	0
1	10	10	10
2	10	20	40
3	10	30	90
4	10	40	160
5	10	50	250
6	10	60	360
7	10	70	490
8	10	80	640
9	10	90	810
Total	100	450	2850

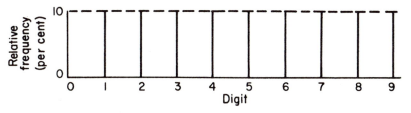

Fig. 7.1.   Rectangular Distribution of Digits in a Perfect Table of Random Digits

Now, applying formula 3.7 to the distribution in Table 7.1, we find that the mean of the digits in a perfect table of random digits is $\mu_x = 450/100 = 4.50$. Applying formula 3.8 with $N = \Sigma f = 100$, we find that the standard deviation of the distribution of digits in a perfect table of random digits is

$$\sigma_x = (1/100)\sqrt{100(2850) - (450)^2} = (1/100)\sqrt{82500} = 2.87$$

However, even if the mean of all the digits in Table A.15 is 4.50, it is not very likely that the mean of the digits in each of your samples will be exactly 4.50. Some of your samples may contain more than 10 per cent of sevens and less than 10 per cent of fours, for example. Other samples may contain too many fives and too few eights. And so on. These variations from sample to sample, these deviations from the true distribution for the whole universe as a result of chance in random sampling, are called *sampling errors*. Consequently, do not be surprised if in none of your four samples of 30 digits is the mean exactly 4.50 or the standard deviation exactly 2.87.

7.2 SHAPE OF THE DISTRIBUTION OF MEANS

Tabulate the means of the 100 or more samples, each of which contained 30 random digits. Arrange this tabulation in the form of a frequency distribution. Then construct a histogram for the frequency distribution of means. What is the shape of this histogram? Is it a rectangular distribution? It ought to be the kind of histogram for which the best approximating smooth curve would be a bell-shaped, nearly normal curve.

Construct another frequency distribution and another histogram for the 100 or more means of the samples each of which contained 120 random digits. In what way does this second histogram differ from the first histogram? The histogram for the means of the samples of size 120 ought to be more concentrated near the middle of the range of the means than the histogram for the means of the samples of size 30. In other words, increasing the sample size from 30 to 120 decreases the variability and the standard deviation of the distribution of the means of samples.

Compute the mean and the standard deviation for each of the two distributions of means of samples. Compare these two means with the mean of the universe from which the samples were drawn, namely, $\mu_x = 4.50$. Compare the two standard deviations with the standard deviation of the universe from which the samples were drawn, namely, $\sigma_x = 2.87$. The standard deviation of your distribution of 100 means of samples, each sample containing 30 digits, ought to be approximately equal to $2.87/\sqrt{30} = 0.524$ and the standard deviation of your distribution of 100 means of samples, each sample containing 120 digits, ought to be approximately equal to $2.87/\sqrt{120} = 0.262$.

Only about 5 out of 100 means of samples of size 30 ought to be outside the interval from 3.473 to 5.527. Only about 1 out of 100 means of samples of size 30 ought to be outside the interval from 3.148 to 5.852. Only about 5 out of 100 means of the samples of size 120 ought to be outside the interval from 3.986 to 5.014. Only about 1 out of 100 of the means of the samples of size 120 ought to be outside the interval from 3.824 to 5.176. Verify these statements for yourselves by counting the numbers of means of the samples drawn by the class that do not fall within the above intervals.

This experiment in sampling ought to have demonstrated to you three basic laws of practical statistics. First, the distribution of the means of samples from a universe is approximately a normal distribution, even though the distribution of the variable in the universe from which the samples are drawn is a rectangular distribution. Second, the mean of the distribution of means of samples from a universe is approximately the same as the mean of the variable in the universe from which the samples are drawn. Third, the standard deviation of the distribution of means of samples drawn from a universe is much smaller than the standard deviation of the variable in the universe from which the samples are drawn. Furthermore, the standard deviation of the distribution of means of samples of size 120 is approximately half as large as the standard deviation of the distribution of means of samples of size 30.

We can generalize the results of this sampling experiment as follows:

1. If we draw a large number of large random samples from a very large universe and if we compute the mean of each sample, then all the means of the samples will form a normal or nearly normal frequency distribution. Furthermore, the distribution of means of samples will be approximately normal no matter what the shape of the distribution of the variable in the universe.

2. The mean μ_m of the distribution of means of large random samples theoretically will be equal to, and practically will be at least approximately equal to, the mean μ_x of the variable x in the universe from which the samples are drawn. That is,

$$\mu_m = \mu_x \tag{7.1}$$

3. The standard deviation σ_m of the distribution of means of large random samples theoretically will be equal to, and practically will be at least approximately equal to, $\sigma_x/\sqrt{n}$, where σ_x is the standard deviation of the distribution of the variable x in the universe from which the samples are drawn and n is the size of each sample. That is,

$$\sigma_m = \frac{\sigma_x}{\sqrt{n}} \tag{7.2}$$

4. Only about 5 per cent of the means of large samples will be outside the interval from $\mu_x - 1.96\sigma_m$ to $\mu_x + 1.96\sigma_m$. Only about 1 per cent of

the means of large samples will be outside the interval from $\mu_x - 2.58\sigma_m$ to $\mu_x + 2.58\sigma_m$.

5. The standard deviation of the mean varies inversely as the square root of the size of the sample. This is shown by formula 7.2, which contains the square root of n in the denominator. For example, we have seen that the value of σ_m when $n = 120$ is only half as large as the value of σ_m when $n = 30$. That is, when n is multiplied by 4, σ_m is divided by 2.

In making the above generalizations, it was indicated that they are true for large random samples from a very large universe. Perhaps you are wondering where to draw the line between small samples and large samples and between very large universes and universes that are not very large. You probably will agree that your frequency distributions of the means of samples of size 30 and size 120 appear to be very nearly normal distributions with means and standard deviations approximately equal to the values indicated by formulas 7.1 and 7.2. This suggests that the generalizations stated above are valid if n, the size of the sample, is as large as 30, and if N, the size of the universe, is as large as the number of digits in Table A.15.

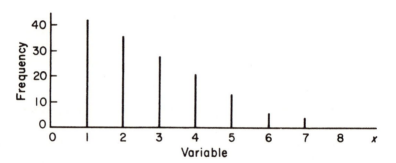

Fig. 7.2.   Right-Triangular Frequency Distribution

Shewhart .demonstrated that the conclusions stated above may be applied, with little risk of serious error, to the means of samples in which n is as small as 4 and the universe is a rectangular universe such as that illustrated in Figure 7.1. He showed also that the distribution of the means of samples is approximately a normal distribution even if n is as small as 4 and the universe is a right-triangular universe such as that illustrated in Figure 7.2.[1] As we proceed, you ought to be impressed by the tremendous power that is given to statisticians by these generalizations about the distributions of means of random samples.

In section 2.8.1, a random sampling procedure called "sampling with

[1]W. A. Shewhart, *Economic Control of Quality of Manufactured Products.* New York: D. Van Nostrand Co., Inc., 1931, pp. 182–184.

replacement after each draw" was described. If the random samples from which the distribution of the means of samples is determined are samples drawn with replacement after each draw, the generalizations stated above are true, regardless of how small the size N of the universe may be.

7.2.1 Adjustment for sampling without replacement.

In section 2.8.2, a random sampling procedure called "sampling without replacement after each draw" was described. The five generalizations stated above are true for random sampling without replacement after each draw provided the finite multiplier is inserted in formula 7.2, giving

$$\sigma_m = \frac{\sigma_x}{\sqrt{n}}\sqrt{\frac{N-n}{N-1}} \qquad (7.3)$$

7.2.2 Effect of sample size on σ_m.

Let us explore the last two generalizations a little further. Table 7.2 shows that if the size of a random sample is increased from 25 to 100 (that is, multiplied by 4) the size of σ_m and the size of the 95 per cent confidence interval are divided by 2 (that is, divided by $\sqrt{4}$). If the size of the sample is multiplied by 4 again to increase n from 100 to 400, the size of σ_m and the size of the 95 per cent confidence interval are divided by 2 again.

TABLE 7.2

The Effect of the Size of a Random Sample on the Reliability of the Sample Mean

Size of Sample	Multiplier	Standard Deviation of Mean	$1.96\sigma_m$	95 Per Cent Limits		Size of 95 Per Cent Interval	Divisor	Cost of a Sample at $3 Per Unit
n	h	σ_m		$4.50 \pm 1.96\sigma_m$			$\sqrt{h}$	
25	—	0.574	1.125	3.375	5.625	2.250	—	$ 75
100	4	0.287	0.562	3.938	5.062	1.125	$\sqrt{4}=2$	300
400	4	0.144	0.281	4.219	4.781	0.562	$\sqrt{4}=2$	1,200
3,600	9	0.048	0.094	4.406	4.594	0.187	$\sqrt{9}=3$	10,800
32,400	9	0.016	0.031	4.469	4.531	0.062	$\sqrt{9}=3$	97,200

Next, the sample size was multiplied by 9, thus increasing n from 400 to 3600. This causes the standard deviation of the mean and the 95 per cent confidence interval around the mean to be divided by 3, that is, divided by $\sqrt{9}$.

It should now be obvious to you that by increasing the size of a random probability sample sufficiently we can reduce the size of the standard deviation of the mean and the size of the confidence interval around the mean as much as we please. But the cost of a statistical investigation may be increased tremendously if we require that the confidence interval around the mean be very small; the cost increases much more rapidly than the confidence interval decreases. In practical statistical situations, we are confronted time after time with the question: Can we afford to increase the sample size enough to enable us to be reasonably certain that the confidence interval for the mean will not exceed some specified amount?

7.3 THE DISTRIBUTIONS OF z AND t

Before we can discuss some of the most interesting applications of the principles developed so far in this chapter, it is necessary to invite your careful attention to a fact that may be easily overlooked. Disregard of this fact might lead to serious errors in application. Consider the values of z obtained by substituting the values of m computed from a large number of random samples of size n from a normal universe in the formula

$$z = \frac{m - \mu_x}{\sigma_x / \sqrt{n}}$$

It can be proved mathematically that the distribution of z is a normal distribution with mean equal to zero and standard deviation equal to unity, that is, the distribution in Tables 6.6 and 6.7 and Figure 6.11.

The standard deviation σ_x that appears in the denominator of the formula for z is a constant, a universe parameter. In practical applications of statistical methods, we usually do not know the exact value of the constant σ_x. If we substitute the sample value of s_x for σ_x in the formula for z for each of a large number of random samples of size n drawn from a normal universe, the value of s_x is not constant but changes from sample to sample. For this reason, let us define t as

$$t = \frac{m - \mu_x}{s_x / \sqrt{n - 1}} \tag{7.4}$$

An investigator who signed himself "Student" showed that the distribution of the values of t computed by this formula for a large number of random samples of size n drawn from a normal universe is not a normal distribution but that it approximates a normal distribution more and more closely as n, the size of the samples, increases.[2]

Figure 7.3 shows graphically the difference between the normal distribution of z and "Student's" distribution of t for two small values of n. The three distributions are bell-shaped and symmetrical, with mean at zero, but the distributions of t are less peaked, that is, more flat-topped, than the normal distribution of z, and the smaller n is, the more flat-topped is the distribution of t. Also, the tails of the distribution of t beyond two standard units on each side of the mean tend to be higher than the tails of the normal distribution of z.

[2] "The Probable Error of a Mean," *Biometrika* 6, No. 1 (1908), pp. 1–25. The author's name was William Sealy Gosset.

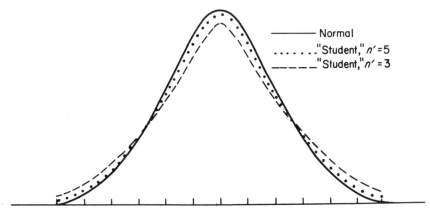

Fig. 7.3. The Normal Distribution and Two "Student" Distributions

It may help you to grasp the idea of the difference between the normal distribution of z and "Student's" distribution of t if you examine the values of the probabilities or areas in the tails of the curves. For example, Table 6.7 shows that in the normal distribution of z a tail containing an area of 0.025 begins at $z = 1.96$ for all values of n. Table 7.3 shows that in the t distribution a tail containing an area of 0.025 begins at $t = 2.228$ if $n' = 10$. Similarly, Table 6.7 shows that for the normal distribution a tail containing an area of 0.005 begins at $z = 2.58$ for all values of n. Table 7.3 shows that for the t distribution a tail containing an area of 0.005 begins at $t = 3.169$ if $n' = 10$.

In other words, it is necessary to go farther than $t = \pm 1.96$ in order to enclose 95 per cent of the area in the t distribution, especially if the value of n is small. And it is necessary to go farther than $t = \pm 2.58$ in order to enclose 99 per cent of the area in the t distribution, especially if the value of n is small. The t distribution is different for each value of n, and the smaller n is, the more the t distribution departs from the normal distribution of z.

For practical purposes, the difference between the normal distribution of z and "Student's" distribution of t may be disregarded as negligible if n is not less than 30 or 32. In other words, if n is not less than 30 or 32, it is permissible to use normal probabilities instead of "Student" probabilities. That the discrepancy between the normal probabilities and the t distribution for which $n' = 30$ is fairly small may be seen by comparing the figures in the last two lines of Table 7.3, the figures in the last line being the same as those for the normal distribution of z.

7.3.1 Confidence limits for the mean of a universe. If we multiply

TABLE 7.3

The *t* Distribution of "Student"*

n'	$P = 0.50$	$P = 0.10$	$P = 0.05$	$P = 0.01$
	Value of *t* for a two-tails test†			
1	1.000	6.314	12.706	63.657
2	0.816	2.920	4.303	9.925
3	0.765	2.353	3.182	5.841
4	0.741	2.132	2.776	4.604
5	0.727	2.015	2.571	4.032
6	0.718	1.943	2.447	3.707
7	0.711	1.895	2.365	3.499
8	0.706	1.860	2.306	3.355
9	0.703	1.833	2.262	3.250
10	0.700	1.812	2.228	3.169
11	0.697	1.796	2.201	3.106
12	0.695	1.782	2.179	3.055
13	0.694	1.771	2.160	3.012
14	0.692	1.761	2.145	2.977
15	0.691	1.753	2.131	2.947
16	0.690	1.746	2.120	2.921
17	0.689	1.740	2.110	2.898
18	0.688	1.734	2.101	2.878
19	0.688	1.729	2.093	2.861
20	0.687	1.725	2.086	2.845
21	0.686	1.721	2.080	2.831
22	0.686	1.717	2.074	2.819
23	0.685	1.714	2.069	2.807
24	0.685	1.711	2.064	2.797
25	0.684	1.708	2.060	2.787
26	0.684	1.706	2.056	2.779
27	0.684	1.703	2.052	2.771
28	0.683	1.701	2.048	2.763
29	0.683	1.699	2.045	2.756
30	0.683	1.697	2.042	2.750
∞	0.674	1.645	1.960	2.576

*Adapted from Table IV of R. A. Fisher, *Statistical Methods for Research Workers*, published by Oliver & Boyd Ltd., Edinburgh, by permission of the author and publishers.

†For a one-tail test, divide the value of *P* by 2.

To determine confidence limits to be associated with the mean of a sample of size *n*, use the values of *t* on the line for $n' = n - 1$.

To test the significance of the difference between two sample means, use the line for $n' = n_1 + n_2 - 2$.

To test the significance of a Pearson linear correlation coefficient obtained from *n* pairs of values, use the line for $n' = n - 2$.

both sides of the formula for z in section 7.3 by $\sigma_x/\sqrt{n}$, we obtain $z\sigma_x/\sqrt{n} = m - \mu_x$. If the value of σ_x is known, then correct confidence limits for the mean μ_x of the universe are

$$m - \frac{z\sigma_x}{\sqrt{n}} < \mu_x < m + \frac{z\sigma_x}{\sqrt{n}}$$

where the value of z is obtained from Table 6.7, corresponding to the desired confidence level.

Similarly, if we multiply both sides of formula 7.4 by $s_x/\sqrt{n-1}$, we obtain $ts_x/\sqrt{n-1} = m - \mu_x$. If the value of σ_x is unknown, correct confidence limits for the mean μ_x of the universe are

$$m - \frac{ts_x}{\sqrt{n-1}} < \mu_x < m + \frac{ts_x}{\sqrt{n-1}} \tag{7.5}$$

where the value of t is obtained from Table 7.3 under the desired confidence level and on the line for $n' = n - 1$.

For the sample in Table 3.8, we know that $n = 242$, $m_x = 28.75$ years, and $s_x = 6.27$ years. To find 95 per cent confidence limits for the mean age at time of marriage of all the men in the universe from which the sample was drawn, we use the value of t on the last line of Table 7.3 under $P = 0.05$, namely, $t = 1.960$. Consequently, 95 per cent confidence limits here are

$$28.75 - 1.960\left(\frac{6.27}{\sqrt{241}}\right) < \mu_x < 28.75 + 1.960\left(\frac{6.27}{\sqrt{241}}\right)$$
$$28.75 - 0.78 < \mu_x < 28.75 + 0.78$$
$$27.97 \text{ years} < \mu_x < 29.53 \text{ years}$$

Let us find 99 per cent confidence limits for the mean μ_x of the universe of truck drivers from which the 12 truck drivers in Table 3.1 were drawn. Here $n = 12$, $m_x = \$1.79$, and $s_x = \$0.28$. In Table 7.3 on the line for $n' = 12 - 1 = 11$ we find $t = 3.106$ under $P = 0.01$. Consequently, 99 per cent confidence limits here are

$$1.79 - 3.106\left(\frac{0.28}{\sqrt{11}}\right) < \mu_x < 1.79 + 3.106\left(\frac{0.28}{\sqrt{11}}\right)$$
$$1.79 - 0.26 < \mu_x < 1.79 + 0.26$$
$$\$1.53 < \mu_x < \$2.05$$

This confidence interval is very wide. The principal reason for the low reliability of the estimate m_x of μ_x is the small size of the sample, $n = 12$. The reason for using a sample of only 12 truck drivers in Chapter 3 was to demonstrate briefly the nature of the computations required for finding the mean and the standard deviation of a serial distribution. In a practical statistical investigation of the rate of pay of truck drivers in a particular city, one would try to have a much larger sample. For example, in its investigation of the rate of pay of truck drivers in the Boston Metropolitan Area in 1962, the Bureau of Labor Statistics collected data for 551 drivers of light trucks. (See Tables 3.15 and 3.16.)

7.3.2 Confidence limits for a universe total. Let m be the mean of a random sample containing n elementary units drawn from a universe containing N elementary units. Then Nm is an estimate of the value of $N\mu$, which is the total of all the values of the variable for all the elementary units in the universe.

It is easy to prove (see exercise 12 on p. 357) that if every value of the variable in a distribution is multiplied by a positive constant, the value of the standard deviation of the distribution also is multiplied by the constant. Therefore, because N is a positive constant, confidence limits for $N\mu$ are $Nm \pm tNs_x/\sqrt{n-1}$, where the value of t is obtained from Table 7.3 on the line for $n' = n - 1$.

For example, suppose that the mean of a variable in a random sample of 100 elementary units drawn from a universe that contains 11,783 elementary units is $m = 1.28$ and that the standard deviation of the sample is $s_x = 0.43$. Then an estimate of the universe total is $11,783(1.28) = 15,082.24$, and 95 per cent confidence limits for the true total are $15,082.24 \pm (1.960)(11,783)(0.43)/\sqrt{99} = 15,082.24 \pm 998.31 = 14,083.93$ and $16,080.55$ which probably would be rounded off to 14,000 and 16,100.

7.3.3 Quality control chart for the mean. A manufacturer of television sets uses many small parts in assembling the chassis of the sets. One of these parts has a desirable length of 2.500 inches, but the part can be used if it varies slightly from the desired length. However, a part that is more than a twentieth of an inch longer or shorter than 2.500 inches must be reworked or scrapped. In other words, the manufacturing specifications require that the length be between 2.450 inches and 2.550 inches.

Let us suppose that the automatic machine that produces this part has been adjusted and stabilized so that when it is operating properly it produces parts with average length 2.500 inches and standard deviation 0.020 inches. So long as the machine is operating properly, it is neither necessary nor economical to inspect each part that comes from the machine. It is only necessary to inspect periodically a sample of the parts produced by the machine. For example, if the machine produces several hundred parts per hour, it might be satisfactory to select at random during each hour a sample of five parts and inspect them by measuring their lengths. Five is the most common sample size for this kind of quality control in American industry today.

The most commonly used confidence limits in statistical quality control work in American industry are the 3-sigma control limits. In our illustration, $\mu_x = 2.500$, $\sigma_x = 0.020$, and $n = 5$. By formula 7.2, $\sigma_m = 0.020/\sqrt{5} = 0.009$. Consequently, the 3-sigma control limits here are $2.500 \pm 3(0.009) = 2.500 \pm 0.027 = 2.473$ inches and 2.527 inches.

If we draw a random sample of five parts from the production for each hour, then compute the average length of the five parts in each sample, and compare this average for each sample with the confidence interval, the samples will indicate whether or not the machine is operating satisfactorily. Table 7.4 shows the lengths of the five parts in each of 17 samples and the average length for each sample. You will notice that the mean of the seventeenth sample is outside of the 3-sigma confidence interval. This is interpreted as a signal that the production process is probably no longer under statistical control and production will be stopped until the cause of the trouble is found and corrected.

TABLE 7.4

**Statistical Quality Control Samples in the
Manufacture of Television Sets**

Sample No.	Length of Part in the Sample (Inches)					Mean Length (Inches)
1	2.513	2.475	2.506	2.495	2.466	2.491
2	2.529	2.483	2.511	2.496	2.516	2.507
3	2.510	2.502	2.479	2.485	2.529	2.501
4	2.475	2.481	2.513	2.518	2.503	2.498
5	2.495	2.525	2.504	2.516	2.560	2.520
6	2.495	2.470	2.496	2.501	2.483	2.489
7	2.503	2.492	2.485	2.512	2.473	2.493
8	2.497	2.507	2.516	2.510	2.565	2.519
9	2.505	2.490	2.486	2.512	2.492	2.497
10	2.512	2.488	2.493	2.508	2.494	2.499
11	2.500	2.491	2.511	2.516	2.492	2.502
12	2.485	2.515	2.497	2.499	2.589	2.517
13	2.471	2.521	2.518	2.483	2.497	2.489
14	2.504	2.530	2.485	2.487	2.469	2.495
15	2.514	2.499	2.482	2.509	2.586	2.518
16	2.486	2.495	2.522	2.507	2.460	2.494
17	2.534	2.525	2.539	2.519	2.528	2.529

After the average length for the five parts in a sample has been computed, this average is marked on a control chart by a dot or a cross directly above the point on the horizontal scale that indicates the number of the sample or the time period represented by the sample. The means of the 17 samples in Table 7.4 have been plotted on Figure 7.4.

That a process no longer is operating under statistical control may be indicated by other features of the control chart than the plotting of a

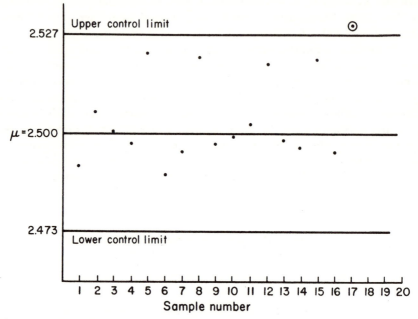

Fig. 7.4.   Statistical Quality Control Chart for the Mean in the Manufacture of Television Sets

point outside of the control limits. For example, a "run," that is, a long series of points in succession on one side of the average line, with none of the points falling outside the control limits, may be an indication that the process is not in statistical control. You learned how to interpret "runs" in section 6.7.

7.4 TESTS OF HYPOTHESES ABOUT MEANS

Sometimes, instead of desiring to attach confidence limits to an estimate, we wish to test the sample mean against some hypothetical value of the true mean. Suppose, for example, that someone states that he has a suspicion that the mean age at time of marriage of the universe of men from which the sample in Table 3.8 (p. 81) was drawn is not greater than 27 years. The test consists of finding the probability that a random sample of 242 men drawn from a universe in which the true mean is not greater than 27 years would produce a value of $m = 28.75$ years. We must notice that the hypothesis being tested here and its logical alternative, namely, that μ_x is greater than 27 years, require a one-tail test. Suppose that we decide to make the test at the $\alpha = 0.05$ significance level. We find on the last line of Table 7.3 that $t = 1.645$ for a one-tail test at the $\alpha = 0.05$ level when $n' = n - 1$ is greater than 30. Consequently, $t = 1.645$ is the critical value of t for our test.

Next, we compute $t = (m - \mu_x)/(s_x/\sqrt{n-1}) = (28.75 - 27)/(6.27/\sqrt{241}) = +4.375$. Because the computed $t = +4.375$ is greater than the critical value $t = 1.645$, we reject the hypothesis that μ_x is not greater than 27 years. Incidentally, Table 6.19 (p. 199) shows that the risk of our having made the wrong decision here is less than 0.00003.

Suppose that we wish to test the hypothesis that the true mean hourly rate of pay of the universe of truck drivers from which the sample of 12 truck drivers in Table 3.2 was drawn is not less than \$2.00. The alternative here is that μ_x is less than \$2.00. This pair of hypotheses requires a one-tail test. Let us make the test at the $\alpha = 0.025$ significance level. Table 7.3 shows that $t = 2.201$ for a one-tail test at the $\alpha = 0.025$ level when $n' = n - 1 = 12 - 1 = 11$. The critical value of t for our test is $t = -2.201$, because the hypotheses being tested here involve the left-hand tail of the t distribution.

Next, we compute $t = (1.79 - 2.00)/(0.28/\sqrt{11}) = -0.21/0.081 = -2.593$. Because the computed $t = -2.593$ is less than the critical value $t = -2.201$, we reject the null hypothesis that the true mean of the universe is not less than \$2.00 and we accept the alternative that the true mean of the universe is less than \$2.00.

EXERCISES

1. Compute 95 per cent confidence limits and 99 per cent confidence limits for the mean of the universe from which the sample in Table 3.7 was drawn. See exercise 3 on page 80. (*Ans.* 53.7 and 57.1; 53.1 and 57.7.)

2. How large would the sample in Table 3.7 need to be to reduce the 95 per cent confidence interval for the mean to one score unit? Hint: In order to divide the confidence interval by 3.4 it is necessary to multiply the sample size by $(3.4)^2$. (*Ans.* 751.)

3. How large would the sample in Table 3.7 need to be to reduce the 99 per cent confidence interval for the mean to one score unit? (*Ans.* 1375.)

4. Find 99 per cent confidence limits for the means that you computed in exercise 13 on page 99. (*Ans.* 28.7 and 30.7; 26.4 and 28.2; 26.3 and 28.1; 23.9 and 25.5.)

5. Find 50 per cent confidence limits and 95 per cent confidence limits for the means computed in exercises 3 to 7 on page 385.

6. Find 90 per cent confidence limits and 99 per cent confidence limits for the mean computed in exercise 12 on page 99.

7. Test the null hypothesis that the true mean of the universe of men from which the sample in Table 3.8 was drawn is greater than 29.5 years. Can you reject this hypothesis at the $\alpha = 0.05$ level? (*Ans.* Yes, because $t = -1.861$ is less than -1.645.)

8. Test the null hypothesis that the true mean of the universe of truck drivers from which the sample in Table 3.2 was drawn is less than $1.70. Can you reject this hypothesis at the $\alpha = 0.05$ probability level? (*Ans.* No.)

9. Test the null hypothesis that the true mean of the universe of men from which the sample in Table 3.8 was drawn is 29.5 years. Make the test at the $\alpha = 0.05$ significance level. Hint: This requires a two-tails test. Reject the null hypothesis and accept the alternative if the computed value of t is less than or equal to -1.960 or equal to or greater than $+1.960$.

10. Compute 95 per cent confidence limits and 99 per cent confidence limits for each of the four or five random samples, each containing 30 random digits, that you drew. Use $t = 2.045$ when $P = 0.05$ and $t = 2.756$ when $P = 0.01$, because for Table 7.3 here $n' = n - 1 = 30 - 1 = 29$. Then pool all the results of the members of the class and determine the percentage of the 95 per cent confidence intervals that contain the mean $\mu = 4.50$ of the universe of random digits. Similarly, determine the percentage of the 99 per cent confidence intervals that contain the mean $\mu = 4.50$ of the universe of random digits.

11. By how much of an error would we have deceived ourselves in the second last paragraph of section 7.3.1 if we had determined 99 per cent confidence limits by substituting $s_x = \$0.28$ for σ_x and using $z = 2.576$? (*Ans.* The confidence interval would have been $0.10 too small, which is 19.2 per cent of the correct confidence interval.)

12. Duplicate Table 7.2 using 99 per cent confidence intervals instead of 95 per cent confidence intervals.

13. Compute 95 per cent confidence limits for the mean of the universe from which the sample in Table 3.16 was drawn. See exercise 12 on page 99. (*Ans.* $2.08 and $2.16.)

14. In seven measurements of the proportion of cobalt in a piece of nickel–chrome alloy, a chemical analyst obtained the following results: 0.45, 0.47, 0.50, 0.45, 0.50, 0.49. (Darnell Salyer and Thomas R. Sweet, "Determination of Small Amounts of Cobalt in Steels and Nickel Alloys," *Analytical Chemistry* 29, No. 1 (January 1957), p. 2.) Determine 95 per cent confidence limits for the true proportion of cobalt in the piece of nickel–chrome alloy. (*Ans.* 0.453 and 0.493.)

15. Consider the universe that consists of the four elementary units whose measurements are $x_1 = 1$ inch, $x_2 = 3$ inches, $x_3 = 5$ inches, and $x_4 = 7$ inches, the measurements having been recorded to the nearest inch. (a) Compute μ_x. (*Ans.* $\mu_x = 4$ in.) (b) Compute σ_x^2. (*Ans.* $\sigma_x^2 = 5$ sq. in.) (c) Write down all the different possible random samples containing three elementary units that can be drawn, without replacement, from this universe. (*Ans.* 1, 3, 5; 1,3, 7; 1,5, 7; 3, 5, 7) (d) Write in serial form the distribution of the means of the samples. (*Ans.* 3, 3⅔, 4⅓, 5) (e) Compute the mean of the sampling distribution of means. (*Ans.* $\mu_m = 4$ in. $= \mu_x$) (f) Compute the variance of the sampling distribution of means. Compare the result with the expression obtained by substituting in formula 7.3 and squaring both sides.

(*Ans.* $\sigma_m^2 = \dfrac{5}{9}$ sq. in. $= \dfrac{5}{3} \cdot \dfrac{4-3}{4-1}$ sq. in.)

7.5 SAMPLE SIZE FOR ESTIMATING A MEAN

Let us suppose that we want to use a random probability sample of the women (or men) in a school to obtain an estimate m_x of the mean height μ_x of all the women (or men) in the school. Suppose, further, that we want the probability to be 0.95 that the amount of sampling error in the estimate m_x will not be greater than $\frac{1}{2}$ inch in absolute value. That is, we want the probability to be 0.95 that $|\, m_x - \mu_x\,| \leq \frac{1}{2}$ inch.

The 95 per cent confidence interval around the mean μ_m of the normal distribution of means of samples of size n from the universe extends $1.96\sigma_m$ above μ_m and $1.96\sigma_m$ below μ_m. Consequently, we want the following equation to be satisfied: $1.96\sigma_m = \frac{1}{2}$ inch. That is, $1.96\sigma_x/\sqrt{n} = \frac{1}{2}$ inch. In order to solve this equation for n, we need to know a suitable value to substitute for σ_x. In most practical sampling situations, we do not know the exact value of σ_x. However, in many practical situations, it is possible to use either our own previous experience with similar investigations or published results of other investigations to find at least a rough estimate of the σ_x that is needed. For example, in many reports of samples of heights of young women and young men the standard deviation was around 2.25 to 2.5 inches. We need not hesitate to use 2.5 inches as the value of σ_x for the purpose of determining an appropriate sample size to use for our investigation here.

First, we may notice that if we do not know a highly accurate value to substitute for σ_x, there is no need to carry two decimal places in the number $z = 1.96$. Let us use $z = 2$. Changing 1.96 to 2 and substituting 2.5 for σ_x, the last equation above becomes $2(2.5)/\sqrt{n} = \frac{1}{2}$. To solve this equation for n, multiply both sides of the equation by $2\sqrt{n}$. The result is $10 = \sqrt{n}$. Now square both sides of this equation to obtain $n = 100$. This tells us that we ought to use a sample containing 100 women (or men).[3]

To illustrate for you the effect that not knowing the exact value of σ_x has in estimating the appropriate sample, size let us consider the following situation. Suppose we knew only that the standard deviation of the height of women is likely to be somewhere between 2 inches and 3 inches. Substituting 2 inches for σ_x and solving the equation produces $n = 64$. Substituting 3 inches for σ_x and solving the equation produces $n = 144$. Consequently, in such a situation, all we could say is that the appropriate sample size is somewhere between 64 and 144, and we probably would use $n = 144$.

In general, this process for estimating the sample size n to use for obtaining an estimate m of the mean μ of a universe, with probability P that the amount of sampling error in m is not more than k units in absolute value, consists in solving for n the equation $z\sigma_m = k$, that is $z\sigma_x/\sqrt{n} = k$. The result is

$$ n = \left[\frac{z\sigma_x}{k}\right]^2 \tag{7.6} $$

[3]If we had used $z = 1.96$ instead of $z = 2$ in solving the equation for n, we would have obtained $n = 96$ instead of $n = 100$.

where the best available estimate of the standard deviation of the variable in the universe is substituted for σ_x.

The appropriate value of z to be substituted into formula 7.6 may be determined in the way that was given in section 4.8. That is, for estimating a parameter and computing a pair of confidence limits, or for a two-tails test of significance, or for a two-tails test of a hypothesis, you may use either one of the two values of z for which $\alpha/2$ is in each tail of the normal curve. (See Table 6.18.) For a one-tail test of significance or a one-tail test of a hypothesis, you may use a value of z that puts all of α in one tail of the normal curve. (See Table 6.19.)

If we desire to take into account the relative sizes of the universe and the sample, when the sample is to be drawn without replacement from a finite universe, we use the formula for σ_m including the finite multiplier. The result is

$$n_1 = n \left[\frac{N}{N + n - 1} \right] \tag{7.7}$$

where n is the value obtained from formula 7.6. The value of n_1 that is obtained from formula 7.7 is smaller than the value of n obtained by formula 7.6. For example, in the illustration that was used at the beginning of this section, let us suppose that there were 1001 women (or men) in the school. Then $N = 1001$ and $n_1 = 100[1001/(1001 + 100 - 1)] = 91$.

This tells us that the mean of a random sample containing 91 elementary units drawn without replacement from a universe of 1001 elementary units has the same degree of reliability, that is, the same size of confidence interval, as the mean of a random sample containing 100 elementary units drawn with replacement from a universe of 1001 elementary units. Also the mean of a random sample containing 91 elementary units drawn without replacement from a universe of 1001 elementary units has the same degree of reliability as the mean of a random sample containing 100 elementary units drawn without replacement from an indefinitely large universe.

One practical situation in which it is usually advisable to use formula 7.7 is the case in which the value of n computed by formula 7.6 is more than 10 per cent of the size of the universe.

Here is an example of a simple method that often provides a satisfactory idea of the value of σ_x in a distribution. Write down the largest and the smallest values of x that you would expect to find in a fairly large sample, say, a sample of 1000 elementary units. These two values give us a rough estimate of the range of the variable in the distribution. In such large samples drawn from a normal distribution, the range of the variable often is somewhere in the neighborhood of six times the standard deviation of the distribution (see Table 6.8). Consequently, one-sixth of the range sometimes is a satisfactory estimate of the value of σ_x when we are determining sample size.

Let us see how this would work in our illustrative example. In a group of about 1000 college women, there probably would not be anyone more than 6 feet tall. It would not be surprising to find a few women less than 5 feet tall. Perhaps 4 feet 10 inches would be a fair guess for the lower end of the range of height of the women.

Using 6 feet and 4 feet 10 inches as rough estimates of the ends of the range of height for the women, we estimate the range of height for the women to be about 14 inches. Now, 14 inches divided by 6 gives 2.33 inches as a rough estimate of the standard deviation of the height of the women. That is in the range of values of σ_x that available reports dealing with distributions of heights of women and men suggested.

Another method may sometimes be used to obtain a rough estimate of the value of σ_x in the universe. This method is called a "pilot study" and it consists in examining a small number of elementary units and computing the standard deviation for them.

EXERCISES

1. Determine the size of a random sample that would provide an estimate of the average height of the women (or men) in a school if the probability is to be 0.99 that the amount of sampling error in the estimate m obtained from the sample is not greater than ½ inch in absolute value. Use 2.5 inches as a rough estimate of σ_x. (*Ans. n* = 169 using *z* = 2.6.)

2. Repeat exercise 1, except that the probability is to be 0.99 that the amount of sampling error in m is not greater than ¼ inch in absolute value. (*Ans. n* = 676.)

3. Repeat exercise 2, except that the probability is to be 0.95. (*Ans. n* = 400.)

4. Summarize and generalize the results obtained in exercises 1, 2, and 3 by stating the relationship between sample size and the absolute value of the amount of sampling error in an estimate of a mean obtained by using a random sample. In other words, what effect does the sample size have on the amount of sampling error to be expected in a mean obtained from a random sample?

5. What assumptions, if any, must be made concerning the shape of the distribution of the variable in the universe when we use formula 7.2 for the standard deviation of the mean of a sample to determine confidence limits for the mean of the universe?

6. Repeat exercises 1, 2, and 3 using equation 7.7 instead of equation 7.6 and assuming that the number of women (or men) in the school is $N = 1001$. (*Ans.* 145, 404, 286.)

7. Rephrase some of exercises 1 to 3 so as to require one-tail tests of significance or one-tail tests of hypotheses. Then compute appropriate sample sizes.

8. Derive formula 7.7.

7.6 SIGNIFICANCE OF DIFFERENCE IN MEANS OF TWO INDEPENDENT SAMPLES

In many situations two samples are necessary in order to provide answers to certain questions about two separate universes, such as, for example, a universe of men and a universe of women, or about one universe that has been subjected to a kind of treatment or condition that might change it into a different universe.

It will be assumed in sections 7.6.1 and 7.6.2 that the two samples are independent of each other. In practice, this means that the selection of the elementary units for one sample must not influence in any way the selection of the elementary units for the other sample.

7.6.1 A parametric test of significance for two means. In a certain school, a sample survey was made for the purpose of finding out whether or not there was a significant difference between the number of movies attended by boys and the number of movies attended by girls in a particular month. The sample of boys contained 103 boys. The average number of movies attended by the 103 boys during the month of January was $m_1 = 2.26$ per boy. The standard deviation of this distribution was $s_1 = 1.34$ movies. There were 98 girls in the sample of girls. The average number of movies attended by the 98 girls during the month of January was $m_2 = 2.63$ per girl. The standard deviation of this distribution was $s_2 = 1.21$ movies. The difference between the two means is $m_2 - m_1 = 2.63 - 2.26 = +0.37$.

Let us choose $\mu_1 = \mu_2$ as the appropriate null hypothesis to be tested here, and as the alternative the hypothesis that $\mu_1 \neq \mu_2$. This pair of hypotheses requires a two-tails test. Let us make the test at the $\alpha = 0.05$ significance level. In order to test the significance of the difference $+0.37$ movies, we must change it into standard units so that we can use our probability tables. We can use formula 7.9 for this purpose. The result is

$$s = \sqrt{\frac{103(1.34)^2 + 98(1.21)^2}{103 + 98 - 2}} = 1.28, \qquad t = \frac{(2.63 - 2.26) - 0}{1.28\sqrt{\frac{1}{103} + \frac{1}{98}}} = \frac{+0.37}{0.18} = +2.06$$

The critical values of t for this two-tails test are ± 1.960, the values given by the last line of Table 7.3 under $P = 0.05$. Because the computed value of t, $+2.06$, is greater than $+1.960$, we reject the null hypothesis that $\mu_1 = \mu_2$ and accept the alternative that $\mu_1 \neq \mu_2$. Because m_2 is greater than m_1 in this example, we conclude further that the universe of girls in that school attended more movies per girl than the universe of boys attended per boy in January.

Here is the basis for this test of significance. Let us imagine that we have drawn a large number of pairs of independent random samples of sizes

n_1 and n_2 from two large, normally distributed universes with means μ_1 and μ_2 and standard deviations σ_1 and σ_2. Imagine also that we have computed the mean of each of the samples and the difference $m_2 - m_1$ between the two means in each pair of samples. These differences could be tabulated in the form of a frequency distribution and graphed in the form of a histogram.

We need to know the properties of this frequency distribution of the differences $m_2 - m_1$. Here they are: (1) The distribution of the differences is a normal distribution. (2) The mean of the distribution of differences is equal to the difference between the means of the two universes; that is $\mu_{m_2-m_1} = \mu_2 - \mu_1$. (3) The standard deviation of the distribution of differences is $\sigma_{m_2-m_1} = \sqrt{\sigma_1^2/n_1 + \sigma_2^2/n_2}$. The difference $m_2 - m_1$ between a specific pair of sample means can be changed into standard units by using formula 3.5 which yields in this case

$$z = \frac{(m_2 - m_1) - (\mu_2 - \mu_1)}{\sqrt{\dfrac{\sigma_1^2}{n_1} + \dfrac{\sigma_2^2}{n_2}}} \tag{7.8}$$

If the values of σ_1 and σ_2 are known, Table 6.7 may be used to determine the probability of obtaining, by chance in random sampling from a normal distribution, a value of z as large as or larger than the absolute value of the z computed by formula 7.8; this is a valid test for small samples as well as for large samples, provided the two universes are normal.

If the values of σ_1 and σ_2 are unknown, which usually is the fact in practical investigations, a test that is valid for small samples as well as for large samples is available if we know that or are willing to assume that $\sigma_1 = \sigma_2$. In this case, we may let σ represent the two equal but unknown numbers σ_1 and σ_2. Then we consider s_1^2 and s_2^2 to be two independent estimates of σ^2. If we define s and t by the formulas

$$s = \sqrt{\frac{n_1 s_1^2 + n_2 s_2^2}{n_1 + n_2 - 2}} \quad \text{and} \quad t = \frac{(m_2 - m_1) - (\mu_2 - \mu_1)}{s\sqrt{\dfrac{1}{n_1} + \dfrac{1}{n_2}}} \tag{7.9}$$

then Table 7.3 may be used to determine the approximate probability P to be associated with the value of t computed by formula 7.9. Use $n' = n_1 + n_2 - 2$. Of course, if $n_1 + n_2$ is greater than 32, we may use Table 6.7 to find a good approximation to the desired probability.

The hypothesis most frequently tested here is the null hypothesis that the two universes are normal distributions with equal means, that is, that $\mu_1 = \mu_2$ or that $\mu_2 - \mu_1 = 0$.

If the size of either of the two universes from which the two samples were drawn is relatively small compared with the size of the sample, and if the sample was drawn without replacement, the finite multiplier should be incorporated in the formula for $\sigma_{m_2 - m_1}$.

In the example dealing with the numbers of movies attended by the boys and girls in a school during a particular month, suppose that the girls had been selected so that each one in the sample was the sister of a boy in the sample of boys. Then the two samples would not have been considered to be independent; they would have been considered to be correlated samples. The formulas given in this section are not appropriate for use in such situations. However, appropriate formulas for testing the significance of the difference between the means of two correlated samples can be obtained by using formula A.4 (p. 395) in evaluating $\sigma_{m_2 - m_1}$. Also, a good test of significance for differences in related samples is given in section 5.1.1.

In actual practice, it is not necessary to limit the use of this test of significance to situations in which the two universes are known to be perfect normal distributions. The test generally is satisfactory even for universes considerably different from normal universes. For example, the universes might be rectangular distributions such as universes of random digits or they might be bell-shaped but moderately skewed distributions. However, if either of the two universes is extremely unsymmetrical, this procedure for testing the significance of the difference between two means may be unsatisfactory, especially if the sample sizes are small. If the sample sizes are so small that $n_1 + n_2$ is not greater than 32 and if there is danger that either of the universes is extremely unsymmetrical, then you ought to use the test of significance to be described in section 7.6.2.

Perhaps it has occurred to some readers that in order to test completely for differences between two normal universes we should test the significance of the difference between the two standard deviations (or the two variances) of the two samples in addition to testing the significance of the difference between the two means. For even if the two normal universes have the same mean, they might have different standard deviations (and variances). A procedure for testing the significance of the difference between the variances of two samples can be found in more advanced books on statistics. But the test given in this section and the test to be given in the next section probably will suffice for almost all the practical situations that you are likely to encounter during a first course in statistics.

EXERCISES

1. Test the significance of the difference between the mean ages at marriage of the 433 skilled men and the 420 unskilled men in Table 3.17. Use the results obtained in exercise 13 on page 99. Use a two-tails test with $\alpha = 0.01$. What is the null hypothesis that you are testing here? What is the alternative? (*Ans.* $t = +4.90$ and the difference is highly significant.)

2. Test the significance of the difference between the mean ages at marriage of the 433 women in the skilled worker classification and the 420 women in the unskilled worker classification in Table 3.17. Use the results obtained in exercise 13 on page 99. Use a two-tails test at the $\alpha = 0.01$ significance level. What is the null hypothesis that you are testing here? What is the alternative? What are the critical values of t for this test? (*Ans.* $t = +5.65$ and the difference is highly significant.)

3. Test the hypothesis that the average age of the skilled men in York at time of marriage was more than two years greater than the average age of the women in the skilled worker classification at time of marriage. Use the estimates obtained from Table 3.17 in exercise 13 on page 99 and make this one-tail test at the 0.05 significance level. (*Ans.* $+0.79$; do not reject the null hypothesis that $\mu_1 - \mu_2 \leqq 2$.)

4. The purchasing agent for a company has two sources of supply for a machine tool that is used in the factory. A record is kept on 230 randomly selected tools from supplier A and it is found that they wear out in an average of 59 hours of use, the standard deviation of the length of time used being 2.4 hours. A random sample of 185 tools from supplier B last for an average of 55 hours and the standard deviation is 2.5 hours. Supplier A charges a slightly higher price for the tool, but the difference in price is not sufficient reason for buying from supplier B if the tool of supplier A can be depended upon to last 3 hours longer on the average. Test the hypothesis that $\mu_A - \mu_B \leqq 3$ against the alternative that $\mu_A - \mu_B > 3$. What advice would you give to the purchasing agent if he has told you that he is not willing to buy exclusively from supplier A if there is more than a 1 per cent risk that the tool of supplier A does not last at least 3 hours longer on the average than the tool of supplier B? (*Ans.* $t = +4.13$, $P < 0.001$. Reject the null hypothesis and accept the alternative. Advise the purchasing agent that the risk is less than one-tenth of 1 per cent.)

5. In a series of determinations of the amounts of carbon and hydrogen in ephedrine hydrochloride, analyst A made 48 carbon determinations with mean error -0.3 parts per 1000, the standard deviation of his errors being 2.4 parts per 1000. Analyst B made 61 carbon determinations with mean error $+1.1$ parts per 1000, the standard deviation of his errors being 2.3 parts per 1000. Analyst C made 37 hydrogen determinations with mean error -5.0 parts per 1000, the standard deviation of his errors being 20.0 parts per 1000. Analyst D made 77 hydrogen determinations with mean error $+8.0$ parts per 1000, the standard deviation of his errors being 21.0 parts per 1000. (Francis W. Power, "Accuracy and Precision of Microanalytical Determination of Carbon and Hydrogen," *Analytical Chemistry* 11, No. 12 (December 15, 1939), p. 661.) Test the significance of the difference between the mean errors of A and B and between the mean errors of C and D. (*Ans.* Both differences are significant; for A and B, $t = 3.06$; for C and D, $t = 3.12$.)

7.6.2 A nonparametric test of significance for two samples.

The test using formula 7.9 is the best available if all the assumptions on which it is based are valid for the practical situation in which you need to test the significance of the difference between two means. Here is a list of the theoretical requirements for the valid application of the test.

1. The two samples are independent.

2. The elementary units in each sample were randomly and inde-
pendently selected.
3. The underlying variable in the investigation is a continuous variable.
4. The data in the two samples are in the form of truly quantitative
measurements.
5. The two samples were drawn from normally distributed universes.
6. The two universes have the same variance (or in special cases, a
known ratio of variances).

As indicated in section 7.6.1, the test usually is satisfactory even when
some of the theoretical requirements are relaxed a little. But the trouble is
that we often do not know how much the requirements must be relaxed to
fit a practical situation in which we are interested. For example, the data
obtained in many investigations of social, psychological, and economic
phenomena are only partly quantitative rather than truly quantitative
classifications. Again, we may not know how far from being normal distributions are the universes from which the samples were drawn, or we may
not know whether or not the variances of the two universes are even approximately equal. Doubts about the validity of assumptions 5 and 6 in
a particular situation may become especially worrisome if the sizes of the
two samples are very small.

The method that will now be described eliminates the need for requirements 4, 5, and 6. It is a valid test regardless of whether the data are
in the form of truly quantitative measurements or only partly quantitative
classification. It is a valid test no matter what the shapes of the distributions in the two universes from which the samples were drawn. It is a
valid test no matter whether or not the variances of the two universes are
equal. It is a valid test no matter how small or how large the sizes of the
two samples, although the work involved in carrying out the test may become somewhat unwieldy if the sample sizes are very large. Because it eliminates the need for requirements 5 and 6, this test is called a distribution-
free test and a nonparametric test.

Suppose that we have two independent random samples in which the
values of the variable are as follows:

Sample A: 18,   13,   10,   7,   3
Sample B: 20,   16,   14,   11,   9,   8,   6,   6,   2

Combine the two samples, arranging the whole batch of values of the
variable in the order of their magnitude beginning with the lowest value of
the variable, but marking each value so as to indicate the sample from
which it came. Assign ranks to the values of the variable in the combined
list by the method explained in section A.5.5 (p. 358). Let ΣR_A = the sum

of the ranks attached to values that came from sample A and let $\Sigma R_B =$ the sum of the ranks attached to values from sample B. The result is

Value:	2	3	6	6	7	8	9	10	11	13	14	16	18	20
Sample:	B	A	B	B	A	B	B	A	B	A	B	B	A	B
Rank:	1	2	3.5	3.5	5	6	7	8	9	10	11	12	13	14

$$\Sigma R_A = \quad 2 \quad\quad + \quad\quad 5 \quad\quad + \quad\quad 8 + 10 \quad\quad + \quad\quad 13 \quad = 38$$
$$\Sigma R_B = 1 \;+\; 3.5 + 3.5 \quad + 6 + 7 \;+\; 9 \;+\; 11 + 12 \;+\; 14 = 67$$

Let $n_1 =$ the number of values of the variable in sample A, and let $n_2 =$ the number of values of the variable in sample B. In our illustration, $n_1 = 5$ and $n_2 = 9$. You can use the fact that $\Sigma R_A + \Sigma R_B = (n_1 + n_2)$ $(n_1 + n_2 + 1)/2$ to check your work. Then evaluate the two expressions

$$n_1 n_2 + \frac{n_1(n_1 + 1)}{2} - \Sigma R_A \quad \text{and} \quad n_1 n_2 + \frac{n_2(n_2 + 1)}{2} - \Sigma R_B$$

Let x be the smaller of the two values obtained from these expressions and let y be the larger of the two values. For our illustration, $x = 22$ and $y = 23$.

The null hypothesis for the test here is that the two universes from which the two samples were drawn have the same distribution. In some practical situations the alternative may be simply that the two universes do not have the same distribution. Then the test is a two-tails test. In other practical situations, we may desire to use as the alternative the statement that the universe from which sample A was drawn has a mean or a median that is larger than the mean or the median of the universe from which sample B was drawn. In that case the test is a one-tail test. In still other practical situations, we may need to use as the alternative the statement that the mean or the median of one universe, say, universe B, exceeds the mean or the median of the other universe by a constant c; in such a situation, subtract c from every value in sample B and then carry out the test.

Tables 7.5, 7.6, 7.7, and 7.8 simplify the work of carrying out the test. Suppose that we desire to make the test at the $\alpha = 0.05$ level of significance for the two-tails test. By looking at Table 7.7 with $n_1 = 5$ and $n_2 = 9$, we see that the critical value of x is 7. This means that if the value of x obtained for the two samples is less than or equal to 7, we should reject the null hypothesis at the $\alpha = 0.05$ level of significance. Because the value of x for our two samples is 22, which is greater than the critical value 7, we do not reject the null hypothesis that the two universes from which sample A and sample B were drawn have the same distribution. If two universes have the same distribution, then, of course, each parameter of one of the universes has exactly the same value as the corresponding parameter of the other universe.

TABLE 7.5

Critical Values of x in the Mann-Whitney Test*

At $\alpha = 0.01$ for a two-tails test and at $\alpha = 0.005$ for a one-tail test

Value of n_1	Value of n_2

	1	2	3	4	5	6	7	8	9	10	11	12	13	14	15	16	17	18	19	20
							Reject the null hypothesis if x is equal to or less than													
1																				
2																			0	0
3									0	0	0	1	1	1	2	2	2	2	3	3
4				0	0	1	1	2	2	3	3	4	5	5	6	6	7	8		
5				0	1	1	2	3	4	5	6	7	7	8	9	10	11	12	13	
6				0	1	2	3	4	5	6	7	9	10	11	12	13	15	16	17	18
7				0	1	3	4	6	7	9	10	12	13	15	16	18	19	21	22	24
8				1	2	4	6	7	9	11	13	15	17	18	20	22	24	26	28	30
9			0	1	3	5	7	9	11	13	16	18	20	22	24	27	29	31	33	36
10			0	2	4	6	9	11	13	16	18	21	24	26	29	31	34	37	39	42
11			0	2	5	7	10	13	16	18	21	24	27	30	33	36	39	42	45	48
12			1	3	6	9	12	15	18	21	24	27	31	34	37	41	44	47	51	54
13			1	3	7	10	13	17	20	24	27	31	34	38	42	45	49	53	56	60
14			1	4	7	11	15	18	22	26	30	34	38	42	46	50	54	58	63	67
15			2	5	8	12	16	20	24	29	33	37	42	46	51	55	60	64	69	73
16			2	5	9	13	18	22	27	31	36	41	45	50	55	60	65	70	74	79
17			2	6	10	15	19	24	29	34	39	44	49	54	60	65	70	75	81	86
18			2	6	11	16	21	26	31	37	42	47	53	58	64	70	75	81	87	92
19		0	3	7	12	17	22	28	33	39	45	51	56	63	69	74	81	87	93	99
20		0	3	8	13	18	24	30	36	42	48	54	60	67	73	79	86	92	99	105

*Adapted from Table 2 of D. Auble, "Extended Tables for the Mann-Whitney Statistic," *Bulletin of the Institute of Educational Research* at Indiana University, Vol. 1, No. 2 (1953), with the permission of the author and the Director of the Institute.

Solomon and Coles trained five rats to follow leaders to secure a food reward.[4] Four other rats were chosen but were not trained to follow leaders to a food reward. Then the nine rats were tested in a maze to see how many trials each rat required to learn how to escape from the maze and obtain a food reward without receiving an electric shock while passing through the maze. The purpose of the investigation was to see whether or not the rats that had been trained to follow leaders to food would transfer this training and learn to escape from the maze without receiving an electric shock in fewer trials than the number of trials required by untrained rats. The five trained rats will be designated the experimental (E) rats, and the other four rats will be designated the control (C) rats. The number of

[4]R. L. Solomon and M. R. Coles, "A Case of Failure of Generalization of Imitation Across Drives and Across Situations," *Journal of Abnormal and Social Psychology* 49 (1954), pp. 7–13.

TABLE 7.6

Critical Values of x in the Mann-Whitney Test*

At $\alpha = 0.02$ for a two-tails test and at $\alpha = 0.01$ for a one-tail test

	Value of n_2																			
Value of n_1	1	2	3	4	5	6	7	8	9	10	11	12	13	14	15	16	17	18	19	20
	Reject the null hypothesis if x is equal to or less than																			
1																				
2													0	0	0	0	0	0	1	1
3						0	0	1	1	1	2	2	2	3	3	4	4	4	5
4				0	1	1	2	3	3	4	5	5	6	7	7	8	9	9	9	10
5				0	1	2	3	4	5	6	7	8	9	10	11	12	13	14	15	16
6				1	2	3	4	6	7	8	9	11	12	13	15	16	18	19	20	22
7			0	1	3	4	6	7	9	11	12	14	16	17	19	21	23	24	26	28
8			0	2	4	6	7	9	11	13	15	17	20	22	24	26	28	30	32	34
9			1	3	5	7	9	11	14	16	18	21	23	26	28	31	33	36	38	40
10			1	3	6	8	11	13	16	19	22	24	27	30	33	36	38	41	44	47
11			1	4	7	9	12	15	18	22	25	28	31	34	37	41	44	47	50	53
12			2	5	8	11	14	17	21	24	28	31	35	38	42	46	49	53	56	60
13		0	2	5	9	12	16	20	23	27	31	35	39	43	47	51	55	59	63	67
14		0	2	6	10	13	17	22	26	30	34	38	43	47	51	56	60	65	69	73
15		0	3	7	11	15	19	24	28	33	37	42	47	51	56	61	66	70	75	80
16		0	3	7	12	16	21	26	31	36	41	46	51	56	61	66	71	76	82	87
17		0	4	8	13	18	23	28	33	38	44	49	55	60	66	71	77	82	88	93
18		0	4	9	14	19	24	30	36	41	47	53	59	65	70	76	82	88	94	100
19		1	4	9	15	20	26	32	38	44	50	56	63	69	75	82	88	94	101	107
20		1	5	10	16	22	28	34	40	47	53	60	67	73	80	87	93	100	107	114

*Adapted from Table 3 of D. Auble, "Extended Tables for the Mann-Whitney Statistic," *Bulletin of the Institute of Educational Research* at Indiana University, Vol. 1, No. 2 (1953), with the permission of the author and the Director of the Institute.

trials that each rat required in learning the maze is shown in the following table.

C rats	51	53	70	110	
E rats	45	64	75	78	82

The null hypothesis here is that the number of trials required by trained rats is not less than the number of trials required by untrained rats. The alternative that would be of interest in such an investigation is, of course, that trained rats require fewer trials than untrained rats to succeed in the shock-avoidance situation. Let us choose the $\alpha = 0.05$ significance level for this one-tail test.

First, we must combine the two samples, arranging the scores for the rats in the order of their magnitude and labeling each score so as to indicate the sample from which it came. Then we assign ranks to the values of the variable.

TABLE 7.7

Critical Values of x in the Mann-Whitney Test*

At $\alpha = 0.05$ for a two-tails test and at $\alpha = 0.025$ for a one-tail test

Value of n_1	\multicolumn{20}{c}{Value of n_2}

Value of n_1	1	2	3	4	5	6	7	8	9	10	11	12	13	14	15	16	17	18	19	20
	\multicolumn{20}{c}{Reject the null hypothesis if x is equal to or less than}
1																				
2								0	0	0	0	1	1	1	1	1	2	2	2	2
3					0	1	1	2	2	3	3	4	4	5	5	6	6	7	7	8
4				0	1	2	3	4	4	5	6	7	8	9	10	11	11	12	13	13
5			0	1	2	3	5	6	7	8	9	11	12	13	14	15	17	18	19	20
6			1	2	3	5	6	8	10	11	13	14	16	17	19	21	22	24	25	27
7			1	3	5	6	8	10	12	14	16	18	20	22	24	26	28	30	32	34
8		0	2	4	6	8	10	13	15	17	19	22	24	26	29	31	34	36	38	41
9		0	2	4	7	10	12	15	17	20	23	26	28	31	34	37	39	42	45	48
10		0	3	5	8	11	14	17	20	23	26	29	33	36	39	42	45	48	52	55
11		0	3	6	9	13	16	19	23	26	30	33	37	40	44	47	51	55	58	62
12		1	4	7	11	14	18	22	26	29	33	37	41	45	49	53	57	61	65	69
13		1	4	8	12	16	20	24	28	33	37	41	45	50	54	59	63	67	72	76
14		1	5	9	13	17	22	26	31	36	40	45	50	55	59	64	67	74	78	83
15		1	5	10	14	19	24	29	34	39	44	49	54	59	64	70	75	80	85	90
16		1	6	11	15	21	26	31	37	42	47	53	59	64	70	75	81	86	92	98
17		2	6	11	17	22	28	34	39	45	51	57	63	67	75	81	87	93	99	105
18		2	7	12	18	24	30	36	42	48	55	61	67	74	80	86	93	99	106	112
19		2	7	13	19	25	32	38	45	52	58	65	72	78	85	92	99	106	113	119
20		2	8	13	20	27	34	41	48	55	62	69	76	83	90	98	105	112	119	127

*Adapted from Table 5 of D. Auble, "Extended Tables for the Mann-Whitney Statistic," *Bulletin of the Institute of Educational Research* at Indiana University, Vol. 1, No. 2 (1953), with the permission of the author and the Director of the Institute.

Value:	45	51	53	64	70	75	78	82	110
Sample:	E	C	C	E	C	E	E	E	C
Rank:	1	2	3	4	5	6	7	8	9

$$\Sigma R_C = \quad 2 + 3 \quad + \quad 5 \qquad\qquad + \qquad\qquad 9 = 19$$
$$\Sigma R_E = 1 \qquad + \qquad 4 \quad + \quad 6 + 7 + 8 \qquad = 26$$

In this illustration, $n_1 = 4$, $n_2 = 5$, $x = 9$, and $y = 11$. Table 7.8 shows that the critical value associated with $n_1 = 4$, $n_2 = 5$ is 2 at the $\alpha = 0.05$ significance level for a one-tail test. Because $x = 9$ is greater than the critical value 2, we do not reject the null hypothesis and, therefore, we do not accept the alternative. In other words, training the rats to follow leaders to food did not have a significant effect on the rats under the other circumstances.

If either n_1 or n_2 is larger than 20, or if both n_1 and n_2 are larger than 20, we carry out the test by using the normal probability as an excellent approximation to the true probability associated with the value of x com-

TABLE 7.8

Critical Values of x in the Mann-Whitney Test*

At $\alpha = 0.10$ for a two-tails test and at $\alpha = 0.05$ for a one-tail test

		Value of n_2																		
	1	2	3	4	5	6	7	8	9	10	11	12	13	14	15	16	17	18	19	20
	Reject the null hypothesis if x is equal to or less than																			
1																			0	0
2				0	0	0	1	1	1	1	2	2	2	3	3	3	4	4	4
3		0	0	1	2	2	3	3	4	5	5	6	7	7	8	9	9	10	11
4		0	1	2	3	4	5	6	7	8	9	10	11	12	14	15	16	17	18
5	0	1	2	4	5	6	8	9	11	12	13	15	16	18	19	20	22	23	25
6	0	2	3	5	7	8	10	12	14	16	17	19	21	23	25	26	28	30	32
7	0	2	4	6	8	11	13	15	17	19	21	24	26	28	30	33	35	37	39
8	1	3	5	8	10	13	15	18	20	23	26	28	31	33	36	39	41	44	47
9	1	3	6	9	12	15	18	21	24	27	30	33	36	39	42	45	48	51	54
10	1	4	7	11	14	17	20	24	27	31	34	37	41	44	48	51	55	58	62
11	1	5	8	12	16	19	23	27	31	34	38	42	46	50	54	57	61	65	69
12	2	5	9	13	17	21	26	30	34	38	42	47	51	55	60	64	68	72	77
13	2	6	10	15	19	24	28	33	37	42	47	51	56	61	65	70	75	80	84
14	2	7	11	16	21	26	31	36	41	46	51	56	61	66	71	77	82	87	92
15	3	7	12	18	23	28	33	39	44	50	55	61	66	72	77	83	88	94	100
16	3	8	14	19	25	30	36	42	48	54	60	65	71	77	83	89	95	101	107
17	3	9	15	20	26	33	39	45	51	57	64	70	77	83	89	96	102	109	115
18	4	9	16	22	28	35	41	48	55	61	68	75	82	88	95	102	109	116	123
19	0	4	10	17	23	30	37	44	51	58	65	72	80	87	94	101	109	116	123	130
20	0	4	11	18	25	32	39	47	54	62	69	77	84	92	100	107	115	123	130	138

(Row label: Value of n_1)

*Adapted from Table 7 of D. Auble, "Extended Tables for the Mann-Whitney Statistic," *Bulletin of the Institute of Educational Research* at Indiana University, Vol. 1, No. 2 (1953), with the permission of the author and the Director of the Institute.

puted from the two samples. In fact, the normal probability is a fairly good approximation to the true probability even if the two sample sizes are as small as $n_1 = 8$ and $n_2 = 8$. To find the normal probability approximation, we compute

$$\mu = \frac{n_1 n_2}{2}, \qquad \sigma = \sqrt{\frac{n_1 n_2 (n_1 + n_2 + 1)}{12}}, \quad \text{and} \quad z = \frac{x - \mu}{\sigma}$$

and then look for the value of P associated with this value of z in Table 6.7.

The basic principle of the test is simple. It involves the analysis of the $(n_1 + n_2)!/(n_1! n_2!)$ equally likely sequences that it is possible to obtain, under the null hypothesis, with n_1 values from sample A and n_2 values from sample B, in order to find the probability to be associated with a sequence in which the number of times that a value from sample A precedes a value from sample B is x. This test usually is called the Mann-Whitney test.[5]

[5] H. B. Mann and D. R. Whitney, "On a Test of Whether One of Two Random Variables is Stochastically Larger than the Other," *Annals of Mathematical Statistics* 18 (1947), pp. 50–60.

The method given above for applying the Mann-Whitney test to two samples is exact only if none of the values of the variable in sample A is tied with any of the values of the variable in sample B. A tie that is limited to the values of the variable in one of the samples does not affect the value of x. You can demonstrate this easily for yourself by changing one of the two tied values in sample B to 5 and recomputing x. But a tie that involves values of the variable in both samples does affect the value of x slightly. However, unless a large number of the values of the variable in one sample are tied with values of the variable in the other sample, it is not worth while to make an adjustment in the formulas to take care of the ties.

If both sample sizes are larger than 8 and if the number of values of the variable in one sample that are tied with values of the variable in the other sample is relatively large, an adjustment can be made in the formula for σ and the adjusted value of σ may be used to compute the value of z. The (seldom needed) adjusted formula for σ is

$$\sigma = \sqrt{\frac{n_1 n_2}{n(n-1)}\left(\frac{n^3 - n}{12} - \Sigma T\right)}$$

where $n = n_1 + n_2$, $T = (k^3 - k)/12$ is the adjustment for a group of k tied values, and ΣT is the sum of all the adjustments for ties. This adjustment to the formula for σ to take care of ties is similar to the adjustment for ties that will be described more fully in section 8.3.8.

The effect of making the adjustment for ties is to increase slightly the value of z, thereby making it slightly more significant by reducing the associated value of P. However, the increase in the value of z usually is negligible, that is, it usually is not large enough to change our decision about the null hypothesis unless it happens that the value of z computed without making the adjustment for ties is very slightly less than the critical value of z for the test. Therefore, you may adopt this rule: Compute the value of z without making the adjustment for ties and find the value of P associated with the computed value of z. If this value of P is just larger than the chosen significance level, α, recompute the value of z by using the adjusted formula for σ and base your decision about the null hypothesis on the value of P associated with the new value of z.

EXERCISES

1. Two independent random samples of trainees made the same 100 readings under identical poor viewing conditions on two dials that were the same except for the forms of the digits. The digits on one dial were of the standard or conventional form. The digits on the other dial were in the form of angular blocks designed for easy reading. The scores of seven trainees on the dial with conventional digits were 45, 47, 49, 51, 53, 55, and 61. The scores of eight trainees on the dial with angular

block digits were 54, 60, 66, 67, 68, 70, 73, and 75. The scores are the numbers of correct readings made by the trainees. Use the Mann-Whitney test of significance for the difference between these two samples. What is your conclusion about the comparative legibility of the two types of digits under the particular viewing condition? (*Ans.* $n_1 = 7$, $n_2 = 8$, $x = 3$. The critical value of x at the $\alpha = 0.005$ level for a one-tail test is 6. The angular block digits are significantly easier to read under the particular viewing condition.)

2. When a strike was being considered by the employees of a factory, an attitude scale dealing with the situation was administered to a random sample of 25 men employed at the factory and a random sample of 20 wives of men employed at the factory. The scores of the 25 men on the attitude test were 18, 23, 29, 30, 34, 34, 37, 39, 42, 44, 45, 47, 47, 47, 49, 49, 50, 50, 52, 52, 52, 52, 55, 58, 64. The scores of the 20 wives were 19, 20, 21, 24, 24, 27, 28, 28, 28, 31, 32, 32, 35, 36, 38, 41, 41, 43, 48, 54. Consider that these scores are partly quantitative data and that the higher the person's score the more favorable his or her attitude toward going on strike. Is there sufficient evidence to conclude that the wives of the men employed at the factory were less favorable toward the strike than the men were? (*Ans.* Yes, $z = -3.27$ and $P = 0.001$ for this one-tail test.)

3. Use the Mann-Whitney test to test the significance of the difference between the sample of measurements by chemical analyst A and the sample of measurements by chemical analyst B given in section 3.7 (p. 78). Make the one-tail test at the $\alpha = 0.01$ significance level. (*Ans.* $x = 0$, $n_1 = 4$, $n_2 = 6$. The measurements by B are significantly greater than the measurements by A. The difference is significant even at the $\alpha = 0.005$ level.)

4. Use the Mann-Whitney test to test the significance of the difference between the two samples supplied by subcontractors A and B in exercise 9 on page 81. (*Ans.* $x = 10$, $n_1 = 5$, $n_2 = 5$, for a two-tails test; do not reject the hypothesis that the two samples were drawn from the same universe or from two universes with the same distribution.)

REFERENCES

Robert V. Hogg and Allen T. Craig, *Introduction to Mathematical Statistics*. New York: The Macmillan Company, 1959.

William H. Kruskal, "Historical Notes on the Wilcoxon Unpaired Two-Sample Test," *Journal of the American Statistical Association* 52 (1957), pp. 356–60. (This article is about the Mann-Whitney test.)

Chapter 8

CORRELATION

8.1 INTRODUCTION

We have seen that statistics may be thought of as the study of variations. The study of the simultaneous variation of two or more characteristics of the elementary units in a sample drawn from a universe is one of the most frequently applied parts of the statistical method. Uusally it is called the study of *correlation*, although a more descriptive word might be covariation. Sometimes it is called the study of *association of variations*.

The purpose of investigation of the correlation of two or more characteristics of the elementary units in a sample usually is, of course, to obtain reliable estimates of the correlation of the same characteristics of the elementary units in the universe from which the sample was drawn.

Man has always wanted to be able to prophesy. He forecasts the weather and the movements of the stars. He tries to predict the cost of living, the size of the population, stock market prices, farm product prices, the incidence of crime, and a host of other things, for the coming months or years. Nearly all the formulas by which statisticians make predictions of things to come are developed by studying the associations or relationships between two or more variables.

8.2 TEST OF RELATIONSHIP IN A 2 × 2 TABLE

Do the data in Table 8.1 lead you to think that a child is more likely to be left-handed if at least one of its parents is left-handed than if both parents are right-handed?

Consider the 1987 right-handed children to be a random sample of children from a universe of right-handed children. The proportion of these children who come from families in which at least one of the parents is left-handed is $p_1 = 145/1987 = 0.073$. Consider the 191 left-handed children to be a random sample of children from a universe of left-handed children. The proportion of these children who come from families in which at least one of the parents is left-handed is $p_2 = 40/191 = 0.209$.

TABLE 8.1

Handedness of Parents and Their Children*

	Both Parents Right-Handed	At Least One Parent Left-Handed	Total
Right-Handed Children	1842	145	1987
Left-Handed Children	151	40	191
Total	1993	185	2178

**Source:* D. C. Rife, "Handedness, With Special Reference to Twins," *Genetics* XXV (1940), Table 3, p. 182 and Table 6, p. 184. Rife defined "handedness" as the hand preferred in the performance of the following one-hand operations: throwing, bowling, marbles, knife, spoon, hammer, saw, sewing, writing, scissors. He classified as right-handed only those who used the right hand for all ten operations, and as left-handed those who used the left hand or either hand with equal ease in one or more of the ten operations.

Is the difference between p_1 and p_2 small enough to be considered as due merely to chance in drawing two random samples of children from two universes of children in which $\pi_1 = \pi_2$? If the difference is small enough to be attributed to chance in random sampling, then there is not sufficient evidence in Table 8.1 to indicate that the handedness of children is related to the handedness of their parents. In that case, we would be justified in thinking that, insofar as the data in Table 8.1 are concerned, the handedness of children is independent of the handedness of their parents.

Consequently, all that we need to do here is use the method of section 5.2 for testing the significance of the difference in two proportions. Probably Professor Rife had in mind the idea that a child is more likely to be left-handed if at least one of the parents is left-handed than if both parents are right-handed. In other words, that probably is the alternative that he desired to accept. Consequently, the null hypothesis here is $\pi_2 \leqq \pi_1$ and the alternative is $\pi_2 > \pi_1$. This pair of hypotheses requires a one-tail test. Formula 5.2 can be used for this test. Let us suppose that Professor Rife did not want to take more than a 1 per cent risk of drawing the wrong conclusion; that is, $\alpha = 0.01$ for the test.

For $p_1 = 0.073$ and $p_2 = 0.209$, formula 5.2 shows that $z = +6.45$. Table 6.19 shows that the value of z at the 0.01 probability level is 2.33. Consequently, the probability of obtaining, by chance in random sampling, a value of z as large as or larger than $z = +6.45$ is far smaller than 0.01. Therefore, we reject the null hypothesis and accept the alternative.

There is strong evidence in Table 8.1 of positive association, that is, of association of left-handedness in children with left-handedness in their parents and of right-handedness in children with right-handedness in their parents.

Here is how a problem of this kind might have its beginning. Through your experience you become aware of an unexpected proportion of happenings of a certain kind. This observation then leads to the proposal of a hypothesis to the effect that the proportions of these happenings is unusually large (or small), that is, that these events happen more often (or less often) than would be expected on the basis of pure random chance or on some other specific basis. This surmise of yours then leads to the idea of carrying out a statistical investigation for the purpose of reaching a conclusion with respect to the acceptability of the proposed hypothesis.

Table 8.1 is called a 2×2 table because there are only two essential columns of numbers and two essential rows of numbers in the table. The marginal totals at the right-hand edge of the table and along the bottom of the table can be omitted without the loss of any essential information. However, it often is considered desirable to include the marginal totals as a convenience for the reader.

The number in each essential cell of a 2×2 table is a frequency corresponding to a pair of characteristics of an elementary unit. For example, in Table 8.1, the number 1842 is a frequency, namely, the number of families in which the child was right-handed and both parents were right-handed. The number 151 is a frequency, namely, the number of families in which the child was left-handed and both parents were right-handed.

8.2.1 Caution. It would be far from safe to conclude from the data in Table 8.1 that heredity is the sole cause of the greater proportion of left-handed children among families in which at least one of the parents is left-handed. Some of the left-handedness among the children might be due to imitation. For example, some children may be left-handed writers simply because the first time they thought of writing they noticed one of the parents holding a pencil in the left hand.

TABLE 8.2

Handedness of Identical Twins and Their Relatives*

	Both Twins Right-Handed	One Twin Right-Handed, Other Twin Left-Handed	Total
Without Left-Handed Relatives	105	25	130
With Left-Handed Relatives	26	22	48
Total	131	47	178

*Source: Same as Table 8.1

It seems so natural for the human mind to seek principles of causation for observed events that sometimes we are tempted to accept a cause-and-effect relationship when all that analysis of the data has demonstrated is an association or coincidence. Robert Garland of Pennsylvania worked so hard promoting the idea of daylight-saving time that he has been called "the father of daylight saving." One of his arguments for daylight saving was that he had observed that boys in daylight-saving areas have larger feet than boys elsewhere. He said "having bigger feet makes better men."

In statistical work, it is not necessary to prove that there are cause-and-effect relationships in every situation. Often, the establishment of the fact that there exists an association, and the measurement of its degree or intensity, are all that the investigator needs for his purposes of estimation and prediction. The statistician often is more interested in what he can accomplish by using observed associations or relationships than he is in assigning cause-and-effect or other explanations to them.

TABLE 8.3
Handedness of Fraternal Twins and Their Relatives*

	Both Twins Right-Handed	One Twin Right-Handed, Other Twin Left-Handed	Total
Without Left-Handed Relatives	84	12	96
With Left-Handed Relatives	16	15	31
Total	100	27	127

*Source: Same as Table 8.1

Some additional evidence in this matter of handedness of children and their parents can be found in Tables 8.2 and 8.3. You may apply the test of relationship to those tables.

8.2.2 A wartime problem solved. Here is an illustration of the many ways in which ideas of elementary practical statistics were used to help save lives and win a war. During World War II, when the Japanese suicide planes (*Kamikaze*) began to attack U. S. Navy ships, it was necessary to determine the best tactics for ships to use in defending themselves against the *Kamikaze*. The two best methods of protection seemed to be antiaircraft gunfire and violent maneuvers such as a series of sharp turns.

The purpose of the violent maneuvers was to spoil the aim of the diving suicide plane. But the radical maneuvers tended also to spoil the accuracy of the antiaircraft fire, especially for small ships such as destroyers and auxiliaries. The accuracy of antiaircraft gunfire of large ships such as

battleships, cruisers, and aircraft carriers was less affected by violent maneuvers than that of the small ships.[1]

Consider first the large ships. We might choose as the null hypothesis
to be tested the statement that the two kinds of tactics are equally effective;
that is, $\pi_1 = \pi_2$. The alternative is that $\pi_1 \neq \pi_2$. This pair of hypotheses
requires a two-tails test. Let us use the $\alpha = 0.01$ significance level.

For the large ships, $n_1 = 8 + 28 = 36$, $n_2 = 30 + 31 = 61$, $p_1 = 8/36 =$
0.22, $p_2 = 30/61 = 0.49$, $p_2 - p_1 = 0.49 - 0.22$, and formula 5.2 shows
that $z = +2.62$. Table 6.7 shows that the critical values of z at the 0.01
significance level are ± 2.58. Consequently, the probability corresponding
to $z = +2.62$ is less than 0.01, and we reject the null hypothesis. We not
only accept the alternative that the two kinds of tactics are not equally
effective but, because p_2 is greater than p_1, we conclude that violent maneuver
is a significantly more effective form of protection for large ships.

Proceeding in a similar way for the small ships, we find that $z = -1.81$;
in this case we would not reject the null hypothesis at the 0.01 significance
level. Nor would we be able to reject the null hypothesis even if we had
decided to use the $\alpha = 0.05$ significance level. We could reject the null
hypothesis at the $\alpha = 0.10$ significance level. This may be interpreted as
tending to indicate that the small ships ought not employ violent maneuvers
when attacked by *Kamikaze* but should concentrate on antiaircraft gunfire.

| | *Large Ships* | | | *Small Ships* | |
	Hit by Kamikaze	Not hit by Kamikaze		Hit by Kamikaze	Not hit by Kamikaze
Maneuvering	8	28	Maneuvering	52	92
Not maneuvering	30	31	Not maneuvering	32	92

EXERCISES

1. Apply the test of the significance of the difference between two proportions to
the data in Tables 8.2 and 8.3. (*Ans.* $z = +3.57$, $z = +4.25$) What conclusions do
you draw? Do these conclusions agree with the conclusion that we drew from the
data in Table 8.1? (*Ans.* Yes.)

2. State the null hypothesis and the alternative for each of the tests in exercise 1.
What conclusions do you reach about these null hypotheses? In other words, do you
reject these null hypotheses? (*Ans.* Yes.) At what significance level?

3. Show that the same results are obtained from the test of significance of the
difference between two proportions for the data in Table 8.1 if you use the data in
columns two and four of the table to determine the two proportions to be tested instead of using the data in columns three and four as we did in the illustration given at
the beginning of this chapter.

[1]For this and other applications of statistical methods in World War II, see Philip
M. Morse and George E. Kimball, *Methods of Operations Research.* New York: John Wiley
and Sons, Inc., 1951.

4. State an appropriate null hypothesis and alternative concerning the attitudes of the boys and the girls in the following tabulation with respect to a particular television program. Choose an appropriate significance level. Apply the test of the significance of the difference between two proportions and then decide whether or not to reject the null hypothesis. (*Ans.* $z = +2.11$)

	Liked the Program	*Did Not Like the Program*
Boys	216	234
Girls	196	157

5. Here are some of the data that resulted from a sample survey of the attitudes of clerical employees toward their jobs in a very large government organization. The sample contained 510 employees. Study these 2 × 2 tables and use the method for testing the significance of the difference between two proportions to help you to decide for yourself what significant relationships, if any, are indicated by the data.

For example, do the data indicate that there is a strong association between job satisfaction and one's feeling about his chance of being promoted? Is there a definite relationship between attitude toward one's supervisor and feeling about chance of being promoted? Is there any relationship between feeling of job security and job satisfaction?

	Feel That Chance to Be Promoted Is Satisfactory	*Do Not Feel That Chance to Be Promoted Is Satisfactory*
High job satisfaction	351	57
Low job satisfaction	50	52

	Feel That Chance to Be Promoted Is Satisfactory	*Do Not Feel That Chance to Be Promoted is Satisfactory*
Like their supervisor	354	106
Do not like their supervisor	24	26

	Think That Their Jobs Are Reasonably Secure	*Do Not Think That Their Jobs Are Reasonably Secure*
High job satisfaction	306	102
Low job satisfaction	55	47

	Want to Change Their Job Channels	*Do Not Want to Change Their Job Channels*
High job satisfaction	135	273
Low job satisfaction	88	14

	Want to Change Their Job Channels	*Do Not Want to Change Their Job Channels*
Like their supervisor	216	244
Do not like their supervisor	37	13

6. Suppose that you are the person who had been hired by the managers of the government organization to make the attitude survey described in exercise 5. If you were in the stage now of preparing your final report to the managers of the organization, which of the following twenty statements would you feel justified in including in your report?

Construct a table containing three columns. Mark the first column "Acceptable," and list in that column the letters designating the statements that you consider to be acceptable on the basis of the data given in exercise 5 and your computations using the method of testing the significance of the difference between two proportions. Mark the second column "Rejected or Contradicted," and list in that column the letters designating the statements that you consider to be contradicted or rejected by the evidence. Mark the third column "Omit," and list in that column the letters designating the statements that you consider to be irrelevant or to be neither clearly supported nor clearly contradicted by the evidence given in exercise 5.

Bear in mind that the managers want you to report to them conclusions that can be used to describe attitudes of the whole universe of several thousand clerical workers in the organization rather than statements that can be applied only to the 510 clerical workers who happened to be drawn into the sample.

(a) The majority of the employees like working for this organization.

(b) About 20 per cent of the employees do not like their jobs.

(c) About 10 per cent of the employees do not like their supervisors.

(d) The majority of the employees feel that they are in the wrong job channels.

(e) About one-third of the employees are not satisfied with their chances of promotion.

(f) Employees who rate high in job satisfaction tend to like their supervisors.

(g) Desire to change job channels is likely to be associated with low job satisfaction.

(h) An employee who wants a different kind of job channel is more likely to be found in the group that does not like the supervisors.

(i) High job satisfaction and the feeling that the outlook for promotion is good go hand in hand.

(j) Those who feel most secure in their jobs tend to be in the high job-satisfaction group.

(k) Employees who have permanent Civil Service status are more likely to be satisfied with their jobs than nonpermanent employees are.

(l) Employees who have low attitudes toward their chances of promotion are likely also to have negative attitudes toward their supervisors.

(m) The tables can not tell us whether the employees want to change their job channels because they dislike their supervisors, or whether they tend to react unfavorably toward their supervisors because they dislike the kind of work they are doing. Quite likely the relation is a circular one.

(n) Surprisingly enough, there does not seem to be a strong relationship between job satisfaction and job security in this organization.

(o) Although good supervision is an important element in high job satisfaction, it is clear that this factor alone is not sufficient to account for high job satisfaction.

(p) High job satisfaction is somewhat more likely to be found among males, older employees, employees with greater seniority, the less well educated, and Negroes.

(q) There are employees who do not think very highly of their supervisors but still feel that this is a good place to work.

(r) Liking the supervisor is important, but it is not sufficient to cause an employee to be highly satisfied with his job.

(s) Dislike for the supervisor is likely to be tied to the feeling of poor promotion possibilities, even though it is quite possible that the supervisor can do nothing about getting promotions.

(t) A supervisor in charge of employees who cannot hope for promotions may expect to be the target of criticisms about himself and his supervisory tactics, even though the heart of the matter may be simply the employees' frustration over promotion opportunities.

8.3 RANK CORRELATION ANALYSIS

One day late in September a few years ago, seven friends were chatting about college football and making predictions as to the teams that would be the best of the year. In the course of the conversation, it was suggested that each member of the group write down the rank order in which he or she predicted that some of the teams in which they were interested would stand at the end of the football season. At the end of the season, they would compare their predictions with the final standings of the teams and give a prize of six dollars to the one who made the best set of predictions. The one who made the worst set of predictions would have to give a little party for the group.

Fifteen teams were selected for rating. There was some discussion about how to decide who made the best set of predictions and who made the worst set of predictions. One member of the group said that there was a statistical method called "correlation" that could be used for the purpose. It was decided that the final standings as reported in the newspapers by the Associated Press would be accepted as the criterion and that the method of rank correlation analysis would be used to decide who made the set of predictions most nearly in agreement with the final Associated Press standings and who made the set of predictions least in agreement with the final Associated Press standings.

On December 3, newspapers published the final Associated Press ratings for a long list of college football teams, including the fifteen teams for which the members of the group had made predictions. The seven rank correlation coefficients were computed and they are shown in Table 8.4. William F. won the prize of six dollars and James R. took the group out for an evening of bowling.

You may be interested in knowing that although Mary C. and Allan O. obtained the same rank correlation coefficient, namely, 0.70, their predictions were not identical. Table 8.5 shows the final Associated Press ratings and the predictions made by Mary C. and Allan O. This illustrates an important principle of the correlation method. The fact that two,

TABLE 8.4

Ability to Predict Winning Football Teams

Person Who Made the Predictions	Rank Correlation of the Predictions with the Final A. P. Standings
William F.	.94
Sterling W.	.91
Carlysle C.	.78
Mary C.	.70
Allan O.	.70
Samuel D.	.50
James R.	.47

members of this group had the same rank correlation coefficient means that Mary C.'s predictions, as a whole, were in agreement with the A. P. ratings as much as Allan O.'s predictions, on the whole. Correlation analysis measures the strength of the relationship, if any, between two sets of values, taking into consideration not the pairs of values one pair at a time, but the whole set of pairs simultaneously.

TABLE 8.5

Football Predictions Made by Mary C. and Allan O.

Team	Final A. P. Rank	Rank Predicted by Mary C.	Rank Predicted by Allan O.
Minnesota	1	1	1
Duke	2	10	7
Notre Dame	3	2	9
Texas	4	7	8
Michigan	5	8	3
Fordham	6	3	5
Texas A & M	7	5	2
Navy	8	4	6
Northwestern	9	13	4
Oregon State	10	9	14
Ohio State	11	6	12
Pennsylvania	12	14	10
Mississippi State	13	11	13
Tulane	14	12	11
Clemson	15	15	15

8.3.1 The Spearman rank correlation coefficient. Here is the way in which the rank correlation coefficients were computed for the football predictions. The computations will be shown for the predictions of Mary C. The first step in the process is to construct Table 8.6.

TABLE 8.6

Tabulation for Computing a Rank Correlation Coefficient

Elementary Unit (Team)	Rank by A. P.	Rank by Mary C.	Difference Between Ranks by Mary C. and A. P. d		d^2
Minnesota	1	1	0		0
Duke	2	10		-8	64
Notre Dame	3	2	1		1
Texas	4	7		-3	9
Michigan	5	8		-3	9
Fordham	6	3	3		9
Texas A & M	7	5	2		4
Navy	8	4	4		16
Northwestern	9	13		-4	16
Oregon State	10	9	1		1
Ohio State	11	6	5		25
Pennsylvania	12	14		-2	4
Mississippi State	13	11	2		4
Tulane	14	12	2		4
Clemson	15	15	0		0
			$\Sigma d = 20$ $-20 = 0$		$166 = \Sigma d^2$

The positive differences have been placed at one side of the fourth column and the negative differences at the other side of the column, in order to emphasize the fact that the total of the positive differences is exactly the same as the total of the negative differences except for the signs. Consequently, $\Sigma d = 0$. This is true for all rank correlation problems. Therefore, Σd is of no use to us except as a check on the accuracy of our computations; if Σd turns out to be other than zero, then we know that we have made a mistake in assigning ranks or in the subtraction or addition.

The symbol r', "r prime," will be used in this book to stand for the Spearman rank correlation coefficient for a sample. The value of r' is computed by the formula

$$r' = 1 - \frac{6\Sigma d^2}{n^3 - n} \tag{8.1}$$

Charles Spearman (1863-1945) was an English psychologist who made several contributions to statistical methods.

For the data in Table 8.6, the number of pairs of ranks is $n = 15$ and $\Sigma d^2 = 166$. Consequently,

$$r' = 1 - \frac{6(166)}{15^3 - 15} = 1 - \frac{996}{3375 - 15} = 1 - 0.30 = +0.70$$

The symbol ρ', "rho prime," will be used in this book to stand for the

Spearman rank correlation coefficient of two sets of ranks for all the elementary units in a universe.

8.3.2 Perfect correlation. Consider Table 8.8, which was constructed by assigning ranks to the ten baseball teams listed in Table 8.7. The two sets of ranks are in perfect agreement for all the teams. All the differences are zeros. Consequently, all the numbers in the d^2 column are zeros and, therefore, Σd^2 is zero in this situation. If you substitute zero for Σd^2 in formula 8.1 you find that $\rho' = +1$, and this would be true no matter what the value of N. In other words, perfect positive rank correlation is represented by $\rho' = +1$. The value of ρ' can never be greater than $+1$.

TABLE 8.7

National League Baseball Results in 1962*

Team	No. of Games Won	No. of Games Lost	Proportion of Games Won	Batting Average	No. of Home Runs
San Francisco	103	62	0.624	0.278	204
Los Angeles	102	63	0.618	0.268	140
Cincinnati	98	64	0.605	0.270	167
Pittsburgh	93	68	0.578	0.268	108
Milwaukee	86	76	0.531	0.252	181
St. Louis	84	78	0.519	0.271	137
Philadelphia	81	80	0.503	0.260	142
Houston	64	96	0.400	0.247	105
Chicago	59	103	0.364	0.253	126
New York	40	120	0.250	0.240	139

*Source: The World Almanac 1963, p. 802. Copyright, 1963, by New York World-Telegram Corporation. Reprinted by permission.

TABLE 8.8

An Illustration of Perfect Positive Rank Correlation

Team	Rank by No. of Games Won	Rank by Proportion of Games Won	Difference in Ranks d	d²
San Francisco	1	1	0	0
Los Angeles	2	2	0	0
Cincinnati	3	3	0	0
Pittsburgh	4	4	0	0
Milwaukee	5	5	0	0
St. Louis	6	6	0	0
Philadelphia	7	7	0	0
Houston	8	8	0	0
Chicago	9	9	0	0
New York	10	10	0	0
Total			0	0

8.3.3 Zero correlation. When there is no correlation whatsoever between the two sets of ranks for all the elementary units in a universe, the value of ρ' is zero.

8.3.4 Negative correlation. Sometimes the situation is such that the ranks of the elementary units for one variable characteristic are almost the reverse of their ranks for the other variable characteristic. For example, consider Table 8.7 again. If we assign ranks to the ten teams on the basis of games won and also on the basis of games lost, the two sets of ranks will be as shown in Table 8.9. The value of ρ' computed from the ranks in Table 8.9 by formula 8.1 is $\rho' = -1$. Consequently, perfect negative rank correlation is represented by $\rho' = -1$. The value of ρ' never can be less than -1.

TABLE 8.9

An Illustration of Perfect Negative Rank Correlation

Team	Rank by No. of Games Won	Rank by No. of Games Lost	Difference in Ranks d		d^2
San Francisco	1	10		-9	81
Los Angeles	2	9		-7	49
Cincinnati	3	8		-5	25
Pittsburgh	4	7		-3	9
Milwaukee	5	6		-1	1
St. Louis	6	5	$+1$		1
Philadelphia	7	4	$+3$		9
Houston	8	3	$+5$		25
Chicago	9	2	$+7$		49
New York	10	1	$+9$		81
Total			$+25$	-25	330

8.3.5 Range of the value of ρ'. The Spearman rank correlation coefficient ρ' always has a value in the range from -1 to $+1$. The correlation coefficient is positive if the rank in one variable characteristic tends to increase when the rank in the other variable characteristic increases, that is, when an elementary unit that has high rank in one of the characteristic tends to have high rank in the other characteristic. The correlation coefficient is negative if the rank in one variable characteristic tends to become smaller as the rank in the other variable characteristic becomes larger that is, when an elementary unit that has high rank in one of the characteristics tends to have low rank in the other characteristic, and vice versa.

8.3.6 Reliability of rank correlation. Suppose that a universe consists of 10,000 individuals and that the individuals are ranked on each of two variable characteristics A and B that are completely independent of one another. Then there is no rank-order relationship whatsoever between the values of the two characteristics A and B for all the individuals in the universe. For such a universe, $\rho' = 0$.

We could construct such a universe by using a table of random digits such as Table A.15 to determine the two ranks for each elementary unit in the universe. We could begin by assigning the ranks 1 to 10,000 to the individuals for the characteristic A in any convenient order, for example, the alphabetical order of their surnames. Then we could determine each individual's rank for characteristic B by drawing a random number in the range from 1 to 10,000 from Table A.15, not using the same random number more than once if we prefer that. In this way, an individual's rank in the second characteristic would be completely independent of his rank in the first characteristic.

Now, suppose that we draw a small random sample of, say, eight individuals from this universe and record in a table the two ranks for each of the eight individuals. Then, reduce these ranks on both characteristics to the range from 1 to 8. Next, compute the rank correlation coefficient r' for the sample.

Although the value of the rank correlation coefficient ρ' for this universe is zero, the value computed for the rank correlation coefficient r' for the sample probably will not be exactly zero; it may be positive or it may be negative. The value of r' for the sample is not likely to be exactly zero because of sampling error, that is, because of the chance nature of the sampling process that caused the particular group of eight individuals with their particular combination of ranks to be drawn into the sample rather than any other group of eight individuals in the universe. If a different group of eight individuals had happened to be drawn into the sample, it is likely that they would have had a somewhat different combination of ranks for the two characteristics and that the value of the rank correlation coefficient r' computed for them would have been somewhat different.

You would find it a very laborious task to write down all the 40,320 possible combinations of two ranks for eight individuals, compute the rank correlation coefficient for each of the 40,320 combinations, and construct a frequency distribution for the different values of the rank correlation coefficient; some different combinations of ranks would produce the same value of the rank correlation coefficient. The relative frequency of a possible value of the rank correlation coefficient here would be the probability of that particular value of r' occurring as a result of pure chance in random sampling from a universe in which the rank correlation coefficient ρ' is zero.

Fortunately for all of us, it is not necessary to carry out this very laborious task. E. G. Olds has determined, for all values of n from 4 to 30 the possible values of r' that are most unlikely to occur because of pure chance in random sampling from a universe in which ρ' is zero. These unlikely or critical values of r' are tabulated for you in Table 8.10. By using Table 8.10, you can test the significance of a value of the rank correlation coefficient r' computed for n elementary units, if n is not greater than 30.

If you ever need to test the significance of a rank correlation coefficient derived from a random sample in which the number of elementary units and, consequently, the number of pairs of ranks is greater than 30, you cannot use Table 8.10. However, you can obtain satisfactory approximations by using Table 8.17 or the method to be explained in section 8.4.9.

8.3.7 Test of significance for rank correlation. For any random sample consisting of eight elementary units and their eight pairs of ranks in two variable characteristics, we find on the line for $n = 8$ in Table 8.10 that the critical value of r' under $P = 0.01$ is 0.881. This means that the probability is approximately 0.01 that a value of r' as large as or larger than $+0.881$ or as small as or smaller than -0.881 would occur in a random sample of eight elementary units from a universe in which the correlation between the ranks of the elementary units in the two variables is zero.

Consequently, any observed value of r' as large as or larger than $+0.881$ or as small as or smaller than -0.881 for eight pairs of ranks obtained from a random sample containing eight elementary units of a universe is considered to be significant at the very strong 1 per cent probability level.

Similarly, any observed value of r' as large as or larger than $+0.738$ or as small as or smaller than -0.738 for eight pairs of ranks obtained from a random sample containing eight elementary units of a universe is considered to be significant at the reasonably strong 5 per cent probability level.

Also, any observed value of r' as large as or larger than $+0.643$ or as small as or smaller than -0.643 for eight pairs of ranks obtained from a random sample containing eight elementary units of a universe is considered to be significant at the relatively weak 10 per cent probability level.

If you desire to make a one-tail test of significance for an observed value of r', all that you need do to Table 8.10 is divide the values of P by 2.

In other words, Table 8.10 shows that on the basis of eight pairs of ranks determined by a random sample of eight elementary units drawn from a universe we can feel mildly safe in rejecting the null hypothesis that the value of ρ' for the universe is zero if the absolute value of r' for the sample is as great as 0.643; we can feel reasonably safe in rejecting that null hypothesis if the absolute value of r' for the sample is as great as 0.738; and we can feel very safe in rejecting that null hypothesis if the absolute value of r' for the sample is as great as 0.881.

If the 15 teams listed in Table 8.6 were a random sample of 15 teams drawn from the universe consisting of all the college football teams in the United States that were ranked by the Associated Press at the end of that year (or some other well-defined and suitable universe of football teams)

TABLE 8.10

**Lowest Rank Correlations That Are Significantly Different
from Zero, at Three Probability Levels***

No. of Pairs of Ranks n	P = 0.10	P = 0.05	P = 0.01
	Lowest significant value of r', sign ignored †		
4	1.000		
5	.900	1.000	
6	.829	.886	1.000
7	.714	.786	.929
8	.643	.738	.881
9	.600	.683	.833
10	.564	.648	7.94
11	.520	.620	.785
12	.496	.591	.777
13	.475	.566	.744
14	.456	.544	.715
15	.438	.524	.688
16	.425	.506	.665
17	.411	.490	.644
18	.399	.475	.625
19	.388	.462	.607
20	.377	.450	.591
21	.368	.438	.576
22	.359	.428	.562
23	.351	.418	.549
24	.343	.409	.537
25	.336	.400	.526
26	.329	.392	.515
27	.323	.384	.505
28	.317	.377	.496
29	.311	.370	.487
30	.305	.364	.478

Source: This table was derived from Tables IV and V of E. G. Olds, "Distributions of Sums of Squares of Rank Differences for Small Numbers of Individuals," *Annals of Mathematical Statistics* IX (1938), pp. 133–148, "The 5% Significance Levels for Sums of Squares of Rank Differences and a Correction," *Annals of Mathematical Statistics* XX (1949), pp. 117–118, with the permission of E. G. Olds and the editor of the *Annals.*

†That is, for a two-tails test.

then we could go further and state that Mary C. would have demonstrated at least some competence or talent as a football expert if she had ranked all the football teams in that universe.

However, we are not justified in drawing any conclusions of that kind from the data in Table 8.6 because the 15 teams were not selected by a random process from any well-defined universe of teams; those 15 teams were deliberately selected because they were the teams of greatest interest to that particular group of people.

On the other hand, it was legitimate for that group of people to decide to use the rank correlation method indicated by formula 8.1 for the purpose of determining relatively how closely each person's set of ranks for the 15 teams agreed with the final Associated Press ranks for those 15 teams.

8.3.8 Adjustment of rank correlation for tied ranks. The value of r' obtained by using formula 8.1 is the same as the value that would be obtained by applying to the ranks formula 8.2 for the Pearson linear correlation coefficient (to be described later in this chapter), provided that there are no tied ranks, that is, provided that the two sets of ranks are simply the integers 1, 2, 3, . . . n, arranged in two orders.

If there are only a few tied ranks among a relatively large number of elementary units in a sample, formula 8.1 still is satisfactory. But if there are a relatively great many tied ranks, then we need to adjust formula 8.1 to make allowance for them.

An adjustment is required for each group of tied ranks. For example, if two elementary units are tied for a given rank, then the adjustment for this group of two tied ranks is $A = (2^3 - 2)/12 = 0.5$. If three elementary units are tied for a given rank, then the adjustment for this group of three tied ranks is $A = (3^3 - 3)/12 = 2$. If four elementary units are tied for a given rank, then the adjustment for this group of four tied ranks is $A = (4^3 - 4)/12 = 5$. In general, if k elementary units are tied for a given rank, then the adjustment for this group of k tied ranks is $A = (k^3 - k)/12$.

After all the adjustments have been computed for tied ranks in each variable characteristic, the sum of the adjustments is found for each variable characteristic. Let ΣA_1 be the sum of all the adjustments for tied ranks in the first variable characteristic. Let ΣA_2 be the sum of all the adjustments for tied ranks in the second variable characteristic. Then the adjusted formula for the rank correlation coefficient is:

$$r' = \frac{\dfrac{n^3 - n}{6} - \Sigma d^2 - \Sigma A_1 - \Sigma A_2}{\sqrt{\left(\dfrac{n^3 - n}{6} - 2\Sigma A_1\right)\left(\dfrac{n^3 - n}{6} - 2\Sigma A_2\right)}}$$

If there are no tied ranks among the elementary units in the sample, this formula reduces to formula 8.1, as may be proved easily by substituting zero for ΣA_1 and zero for ΣA_2 in the formula and simplifying the resulting fraction.

Here is an illustration of the way in which adjustments for tied ranks are made. Table 8.11 shows the ranks of 25 girls on a stenographic

test taken when they applied for a position, along with their ranks based
on efficiency ratings assigned to them by their supervisor at the end of
their probationary period on the job in a stenographic pool. The 25 girls
in this illustration were selected by a random sampling process from the
large number of girls hired by the organization during a period of one
year.

TABLE 8.11

**Ranks of Twenty-five Girls on a Stenographic Proficiency
Test and on Efficiency Rating as Stenographers**

Girl	Rank on Stenographic Proficiency Test	Rank on Efficiency on the Job	d	d²
A	2.5	3	−0.5	0.25
B	2.5	9	−6.5	42.25
C	2.5	3	−0.5	0.25
D	2.5	18	−15.5	240.25
E	6.5	3	3.5	12.25
F	6.5	9	−2.5	6.25
G	6.5	9	−2.5	6.25
H	6.5	18	−11.5	132.25
I	9.5	3	6.5	42.25
J	9.5	9	0.5	0.25
K	11.5	3	8.5	72.25
L	11.5	18	−6.5	42.25
M	14.5	9	5.5	30.25
N	14.5	18	−3.5	12.25
O	14.5	9	5.5	30.25
P	14.5	18	−3.5	12.25
Q	18	18	0.0	0.00
R	18	18	0.0	0.00
S	18	18	0.0	0.00
T	20.5	9	11.5	132.25
U	20.5	18	2.5	6.25
V	22	18	4.0	16.00
W	23	18	5.0	25.00
X	24	24.5	−0.5	0.25
Y	25	24.5	0.5	0.25
			0.0	862.00

As indicated by the ranks, many of the girls obtained tie scores on the
proficiency test. Also, because the efficiency ratings were given in the form
of "excellent," "above average," "average," and "below average," there
are a great many tied ranks among the efficiency ranks. If you have forgotten
how ranks are determined for people with tie scores, review section A.5.5.

In order to compute the rank correlation coefficient for the data in Table
8.11 by using the formula, we must compute first the adjustments to be
made on account of the tied ranks. The computation of these adjustments

may be organized as in Table 8.12. The results are $\Sigma A_1 = 18.5$ and $\Sigma A_2 = 148.5$. Here $n = 25$ and, consequently, $(n^3 - n)/6 = (25^3 - 25)/6 = 2600$. Table 8.11 shows that $\Sigma d^2 = 862$.

$$r' = \frac{2600 - 862 - 18.5 - 148.5}{\sqrt{(2600 - 37.0)(2600 - 297.0)}} = \frac{1571}{\sqrt{5902589}} = \frac{1571}{2429.5} = +0.65$$

If you use formula 8.1 to compute the rank correlation coefficient for Table 8.11 you will obtain the value of $+0.67$ for r'. Therefore, the effect of the adjustments for tied ranks is not very great in this illustration.

TABLE 8.12
Computation of Adjustments for Tied Ranks
in Rank Correlation for Table 8.11

For Stenographic Proficiency Test			For Efficiency Rating on the Job		
Tied Rank	No. Tied k	Adjustment $A_1 = \dfrac{k^3 - k}{12}$	Tied Rank	No. Tied k	Adjustment $A_2 = \dfrac{k^3 - k}{12}$
2.5	4	5.0	3	5	10.0
6.5	4	5.0	9	7	28.0
9.5	2	0.5	18	11	110.0
11.5	2	0.5	24.5	2	0.5
14.5	4	5.0			
18	3	2.0			
20.5	2	0.5			
		$\Sigma A_1 = 18.5$			$\Sigma A_2 = 148.5$

Now, compare the rank correlation coefficient $r' = 0.65$ with the critical values of r' for $n = 25$ in Table 8.10. You see that 0.65 is greater than 0.526, the critical value of r' at the 0.01 probability level when $n = 25$. Therefore, it is clear that the correlation between rank on the stenographic proficiency test and rank on efficiency on the job is highly significant. In other words, we can reject at the 1 per cent probability level the null hypothesis that $\rho' = 0$. If the null hypothesis were true, there would be no benefit for the organization in administering the stenographic proficiency test to applicants for jobs.

Having decided that the rank correlation in this illustration is highly significant, let us consider the following pertinent question: Is this stenographic proficiency test a highly accurate predictor of the relative degree of success that a girl will attain on the job in this organization? Or do there appear to be factors not covered by the stenographic proficiency test that play an important part in determining the efficiency rating that a girl receives on the job as a stenographer in this office?'

The answer to the first part of the question is "No," because the value of r' here is only 0.65, which is not very near to 1.00. The answer to the second part of the question is "Yes," because matters such as tardiness, absence from work, and many other factors that may affect the efficiency rating on the job are not taken into consideration by the stenographic proficiency test. Nevertheless, this proficiency test may provide some worthwhile information for both the organization and the applicant in evaluating the prospects of the applicant.

8.3.9 Remarks. The rank correlation coefficient for two variable characteristics of the elementary units in a random sample usually is not as accurate an estimate of the strength of the relationship between the two variable characteristics for all the elementary units in the universe as the estimate provided by the Pearson linear correlation coefficient that is to be studied in the following sections of this chapter. The reason for this is that the rank correlation method takes into account only the rank order of the elementary units with respect to the variable characteristics and disregards the actual magnitudes of the values of the variable characteristics for each elementary unit. In other words, the Spearman rank correlation coefficient requires only partly quantitative data, but the Pearson linear correlation coefficient requires truly quantitative data.

However, the Spearman rank correlation coefficient usually is considered to be easier to compute than the Pearson linear correlation coefficient if n is less than 40. Also, the rank correlation coefficient is often the only available estimate of the strength of the relationship, because nothing is known about the variable characteristics of the elementary units except the ranks of the elementary units in a sample with respect to the two variable characteristics. For example, supervisors of employees usually think that it is much easier to indicate the partly quantitative rank order of the personnel in their departments than to determine truly quantitative numerical values for the efficiency of the employees.

A noteworthy theoretical characteristic of the rank correlation method is that we may apply the test of significance to it and draw conclusions about the characteristics in the universe legitimately without making any assumptions about the shape of the joint distribution of the two variable characteristics in the universe from which the random sample of elementary units was drawn. In other words, valid inferences about the rank correlation ρ' between the two variable characteristics of the elementary units in a universe may be derived from the rank correlation coefficient for a random sample drawn from the universe, regardless of whether or not the joint distribution of the two variable characteristics in the universe is a normal distribution.

EXERCISES

1. According to Table 8.7, how many baseball games were played by each team

in the National League during the year? (*Ans.* 160 to 165.) Show how to determine the "proportion of games won" for each team. Find the value of the rank correlation coefficient ρ' for final team standing and club batting average. (*Ans.* $\rho' = +0.773$.) Find the value of the rank correlation coefficient ρ' for final team standing and number of home runs. (*Ans.* $\rho' = +0.55$.) On the basis of the two values of ρ' obtained here, which of two teams seems likely to have the higher final standing: (a) the team that has the higher club batting average, or (b) the team that hits more home runs? (*Ans.* a.)

2. Compute the value of ρ' for the proportion of games won and team batting average given in Table 8.7, using the adjustment for tied ranks. (*Ans.* $\rho' = +0.772$.)

3. Table 8.13 shows the ranks of a random sample of 20 insurance salesmen on a personality test that a large number of salesmen took when they applied for jobs with a particular insurance company during a particular year, as well as their ranks based on the amount of life insurance each of the 20 men sold during the first year after he was hired. If you have forgotten how ranks are determined for people with tie scores, review section A.5.5. Compute the rank correlation coefficient for these data. Test the significance of the value of r' observed here. (*Ans.* $r' = +0.84$, which is highly significant for $n = 20$, that is, 0.84 is higher than the lowest value of r', namely, 0.591, which is significant at 1 per cent probability level for $n = 20$, according to Table 8.10.)

TABLE 8.13

**Ranks of Twenty Salesmen on a Personality Test and on the
Amount of Insurance Each Sold in His First Year**

Salesman	Rank on Personality Test	Rank on Insurance Sold in First Year
A	1	5
B	2	2.5
C	3	2.5
D	4	7
E	5.5	1
F	5.5	9
G	7	6
H	8	4
I	9	14
J	10	8
K	11	12
L	12	15
M	13	10
N	14	11
O	16	17
P	16	20
Q	16	19
R	18	17
S	19	17
T	20	13

Do you think that this particular personality test gives the insurance company a good indication of the relative success men will have as insurance salesmen for the company? Do you feel that the personality test gives each new applicant a reasonably square deal insofar as indicating his ability as an insurance salesman is concerned?

8.4 LINEAR CORRELATION ANALYSIS

In order to illustrate for you how data suitable for linear correlation analysis may arise, consider Table 8.14. The table shows alphabetical symbols for 42 students (the elementary units) and the values of the intelligence quotient (the variable x) and the grade point average (the variable y) for each of the 42 students.[2] A committee of students decided that their statistical investigation would be to find out how much relationship, if any, there was between the intelligence quotients of the students in their college and the grades they received in their courses.

TABLE 8.14

Intelligence Quotients and Grade Point Averages for 42 College Seniors

Student	IQ x	G.P.A. y	Student	IQ x	G.P.A. y
A	129	2.60	V	111	1.27
B	127	2.18	W	110	0.97
C	125	2.27	X	109	1.61
D	125	2.25	Y	109	0.96
E	124	1.98	Z	108	1.75
F	123	2.38	AA	108	1.12
G	123	1.84	BB	107	2.09
H	123	1.72	CC	107	1.51
I	122	2.26	DD	106	1.16
J	120	1.39	EE	106	1.08
K	119	1.99	FF	105	1.88
L	118	2.17	GG	105	1.39
M	118	1.67	HH	104	1.30
N	117	2.54	II	104	1.30
O	116	2.17	JJ	104	0.99
P	116	2.07	KK	102	1.85
Q	113	2.06	LL	101	1.07
R	112	2.44	MM	100	1.12
S	112	1.50	NN	99	1.01
T	111	2.17	OO	97	1.13
U	111	1.85	PP	96	1.49

[2] Grade point average (G.P.A.) is an average of all the marks of a student during his or her four years in the college. At this particular college, the G.P.A. was based on three points for each semester hour of a course in which the grade A was obtained, two points for each hour of B grade, one point for each hour of C grade, and 0 points for each semester hour of a course in which a grade below C was obtained. Many colleges use four points for A, three points for B, two points for C, one point for D, and 0 points for each semester hour of a course in which a grade below D is obtained.

First, they obtained a list of the names of all the students who were in the senior class during the previous school year. There were 420 names in the list. The students assigned the numbers 1, 2, 3, . . . , 418, 419, 420 to the names. Next they decided to anlayze the data for a sample of only 42 students. They used a table of random digits similar to Table A.15 to select 42 random numbers in the range from 1 to 420. The 42 random numbers determined in this way indicated the 42 names to be used as the sample.

Their teacher then took the list of 42 randomly selected names to the Registrar of the college and asked for the IQ and G.P.A. for each of the 42 students whose names were in the list. The resulting data are shown in Table 8.14, except that letters of the alphabet are used in place of names to designate the students.

How much linear correlation, if any, is there between a student's IQ and his or her academic success in this particular college? If one student has an IQ 10 per cent higher than another student's IQ, is it likely that the first of these two students will obtain a G.P.A. 10 per cent higher than the other student's G.P.A.?

If the linear correlation between IQ and G.P.A. for this universe of 420 students were practically perfect, then it would be possible to advise students in high schools as to how much academic success they could expect in this college. In that case, one could give each high school student an intelligence test, and on the basis of his IQ predict his G.P.A. for this college. Usually, it turns out that the correlation between IQ and G.P.A. is too low to be satisfactory for the purpose of prediction.

In order to shorten and simplify for you the illustration of all the details of the computations that are involved in linear correlation analysis, let us work now with the sample of only 12 students listed in the first column of Table 8.15.

8.4.1 Scatter diagrams. One of the best ways in which to discover and to present the relationship, if any, between two variable characteristics of the elementary units in a sample or in a universe is by making a graph for the data. Use one of the two variables as abscissa and the other variable as ordinate. That is, measure one of the variables along the horizontal scale or axis of the graph and plot the other variable parallel to the vertical scale or axis of the graph. The pair of values of the two variables corresponding to each elementary unit determines a point or dot on the graph. When all the points or dots have been plotted on the graph, you have what usually is called a *scatter diagram* or a dot diagram.

Figure 8.1 is a scatter diagram for the 12 students in Table 8.15. The pattern of the dots indicates that there was some tendency for the students with high intelligence quotients to obtain high grade point averages in the college in question and for the students with low intelligence quo-

TABLE 8.15

**Tabulation for Computing the Pearson Correlation
Coefficient by the Longest Method**

Student	IQ x	G.P.A. y	z_x	z_y	$z_x z_y$
T	111	2.17	0.046	0.671	0.031
N	117	2.54	0.710	1.407	0.999
M	118	1.67	0.821	−0.324	−0.266
BB	107	2.09	−0.397	0.511	−0.203
U	111	1.85	0.046	0.034	0.002
HH	104	1.30	−0.729	−1.061	0.773
A	129	2.60	2.039	1.526	3.112
I	122	2.26	1.264	0.850	1.074
LL	101	1.07	−1.061	−1.518	1.611
PP	96	1.49	−1.614	−0.683	1.102
FF	105	1.88	−0.618	0.094	−0.058
EE	106	1.08	−0.507	−1.499	0.760
Total	1327	22.00			8.937

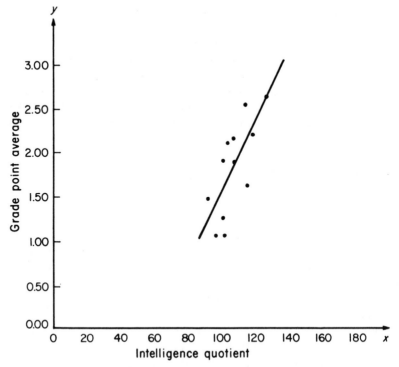

Fig. 8.1.   Scatter Diagram for the Data In Table 8.15

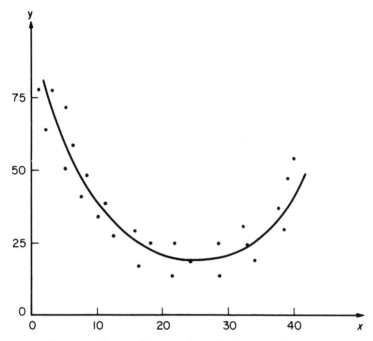

Fig. 8.2. Scatter Diagram for a Nonlinear Relationship

tients to obtain low grade point averages in that college. A straight line has been drawn through the pattern of dots in the scatter diagram to indicate approximately the trend of the relationship between intelligence quotient and grade point average for this sample. Because a straight line appears to be a better representation of the trend of the relationship between the variables x and y in Figure 8.1 than any smooth curve would be, we shall assume that the relationship between the two variables x and y in Figure 8.1 and Table 8.15 is a linear relationship.

Figure 8.2 is a scatter diagram in which the pattern of dots indicates that the trend of the relationship between the two variables is much better represented by a smooth curve than by a straight line. In this case, the relationship between the two variables x and y is said to be a nonlinear relationship or a curvilinear relationship.

Other illustrations of scatter diagrams may be seen in Figures 8.3, 8.4, 8.5, and 8.6.

8.4.2 The Pearson linear correlation coefficient. The scatter diagram in Figure 8.1 shows no evidence of a curved trend line for the relationship between a student's IQ and G.P.A. If the relationship between the two variables x and y in Table 8.15 were such that all the 12 points of the scatter diagram fell exactly on a straight line, we would feel that the linear relationship between the values of x and y in the sample was very definite and very strong. If such a perfect linear relationship existed in the universe, then we could show by using similar triangles that an increase in the value of x always would be accompanied by a proportionate increase in the value of y if the straight line sloped upward to the right. Similarly, an increase in x always would be accompanied by a proportionate decrease in y if the straight line sloped downward toward the right.

Consequently, the strength of a linear relationship corresponds to the degree to which the variation of one variable is accompanied by or associated with proportionate variation of the other variable.

If two variables x and y increase at the same time, their product xy also will increase. If the two variables x and y decrease at the same time, their product xy also will decrease. These facts suggest that a good way to study the simultaneous variation of two variables, that is, to measure their covariation or correlation, is to study the products xy of corresponding pairs of values of the two variables.

Because one of the two variables in a correlation analysis might be measured in one kind of unit, such as inches, and the other variable might be measured in another kind of unit, such as pounds, let us change the values of both variables into standard units before the multiplication takes place. In order to use formula 3.5 for this purpose, we need to know the values of the mean and the standard deviation for each of the two variables. For the values of x and y in columns two and three of Table 8.15, $m_x = 110.583$, $m_y = 1.833$, $s_x = 9.033$, and $s_y = 0.5025$.

Columns four and five of Table 8.15 show the results of changing the sample values of x and y into standard units. The first number in column four was computed as follows: $z_x = (111 - 110.583)/9.033 = 0.417/9.033 = 0.046$. Similarly, the first number in the fifth column was found by $z_y = (2.17 - 1.833)/0.5025 = 0.337/0.5025 = 0.671$. The other numbers in columns four and five were found in the same way.

In the 1890's, Karl Pearson, an English statistician, defined the linear correlation coefficient for a sample as the sum of the pair products of the two variables, in standard units, divided by the number of pairs of values of the two variables in the sample. That is, by definition, $r = (\Sigma z_x z_y)/n$. Consequently, for the 12 pairs of values of x and y in Table 8.15, the value of the Pearson linear correlation coefficient is $r = 8.937/12 = 0.7448 = 0.74$ to two decimal places.

The small letter r is reserved in most statistics books as a symbol for the Pearson linear correlation coefficient computed for a sample. The small Greek letter ρ (rho) will be used as a symbol for the Pearson linear correlation coefficient of two variables for all the elementary units in a universe.

Sometimes more than two variables are of interest. Subscripts may be attached to the symbols r and ρ whenever it is desirable to indicate specifically which two of several variables are involved in a correlation. For example, if three variables x, y, and w were involved in a situation, then it would be possible to compute three correlation coefficients, namely, r_{xy}, r_{xw}, and r_{yw}. Similarly, for the universe we would have the three correlation coefficients ρ_{xy}, ρ_{xw}, and ρ_{yw}.

It is easy to show that the formula given above for r may be written in the equivalent form $r = [\Sigma(x - m_x)(y - m_y)]/(ns_xs_y)$. However, neither of these two formulas is of much practical importance because each of them involves too much computational labor.

8.4.3 A shorter method for computing r. Most of the labor involved in computing r for Table 8.15 occurs in changing the original values of x and y into standard units. It is easy to prove by elementary algebra that the correct value of r is obtained if we disregard the units of measurement in x and y, that is, treat the values of x and y as if they were abstract numbers, and use formula 8.2.

$$r = \frac{n\Sigma xy - \Sigma x\Sigma y}{\sqrt{[n\Sigma x^2 - (\Sigma x)^2][n\Sigma y^2 - (\Sigma y)^2]}} \tag{8.2}$$

TABLE 8.16

Tabulation for Computing the Pearson Correlation Coefficient by the Shorter Method

Student	IQ x	G.P.A. y	x²	y²	xy
T	111	2.17	12321	4.7089	240.87
N	117	2.54	13689	6.4516	297.18
M	118	1.67	13924	2.7889	197.06
BB	107	2.09	11449	4.3681	223.63
U	111	1.85	12321	3.4225	205.35
HH	104	1.30	10816	1.6900	135.20
A	129	2.60	16641	6.7600	335.40
I	122	2.26	14884	5.1076	275.72
LL	101	1.07	10201	1.1449	108.07
PP	96	1.49	9216	2.2201	143.04
FF	105	1.88	11025	3.5344	197.40
EE	106	1.08	11236	1.1664	114.48
Total	1327 $= \Sigma x$	22.00 $= \Sigma y$	147723 $= \Sigma x^2$	43.3634 $= \Sigma y^2$	2473.40 $= \Sigma xy$

Table 8.16 shows the tabulations needed for the shorter method of computing the value of r. The following arrangement is a convenient one for carrying out the calculation of r by formula 8.2.

$$
\begin{array}{llll}
n\Sigma xy & = 12(2473.40) & = 29680.80 \\
\Sigma x\Sigma y & = 1327(22.00) & = 29194.00 \\
\hline
n\Sigma xy - \Sigma x\Sigma y & & = 486.80
\end{array}
$$

$$
\begin{array}{llll}
n\Sigma x^2 & = 12(147723) & = 1772676 \\
(\Sigma x)^2 & = (1327)^2 & = 1760929 \\
\hline
n\Sigma x^2 - (\Sigma x)^2 & & = 11747
\end{array}
$$

$$
\begin{array}{llll}
n\Sigma y^2 & = 12(43.3634) & = 520.3608 \\
(\Sigma y)^2 & = (22.00)^2 & = 484.0000 \\
\hline
n\Sigma y^2 - (\Sigma y)^2 & & = 36.3608
\end{array}
$$

$$(11747)(36.3608) = 427130.3176$$
$$\sqrt{427130.3176} = 653.55$$
$$r = 486.80/653.55 = +0.7449 = +0.74$$

For most of the situations in which we shall deal with the Pearson linear correlation coefficient in this book, two decimal places in the value of r will be sufficient. That is, we will compute the value of r to three or four decimal places and then round off the result to two decimal places.

8.4.4 Other short cuts. There are other short cuts that frequently simplify the work of computing the Pearson linear correlation coefficient. For example, the computation of r for the data in Table 8.15 can be shortened by first subtracting 111 from each value of x and subtracting 1.88 from each value of y. Then construct a table similar to Table 8.16. The numbers in the last three columns of the new table will be considerably smaller and easier to handle in computations than the numbers in the last three columns of Table 8.16. Sometimes the work of computing r can be simplified by first multiplying (or dividing) all the values of x or of y by a positive constant.

8.4.5 A rough indication of the value of r. It is a good idea in many problem situations to guess what would be a reasonable answer before beginning the computations. Then, when the computations have been completed, compare the answer with the guess. If the result of the computations is radically different from the guess, check carefully for errors in the computations.

An easy way to obtain a rough indication of the value of r for a sample is to inspect the scatter diagram for the data. Draw a smooth ellipse, (an oval-shaped curve) so as to enclose all, or nearly all, of the points in the scatter diagram. Do not make the oval any larger than necessary. If the ellipse is long and narrow, as in Figure 8.3, the value of r is high. If the ellipse is nearly as broad as it is long, as in Figure 8.4, the value of r is near zero. If the ellipse is about twice as long as it is wide, as in Figure 8.5, the correlation is of medium strength, that is, the value of *r* probably is somewhere in the range from 0.40 to 0.60.

Furthermore, if the longer axis of the ellipse slopes upward toward the right-hand side of the graph, as in Figures 8.3 and 8.5, the correlation appears to be positive. If the longer axis of the ellipse slopes downward toward the right-hand side of the graph, the correlation will be expected to be negative. If the longer axis of the ellipse is horizontal, that is, parallel to the *x*-axis, the correlation will be expected to be zero or very near zero, as in Figure 8.6.

Why are these statements true? First of all, if the linear correlation

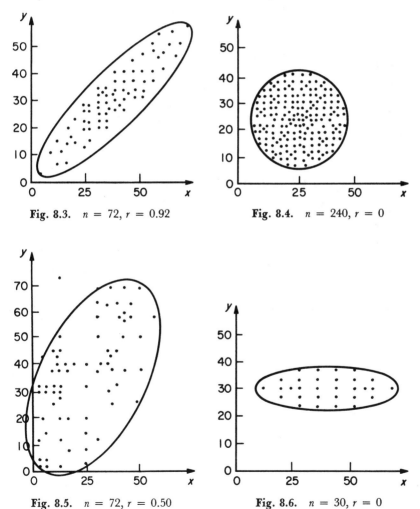

Fig. 8.3. $n = 72, r = 0.92$ **Fig. 8.4.** $n = 240, r = 0$

Fig. 8.5. $n = 72, r = 0.50$ **Fig. 8.6.** $n = 30, r = 0$

between the two variables were perfect, that is, if $r = +1$ or $r = -1$, all the points of the scatter diagram would fall exactly on a straight line and you could draw a straight line through all the points. That would be an ellipse that had collapsed to zero width. If all the points lie very near, but not exactly on, a straight line that is not horizontal and is not vertical, the correlation will be nearly but not exactly $+1$ or -1, and the oval that you can draw around the points will be long and very narrow; the longer axis of the oval will be neither parallel to nor perpendicular to the horizontal axis or scale of the graph.

As the points of the scatter diagram spread out more and more away from a straight line, the value of r becomes lower and lower, that is, nearer to zero, and the ellipse that you can draw around the points becomes wider and wider relative to its length. When an ellipse is as broad as it is long, the curve is a circle. When the curve that seems to be best for surrounding the points of the scatter diagram is approximately a circle, then the value of r is approximately zero.

After you have had some experience with scatter diagrams and correlations, you probably will be able to make fairly good guesses about the amount of correlation in a sample if the number of dots in the scatter diagram is fairly large.

Of course, the scatter diagram serves another very important purpose, namely, indication of whether or not a straight line is a satisfactory representation of the trend of the relationship between the two variables.

8.4.6 Remarks about the value of r. Because perfect positive linear correlation is indicated by $r = +1$, the linear correlation between IQ and G.P.A. in Table 8.15 cannot be said to be very high. The correlation coefficient $r = +0.74$ suggests that there is some relationship between the values of the two variables for the 12 students who happened to be in the sample, but it indicates that the changes in corresponding values of the two variables are not closely proportionate. This is merely another way of saying that the 12 points in the scatter diagram in Figure 8.1 do not fall close to any specific straight line.

The value of r in this illustration is positive. In the long, direct method, the value of r turned out to be positive because $\Sigma z_x z_y$ for Table 8.15 is positive. But notice that some of the products $z_x z_y$ in the last column of Table 8.15 are negative. It is possible for situations to exist in which $\Sigma z_x z_y$ will be negative. Whenever that happens, the value of the correlation coefficient r will be negative. A negative value of r will be obtained whenever the relationship between the two variables is such that one of the variables tends to decrease when the other variable increases. The trend line representing the relationship in the scatter diagram will slope downward toward the right-hand side of the graph when the value of r is negative.

The positive value $r = +0.74$ obtained for the data in Table 8.15 indicates that there was at least a slight tendency for the students with high intelligence quotients to receive high marks in college, and for students with low intelligence quotients to receive low marks in college. But, in spite of this over-all tendency for the sample, there were some students with fairly high intelligence quotients who received rather low grade point averages, and some students in the sample had relatively low intelligence quotients but received fairly good marks in college.

It can be proved by algebra that the value of r never can be less than -1 nor greater than $+1$. The value $r = -1$ indicates perfect negative linear correlation between the values of two variables for the elementary units in a sample. To see for yourself an illustration of perfect negative correlation in a universe, construct a scatter diagram for the data in the two columns "No. of Games Won" and "No. of Games Lost" in Table 8.7. Then compute the Pearson linear correlation coefficient ρ for the pairs of values of the variables in those two columns.

8.4.7 Reliability of r as an estimate. Suppose that (1) the joint distribution of the pairs of values of two variable characteristics of all the elementary units in a universe is a normal distribution, (2) a straight line is a good representation of the trend of the relationship between the two variables in the universe, and (3) a random sample of the elementary units is drawn from the universe by a method in which no control or restriction is placed on the values of either of the two variables that may be in the sample. Then (1) a valid estimate of the strength of the linear correlation between the two variables in the universe can be obtained by computing the value of the Pearson linear correlation coefficient for the sample, and (2) the reliability of the estimate can be determined by computing confidence limits or by applying a test of significance to the estimate.

You must be prepared to allow for some sampling error in any observed value of r as an estimate of the true value of ρ. If you drew a large number of small samples from a large normal universe in which there is fairly high correlation ρ between the two variables, and computed the value of r for each sample, the frequency distribution of the values of r likely would be so unsymmetrical or skewed that you would not be willing to consider it as even approximately a normal distribution. See, for example, the frequency curve for $\rho = 0.80$ in Figure 8.7.

The mathematical equations for these frequency distributions are very complicated and the probabilities cannot be tabulated as simply as those for the normal distribution. Therefore, we usually do not determine confidence limits to be attached to a value of r. Instead, we frequently are content to employ a simple test to see whether or not a value of r obtained from a sample is significantly different from zero.

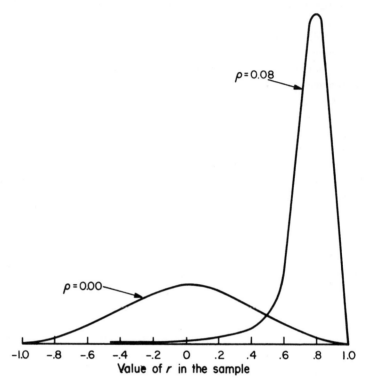

Fig. 8.7.   Sampling Distributions of *r* for Samples of 8 Pairs of Observations Drawn from two Universes Having the Indicated Values of *ρ*

Mathematicians have proved that for random samples from a normally distributed universe in which the value of *ρ* is zero, the distribution of the values of *t* computed by formula 8.3 is the "Student" distribution in Table 7.3.

$$t = r\sqrt{\frac{n-2}{1-r^2}}, \qquad n' = n - 2 \tag{8.3}$$

In section 8.4.3, we obtained *r* = +0.74 for the linear correlation between IQ and G.P.A. for a sample containing 12 students. Our null hypothesis here is that *ρ* = 0 in the normally distributed universe from which the sample was drawn. Let us make the test at the *α* = 0.01 significance level. In Table 7.3 on the line for *n'* = 12 − 2 = 10 under *P* = 0.01 we find that the critical values of *t* for this two-tails test are *t* = ±3.169. The computed value of *t* for the sample is $t = +0.74\sqrt{(12-2)/(1-0.74^2)} = +3.478$ which is greater than *t* = +3.169 and therefore we reject the null hypothesis that *ρ* = 0.

Notice that the null hypothesis here consists of two parts: (1) the joint distribution of the two variables in the universe is a normal distribution; (2) the linear correlation between the two variables for all the elementary units in the universe is zero. Frequently, in practical situations, statisticians assume that the joint distribution in the universe is either normal, or at least approximately normal. Then they reject part (2) of the null hypothesis when the probability corresponding to t computed by formula 8.3 turns out to be small. Failure to accept part (1) of the null hypothesis in any given situation would mean that the test of significance based on formula 8.3 and Table 7.3 is not applicable.

In many practical situations, all we need to know is whether or not it is reasonably safe to reject the null hypothesis that the value of ρ is zero and accept the alternative that ρ is not zero, assuming that the universe is a normal or near normal bivariate distribution. In order to enable us to carry out this test quickly and easily, Table 8.17 has been constructed. It contains the lowest values of r that may be considered significantly different from zero for various probability levels and for various values of n.

Consider again our example in which $r = +0.74$ and $n = 12$. In Table 8.17, on the line for $n = 12$, in the column under $P = 0.01$, we find the number 0.71. The interpretation of the situation here is: There is only 1 chance in 100 that a value of r as large as $+0.71$ or as small as -0.71 would occur by pure chance in drawing a random sample of 12 elementary units from a normal universe in which the value of ρ actually is zero. In other words, 0.71 is the smallest positive value of r that can be considered significant at the 1 per cent probability level for 12 pairs of values of two variables.

Because the observed value $+0.74$ of r for the 12 students in Table 8.15 is greater than $+0.71$, we conclude that the value of r computed for the sample is significantly different from zero at the 1 per cent probability level, and we reject, at the 1 per cent probability level, the null hypothesis that the true linear correlation between IQ and G.P.A. is zero for the universe of students from which the sample of 12 students was drawn. Furthermore, because $r = +0.74$ is positive as well as significant, we conclude that ρ is positive.

Now, suppose that in another problem you obtain $r = +0.63$ for 12 pairs of values. Because this value of r is higher than the value given on the line for $n = 12$ in the column under $P = 0.05$, but lower than the value given on the line for $n = 12$ in the column under $P = 0.01$, you should conclude that the observed value $r = +0.63$ is significantly different from zero at the 5 per cent probability level but not significantly different from zero at the 1 per cent probability level.

Similarly, if for 12 elementary units the value of r turns out to be -0.55, then we should not reject the null hypothesis that $\rho = 0$ at either the 1

TABLE 8.17

**Lowest Linear Correlations That Are Significantly Different
from Zero, at Three Probability Levels***

No. of Pairs of Values of the Variables n	P = 0.10	P = 0.05	P = 0.01
	Lowest significant value of *r*, sign ignored†		
3	.99	1.00	1.00
4	.90	.95	.99
5	.81	.88	.96
6	.73	.81	.92
7	.67	.75	.87
8	.62	.71	.83
9	.58	.67	.80
10	.55	.63	.76
11	.52	.60	.73
12	.50	.58	.71
13	.48	.55	.68
14	.46	.53	.66
15	.44	.51	.64
16	.43	.50	.62
17	.41	.48	.61
18	.40	.47	.59
19	.39	.46	.58
20	.38	.44	.56
21	.37	.43	.55
22	.36	.42	.54
27	.32	.38	.49
32	.30	.35	.45
37	.27	.32	.42
42	.26	.30	.39
47	.24	.29	.37
52	.23	.27	.35
62	.21	.25	.32
72	.20	.23	.30
82	.18	.22	.28
92	.17	.21	.27
102	.16	.19	.25

*Adapted from Table VA of R. A. Fisher, *Statistical Methods for Research Workers*, published by Oliver & Boyd Ltd., Edinburgh, by permission of the author and publishers.

†That is, for a two-tails test.

per cent probability level or the 5 per cent probability level. But we could reject the null hypothesis at the rather weak 10 per cent probability level.

Table 8.17 enables us to decide whether to regard an observed value of *r* as a safe indication that the true value of ρ for the universe is different from zero or as a purely fictitious value that might easily have happened because of sampling error in random sampling from a normal universe in which the true value of ρ is zero.

8.4.8 Some cautions. The procedures given in section 8.4.7 for testing the null hypothesis that $\rho = 0$ cannot be used to test other hypotheses about the value of ρ, for example, the hypothesis that for a particular universe the value of ρ is $+0.5$. Nor can those procedures be used to test the significance of the difference between two observed values of *r*. Tests are available for these purposes, but they will not be discussed in this book.

If the two variables *x* and *y* for all the elementary units in a universe are completely independent of each other, then the value of ρ necessarily is zero. However, if ρ is zero, it does not necessarily follow that the two variables *x* and *y* for all the elementary units in a universe are completely independent of each other; $\rho = 0$ means only that there is no linear relationship between the values of *x* and *y*. It is possible that in a particular universe in which there is no linear relationship between the values of *x* and *y*, a strong curvilinear relationship might exist.

For example, consider the seven pairs of values of *x* and *y* in Table 8.18 and the scatter diagram consisting of the seven corresponding points in Figure 8.8. The scatter diagram shows that the trend of the relationship between the variables *x* and *y* cannot be represented satisfactorily by a straight line. But the trend can be represented exceedingly well by a smooth curve. Consequently, the Pearson linear correlation coefficient should not be computed for such a set of values of *x* and *y*.

If you compute the value of the linear correlation coefficient for the seven pairs of values of *x* and *y* that are given in Table 8.18, you will find that $r = 0$. But the values of *x* and *y* are not independent of each other here. They are related exactly by the equation $y = \sqrt{25 - x^2}$ and the seven points of the scatter diagram lie exactly on the circumference of a semicircle, as indicated in Figure 8.8. In other words, although there is no linear relationship between the values of *x* and *y* in this illustration, there is a perfect circular relationship between them.

The only requirements for legitimate computation of the value of *r* for a sample are (1) that the trend of the relationship between the two variables in the universe is a straight line, and (2) that the values of the variable in the sample are truly quantitative; it is not necessary that the joint distribution of the two variables in the universe be normal. The assumption that the joint distribution in the universe is normal enters into the analysis only at the time that we compute confidence limits or apply a test of significance to the observed value of *r*. It enters the analysis here because the mathematical theory on which the above tests of significance are based

TABLE 8.18

Nonlinear Relationship

Elementary Unit	Values of the Variables	
	x	y
A	−5	0
B	−4	3
C	−3	4
D	0	5
E	3	4
F	4	3
G	5	0

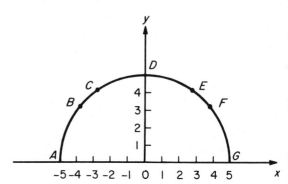

Fig. 8.8.   Nonlinear Relationship

involves the assumption that the universe from which the sample was drawn is a normal distribution.

If all you know about the universe is the information contained in one small random sample, there may be considerable risk involved in making the assumption that the joint distribution of the two variables in the universe is normal. Statisticians who have worked in a particular field for a long time often can decide on the basis of past experience whether or not it is reasonable to assume that the joint distribution of the two variables in the universe they wish to investigate is at least approximately normal.

One of the important properties of the rank correlation method of analysis is that the test of significance given for the rank correlation coefficient does not require any assumption about the shape of the joint distribution of the two variables in the universe. Consequently, if there is danger that the joint distribution may be far from normal and if it is desired to test the significance of the observed correlation coefficient, it may be advisable to use the Spearman rank correlation method rather than the Pearson linear correlation method.

EXERCISES

1. What is the value of the Pearson linear correlation coefficient for the proportion of games won and the number of games won by the baseball teams in Table 8.7? (*Ans.* $\rho = +1$.) What is the value of the Pearson linear correlation coefficient for the proportion of the games won and the number of games lost? (*Ans.* $\rho = -1$.)

2. Subtract 111 from each value of x in Table 8.15 and subtract 1.88 from each value of y. Let $X = x - 111$ and $Y = y - 1.88$. Then treat the numbers in the X column and the Y column as if they were the original data. Construct a new table similar to Table 8.16. Compute the Pearson linear correlation coefficient for X and Y by using formula 8.2 with x replaced by X and y replaced by Y. The resulting value of r here should be exactly the same value of r that was obtained from Table 8.16.

3. Generalize exercise 2. In other words, prove algebraically that the Pearson linear correlation coefficient for two variables x and y is identical in value with the Pearson linear correlation coefficient for the two variables X and Y if $X = x + c$ and $Y = y + k$, where c and k are any two constants. The constants c and k may be positive, negative, or zero.

4. Prove algebraically that the linear correlation coefficient for two variables x and y is identical in value with the linear correlation coefficient for the two variables X and Y if $X = cx$ and $Y = ky$, where c and k are any positive constants. What happens if one of the two constants is positive and the other is negative? What happens if both c and k are negative?

5. Ten pupils selected at random from all the pupils in a very large school took an intelligence test constructed by Otis. A short time later the same ten pupils took an intelligence test constructed by Terman. The scores of the ten pupils on both tests are given in Table 8.19. Construct a scatter diagram for these data. Compute the value of the linear correlation coefficient for these data. (*Ans.* $r = +0.91$.) What conclusion do you draw about the significance of this value of r by comparing it with the critical values of r given in Table 8.17? (*Ans.* Significant at the 1 per cent significance level.) You can reduce considerably the amount of computational labor involved in finding the value of r in this problem if you use some of the short cuts mentioned earlier in these exercises.

TABLE 8.19

Scores of Ten Pupils on Otis and Terman Intelligence Tests

Pupil	Score on Otis Test	Score on Terman Test
A	149	135
B	114	100
C	135	125
D	131	120
E	161	149
F	95	87
G	134	103
H	124	119
I	125	95
J	167	178

Assign ranks to the ten pupils in Table 8.19 for each test. Compute the value of the rank correlation coefficient for the two tests. (*Ans.* $r' = +0.92$.) For how many of these pupils is their rank on the Otis test the same as their rank on the Terman test? (*Ans.* Five.) Do you feel that insofar as ranking pupils in the order of their intelligence is concerned these two tests are practically equivalent?

6. Construct a scatter diagram for the data in Table 8.14. Sketch with a free hand an ellipse or oval enclosing all or almost all of the points in the scatter diagram. By comparing the length of this ellipse or oval with its width, make a rough guess or estimate of the strength of the correlation between a student's intelligence quotient and his or her grade point average in that college. Compute the value of the Pearson linear correlation coefficient for the data in Table 8.14. (*Ans.* $r =$ +0.70.) How closely does the computed value of r agree with your rough guess or estimate? Use Table 8.17 to test the significance of the value of r obtained here.

Compare the value of r computed for the data in Table 8.14 with the value of r obtained in sections 8.4.2 and 8.4.3 for the subsample of 12 students listed in Table 8.15. Does Table 8.17 indicate that the value of r you obtained from the data for the 42 students in Table 8.14 has greater significance than the value of r obtained for the 12 students in Table 8.15?

7. Assuming that the universe is normal and that the n elementary units in each sample are drawn by a strictly random sampling process, what conclusion do you draw about the null hypothesis that $\rho = 0$ in each of the sampling situations described in the following table? Apply Table 8.17.

No. of Pairs of Values n	Observed Correlation r	Conclusion About the Null Hypothesis that $\rho = 0$
10	+0.82	Reject it at the .01 level
21	+0.46	Reject it at the .05 level
37	+0.25	Do not reject it
72	−0.21	Reject it at the .10 level
92	−0.24	Reject it at the .05 level

8. Show that the value of the linear correlation coefficient r obtained by applying formula 8.2 to Table 8.6 with the ranks considered as the values of the variables x and y is the same as the value of the rank correlation coefficient r' that was obtained by applying formula 8.1 to Table 8.6 in section 8.3.1.

9. Show that the value of the linear correlation coefficient r obtained by applying formula 8.2 to Table 8.11 with the ranks considered as the values of the variables x and y is the same as the value of the rank correlation coefficient r' that was obtained in section 8.3.8.

10. Show that the value of the linear correlation coefficient r obtained by applying formula 8.2 to Table 8.13 with the ranks considered as the values of the variables x and y is the same as the value of the rank correlation coefficient r' obtained in exercise 3 on p. 259.

11. Prove that $\dfrac{\Sigma z_x z_y}{n} = \dfrac{n\Sigma xy - \Sigma x \Sigma y}{\sqrt{[n\Sigma x^2 - (\Sigma x)^2][n\Sigma y^2 - (\Sigma y)^2]}}$

12. Prove that an equivalent formula for ρ is

$$\rho = \frac{N\Sigma xy - \Sigma x \Sigma y}{N^2 \sigma_x \sigma_y} \tag{8.4}$$

REFERENCES

Maurice G. Kendall, *Rank Correlation Methods*. London: Charles Griffin & Co., Ltd., 1948.

M. H. Quenouille, *Associated Measurements*. New York: Academic Press, Inc., 1952.

Herbert A. Simon, "Spurious Correlation: A Causal Interpretation," *Journal of the American Statistical Association* 49 (1954), pp. 467–479.

Chapter 9

LINEAR PREDICTION EQUATIONS

9.1 AN ILLUSTRATION

To understand the way in which data for prediction equations often arise and where variations due to chance in random sampling enter into the data in a typical prediction problem, let us examine a specific situation.

For our universe in this example we will use the distribution in Table 6.24 consisting of 120,000 white shirts for men. You may think of the universe as consisting of 120,000 men, each man wearing a white shirt with a particular sleeve length and a particular neck size. The elementary unit in this universe is, then, a white shirt or a man wearing a white shirt.

The principal purposes of the prediction analysis here are (1) to determine whether or not a linear relationship of a particular type exists between two variable characteristics of the elementary units in the universe and, if such a relationship exists, (2) to determine an equation that may be regarded as the best available estimate of the relationship, (3) to use the equation to estimate or predict an average value of one variable characteristic corresponding to a specified value of the other variable characteristic, and (4) to determine confidence limits to be associated with each estimate or prediction.

The variable characteristic whose values are likely to be known in advance or that logically may be specified in advance usually is designated by x; the variable characteristic whose average value will be estimated or predicted usually is designated by y. Here, let x be the sleeve length of a shirt and let y be the neck size of a shirt.

Ordinarily, it is desirable to have very large samples for prediction analysis. However, a sample consisting of 20 shirts is large enough to illustrate the techniques of prediction analysis without requiring more arithmetic computation than you can follow easily.

The values of x in which we are interested here are the five sleeve lengths, namely, $x = 31$, $x = 32$, $x = 33$, $x = 34$, and $x = 35$. For a sample consisting of 20 pairs of values of x and y, we need 20 values of x, but, of course, these values do not need to be 20 different values. In fact, there are only five possible different values of x in this situation. Consequently, we must make up a set of 20 values of x consisting of some of the five numbers 31, 32, 33, 34, and 35. How many of each of these values shall we choose?

Perhaps you may suggest that we use each of the five numbers four times. That method often is used to select the values of x for the sample, especially if almost nothing is known about the shape of the distribution of the values of x in the universe. Here the complete universe is displayed in Table 6.24 and we can see how the variable x is distributed in the universe by examining the totals in the bottom row of the table. Let us assign the 20 values of x for our sample in proportion to the percentages of the various values of x in the universe. Table 9.1 shows that we shall need to use $x = 31$ once, $x = 32$ four times, $x = 33$ eight times, $x = 34$ five times, and $x = 35$ twice.

The next step in the prediction problem is to draw, by a random sampling process, a value of y to be paired with each value of x. Table A.15 (p. 400) was used to draw the 20 values of y. The details of the process are shown in Table 9.2. First, notice that in Table 6.24 there are $3\frac{1}{2}$ hundred dozen shirts with 31-inch sleeve length. It will simplify the discussion if, for the moment, we consider the column totals in Table 6.24 as dozens rather than hundreds of dozens. There are, then, $3\frac{1}{2}$ dozen shirts, that is, 42, with 31-inch sleeve length.

TABLE 9.1

Specification of the Sleeve Lengths for the Sample

Sleeve Length x	Per Cent of Values in the Universe k	No. of Shirts with Sleeve Length x to be in the Sample 20k/100	
31	3.5	0.700	1
32	22.5	4.500	4
33	40.5	8.100	8
34	23	4.600	5
35	10.5	2.100	2
Total	100.0		20

Consequently, because we need only one shirt with 31-inch sleeve length for the sample, we need to draw only one random number in the range from 1 to 42. The random number that was obtained from Table A.15 in this case was 34. Now, starting at the top of column two in Table 6.24, count down through the column until you come to the thirty-fourth shirt among the 42 shirts in that column. There are 12 shirts with neck size 14, 12 shirts with neck size 14½, and 12 shirts with neck size 15. That is a total of 36 shirts, and consequently, the thirty-fourth shirt in the 31-inch sleeve length is in the dozen that have neck size 15. Therefore, the value of y to be paired with $x = 31$ is $y = 15$.

TABLE 9.2

Procedure for Drawing a Random Sample of 20 Shirts with the Specified Sleeve Lengths

Elementary Unit (Shirt)	Specified Value of Sleeve Length x	Shirt Numbers in Distribution	Random Number	Randomly Selected Value of Neck Size y
A	31	1–42	34	15
B	32	1–68	45	14½
C	32	69–135	86	14½
D	32	136–203	184	15
E	32	204–270	251	16
F	33	1–61	57	14½
G	33	62–122	79	14½
H	33	123–182	129	14½
I	33	183–243	202	15
J	33	244–304	247	15
K	33	305–364	333	15½
L	33	365–425	386	15½
M	33	426–486	459	16½
N	34	1–56	51	15
O	34	57–111	83	15
P	34	112–166	132	15½
Q	34	167–221	205	16
R	34	222–276	227	16
S	35	1–63	36	15½
T	35	64–126	111	17

The resulting sample is shown also in the first three columns of Table 9.3. Because the value of y to be paired with each specified value of x was determined by random chance, the 20 values of y may not be a perfect representation of the whole set of values of y in the universe. It is possible that one or two too many of the shirts in the sample have neck size 14½, for example, and that too few of the shirts in the sample have some other neck sizes. You may have noticed, for example, that none of the shirts in the sample in Table 9.3 has neck size 13½ or 14 or 17½ or 18. Variations such as these that are the result of chance in random sampling are the source of what we have called sampling errors in estimates.

TABLE 9.3

Tabulation for Finding the Linear Prediction Equation

Elementary Unit (Shirt)	Sleeve Length (Inches) x	Neck Size (Inches) y	x^2	y^2	xy
A	31	15	961	225	465
B	32	14.5	1024	210.25	464.0
C	32	14.5	1024	210.25	464.0
D	32	15	1024	225	480
E	32	16	1024	256	512
F	33	14.5	1089	210.25	478.5
G	33	14.5	1089	210.25	478.5
H	33	14.5	1089	210.25	478.5
I	33	15	1089	225	495
J	33	15	1089	225	495
K	33	15.5	1089	240.25	511.5
L	33	15.5	1089	240.25	511.5
M	33	16.5	1089	272.25	544.5
N	34	15	1156	225	510
O	34	15	1156	225	510
P	34	15.5	1156	240.25	527.0
Q	34	16	1156	256	544
R	34	16	1156	256	544
S	35	15.5	1225	240.25	542.5
T	35	17	1225	289	595
Total	663	306.0	21999	4691.50	10150.5

9.1.1 The scatter diagram. The 20 pairs of values of *x* and *y* in Table 9.3 determine the 20 points or dots in the scatter diagram in Figure 9.1. Examination of the diagram suggests that there is a slight upward trend from left to right in the pattern of points and that a straight line probably is a satisfactory representation of this trend. Let us assume then that there is a straight-line relationship between the values of *x* and *y* in the universe from which the sample of 20 shirts was drawn.

9.1.2 The linear prediction equation. A linear relationship for making estimates or predictions corresponding to specified values of *x* in the universe must have an equation of the form $\mu'_{y \cdot x} = \gamma + \delta x$ because it can be proved easily that every straight line has an equation of that form and that, conversely, every equation of that form represents a straight line. The two constants γ (gamma) and δ (delta) in the equation are two parameters of the universe. The symbol $\mu'_{y \cdot x}$ will be used in this book to indicate an average value of *y* computed for the universe by substituting a specific value of *x* into the expression $\gamma + \delta x$ when the values of γ and δ for the universe are known.

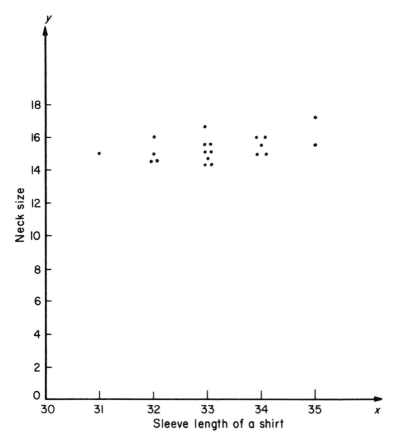

Fig. 9.1. Scatter Diagram for the Data in Table 9.3

Our problem now consists of computing estimates of γ and δ from the data in our sample. We may designate the estimates of the parameters γ and δ by the letters c and d. Consequently, the linear equation determined by the sample will be

$$m'_{y \cdot x} = c + dx \qquad (9.1)$$

We may think of the values of $m'_{y \cdot x}$ computed by equation 9.1 as being estimates of the values of $\mu'_{y \cdot x}$ that could be computed if the values of γ and δ were known.

We need a method for computing the values of c and d. Theoretically, a proper method is "the method of maximum likelihood." Most people use the "method of least squares." There is also a simple method called the "method of average equations." (See section A.10.) The formulas for computing the values of c and d by the latter two methods from the data in a sample are

$$d = \frac{n\Sigma xy - \Sigma x \Sigma y}{n\Sigma x^2 - (\Sigma x)^2} \qquad (9.2)$$

$$c = \frac{\Sigma y - d\Sigma x}{n}$$

where n is the number of pairs of values of x and y used in the computation of c and d. The computation is simpler if we find the value of d before attempting to find the value of c. The "method of maximum likelihood" also produces formulas 9.2 for c and d, if the universe from which the sample is drawn has certain characteristics to be listed in section 9.1.3. This is the theoretical justification for applying formulas 9.2 in connection with the use of a random sample to obtain valid and reliable estimates c and d of the parameters γ and δ.

An inspection of formulas 9.2 shows that we need the following four summations: Σx, Σy, Σx^2, and Σxy. The computation of these four summations or totals for the sample is shown in Table 9.3. For this sample, $n = 20$, $\Sigma x = 663$, $\Sigma y = 306.0$, $\Sigma x^2 = 21,999$, $\Sigma xy = 10,150.5$, and $(\Sigma x)^2 = (663)^2 = 439,569$. Consequently,

$$d = \frac{20(10150.5) - (663)(306.0)}{20(21999) - 439569} = +0.321$$

$$c = \frac{306.0 - 0.321(663)}{20} = +4.659$$

The equation 9.1 for this sample is, therefore,

$$m'_{y \cdot x} = 4.659 + 0.321x$$

The values of y^2 and the total Σy^2 that are included in Table 9.3 are not needed in determining the values of c and d. But Σy^2 will be needed later in the test of the significance of the value of d, and usually it is most convenient to find all the totals given in Table 9.3 at the same time. On some calculating machines it is possible to accumulate the five sums simultaneously.

Now we can draw the straight line whose equation is $m'_{y \cdot x} = 4.659 + 0.321x$ through the scatter diagram in Figure 9.1, obtaining Figure 9.2. The easiest way to determine the position of the straight line on the graph is to determine two points on the line. Substituting 31 for x in the equation, we find that $m'_{y \cdot 31} = 4.659 + 0.321(31) = 14.610$. Consequently, the point corresponding to $x = 31$ and $m'_{y \cdot 31} = 14.610$ is one point on the line. Substituting 35 for x in the equation, we find that $m'_{y \cdot 35} = 15.894$. Consequently, the point for which $x = 35$ and $m'_{y \cdot 35} = 15.894$ is another

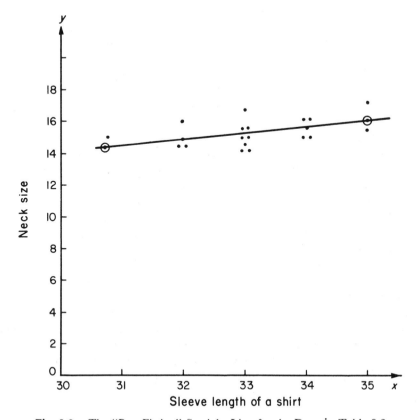

Fig. 9.2.   The "Best Fitting" Straight Line for the Data in Table 9.3

point on the line. These two points have small circles around them in Figure 9.2. After plotting these two points, we can draw the straight line through them.

In the graph of the equation $m'_{y \cdot x} = 4.659 + 0.321x$, the value of the constant c, 4.659, represents the vertical distance from the origin to the point where the straight line crosses the y axis. Notice that the true y axis is 30 units to the left of the position of the vertical axis in Figures 9.1 and 9.2.

If the value of c happens to be negative, then the straight line will cross the true y axis at a point below the origin. If the value of c turns out to be zero, then the straight line passes through the origin, that is, it passes through the point where the x axis and the true y axis intersect.

The constant d represents the rate of change of $m'_{y \cdot x}$ with respect to x. In the equation $m'_{y \cdot x} = 4.659 + 0.321x$ the constant $d = 0.321$ tells us that when the value of x is increased 1 inch, the value of $m'_{y \cdot x}$ is increased only

0.321 inch. This suggests that the average neck size of all the men who wear shirts of one particular sleeve length is 0.321 inch greater than the average neck size of all the men who wear shirts with sleeves 1 inch shorter.

If you have studied trigonometry, you will recognize that d is the tangent of the angle that the straight line makes with the x axis. In other words, the constant d tells us the slope of the line. If the value of d is positive, the straight line slopes upward toward the right-hand side of the graph. If the value of d is negative, the straight line slopes downward toward the right-hand side of the graph. If the value of d turns out to be zero, the straight line is parallel to the x axis.

If the value of d computed for a sample turns out to be zero, then the values of $m'_{y \cdot x}$ computed by substituting values of x into the equation $m'_{y \cdot x} = c + 0x$ are all the same, namely, $m'_{y \cdot x} = c$, no matter what values we give to x. Consequently, when the value of d is zero, the variable x is no help in estimating the value of $\mu'_{y \cdot x}$. This situation arises when there is no linear relationship of the desired type between the values of x and y in the sample.

The equation $m'_{y \cdot x} = 4.659 + 0.321x$ may be thought of as the "best" linear equation for estimating the average value of y, the neck size, for all the shirts or men in the universe in Table 6.24 that have a specified value of x, the sleeve length. It should not be thought of as the "best" linear equation for estimating the average value of x, the sleeve length, for all the shirts or men in the universe that have a specified value of y, the neck size. The "best" linear equation for estimating the average value of x for all the elementary units in the universe that have a specified value of y is

$$m'_{x \cdot y} = e + fy$$

where

$$f = \frac{n\Sigma xy - \Sigma x \Sigma y}{n\Sigma y^2 - (\Sigma y)^2} \quad \text{and} \quad e = \frac{\Sigma x - f\Sigma y}{n}$$

Here is the reason for the distinction. The way in which the straight line whose equation is $m'_{y \cdot x} = 4.659 + 0.321x$ was fitted to the values of x and y for the dots in the scatter diagram minimizes the sum of the squares of the vertical distances between the 20 dots and the straight line. If we desired to find the "best" linear equation for the estimating process when the value of y is specified, then we should fit the straight line in the way that minimizes the sum of the squares of the horizontal distances between the 20 dots and the straight line.

Sir Francis Galton (1822–1911), who was greatly impressed by Charles Darwin's theory of evolution, devoted much of his life to the study of heredity. Galton and Darwin were cousins. During the years 1877–1889, Galton developed his famous "law of regression to type." According to this "law," offspring tend to differ from the mean by only about two-thirds as much as their parents. Prediction equations of the kind discussed in this section usually are called *regression equations*.

9.1.3 Test of significance for d. The slope or direction of the straight line $m'_{y \cdot x} = c + dx$ is its most important geometric characteristic, and that slope or direction is determined entirely by the value of d. Consequently, it is very helpful to have a method for indicating the amount of sampling error that might be expected in the value of d.

If we know that the universe has certain special properties, then we can describe the frequency distribution of the values of d computed from a large number of large samples. Here are the special properties.

1. For each value of x in the universe, there exists a distribution of values of y. For example, in Table 6.24, the column under "sleeve length 33" shows a frequency distribution of the 10 neck sizes, that is, a frequency distribution of the 10 different possible values of y corresponding to $x = 33$. Cover all the columns of Table 6.24 except columns one and four in order to see more clearly this frequency distribution of y for $x = 33$. Then cover all the columns except columns one and two in order to see clearly the frequency distribution of y for $x = 31$. The mean and the standard deviation for each of the five frequency distributions of y are shown in Table 9.4.

TABLE 9.4

**Means and Standard Deviations of the Distributions
of y in Table 6.24**

Sleeve Length	Mean of Neck Size	Standard Deviation of Neck Size
x	$\mu_{y \cdot x}$	$\sigma_{y \cdot x}$
31 in.	14.64 in.	0.52 in.
32 in.	14.94 in.	0.69 in.
33 in.	15.19 in.	0.70 in.
34 in.	15.60 in.	0.79 in.
35 in.	16.00 in.	0.83 in.

2. Each of the distributions of y must be a normal distribution.

3. The means of the series of normal distributions of y are not all identical but, instead, the mean $\mu_{y \cdot x}$ of each distribution of y is exactly equal to the number computed by substituting the value of x for that particular distribution of y into the expression $\gamma + \delta x$. In other words, the means $\mu_{y \cdot x}$ of the distributions of y move along exactly on the straight line $\mu_{y \cdot x} = \gamma + \delta x$ as the value of x changes.

4. The standard deviations of the series of normal distributions of y are all identical, that is, a constant that we may denote by $\sigma_{y \cdot x}$.

It makes no difference here how the values of x are distributed in the universe. However, the method to be given here for testing the significance of d is based on the assumption that in drawing a very large number of samples of n pairs of values of x and y from the universe, the value of s_x, the standard deviation of the values of x in the sample, is the same constant for all the samples. One easy way to insure that s_x remains the same for all repetitions of the sampling process is to specify that the same set of n values of x shall be used in every one of the samples.

Mathematicians have proved that if we drew a large number of large random samples of size n, all of the samples having the same value of s_x, from a universe that has the four special properties listed above, and if we computed the value of d for each of the samples, the frequency distribution of the large number of values of d would be approximated very satisfactorily by a normal distribution whose mean is δ and whose standard deviation is σ_d.

Ordinarily, of course, in a practical situation we do not know the exact value of the parameter δ for the universe of values of x and y and we do not know the exact value of the parameter σ_d for the universe of values of d. But we may use the value of d computed from the sample as the best available estimate of δ, and as an estimate of σ_d we may compute

$$ s_d = \frac{s_y}{s_x} \sqrt{\frac{1 - r^2}{n - 2}} $$

In order to test the significance of the difference between the computed value of d and any proposed value of δ, we find the value of

$$ t = \frac{d - \delta}{s_d} $$

and we use Table 7.3 to determine the probability of obtaining a difference as great as $d - \delta$ solely because of random sampling fluctuations in the values of y.

Here, δ may be any theoretical or hypothetical value of δ that has been proposed for the universe or that is a sensible or logical value of δ to use for the test of significance. The most frequently used value of δ in this test is zero. We usually want to know whether or not the value of δ is different from zero, and for this purpose we use the null hypothesis that $\delta = 0$. If the value of δ is zero, then there is no linear relationship of the desired type between the two variables x and y in that universe.

In our illustration, we found that $d = +0.321$. Let us test the significance of this value of d. That is, let us test the null hypothesis that $\delta = 0$ for the universe of shirts in Table 6.24. First, we must compute s_d. We can find the values of s_x, s_y, and r by using the totals in the bottom line of Table 9.3. We find that $s_x = 1.01$, $s_y = 0.696$, and $r = +0.467$. Consequently,

$$ s_d = \frac{0.696}{1.01} \sqrt{\frac{1 - 0.467^2}{20 - 2}} = +0.144 $$

Therefore,

$$t = \frac{+0.321 - 0}{+0.144} = +2.229$$

Now, turn to Table 7.3 and compare $t = 2.229$ with the values of t on the line for $n' = n - 2 = 20 - 2 = 18$. Our $t = 2.229$ is between $t = 2.101$ for $P = 0.05$ and $t = 2.878$ for $P = 0.01$. Consequently, we conclude that the value of d, namely, $+0.321$, is significant at the 5 per cent probability level, but is not significant at the 1 per cent probability level. There are between 1 and 5 chances in 100 that a value of d as far from zero as 0.321 would occur for a random sample of 20 elementary units if the value of δ is zero.

If we reject the null hypothesis that $\delta = 0$, then we accept the alternative that there is a linear relationship of the form $\mu_{y \cdot x} = \gamma + \delta x$ in the universe, and our best estimate of that linear relationship for the universe in Table 6.24 on the basis of our sample of 20 shirts is the equation $m'_{y \cdot x} = 4.659 + 0.321x$. However, bear in mind that this conclusion is based on the assumption that the universe in Table 6.24 possesses the four special properties listed above.

We have seen already that the universe satisfies the requirements of the first special property. Inspection of the five frequency distributions of y in Table 6.24 shows that although they are bell-shaped, there is some skewness and therefore they are not perfect normal distributions. Figures 9.1 and 9.2 and the five values of the means in the second column of Table 9.4 indicate that the universe satisfies the third requirement fairly well but not perfectly; if the universe satisfied this requirement perfectly, then the values of $\mu'_{y \cdot x}$ and $\mu_{y \cdot x}$ would be identical for any specific value of x; Tables 9.4 and 9.7 show that the values of $\mu'_{y \cdot x}$ and $\mu_{y \cdot x}$ are not exactly identical, but that the differences are small.

The five values of the standard deviation of neck size in the third column of Table 9.4 do not satisfy the requirement of the fourth special property particularly well; there appears to be some tendency for the standard deviation of y to increase slightly as the value of x increases.

We may summarize this review of the universe from Table 6.24 by stating that it does not satisfy perfectly all the requirements that are assumed as a basis for the test of significance of d, but that the universe probably is a sufficiently close approximation to the assumed type of universe so that it is reasonably safe to apply the test of significance for d. The test of significance for d is intended to protect us against accepting an apparent linear relationship in the sample when no linear relationship of the desired type exists between the two variable characteristics of the elementary units in the universe.

9.1.4 Confidence limits for δ. If the test of significance for d shows that the value of d is significant at a satisfactory probability level, confidence limits for the value of δ are $d \pm t s_d$ where the value of t to be used here is obtained from Table 7.3 on the line for $n' = n - 2$ in the column for the probability corresponding to the desired confidence level. In our illustration, the 95 per cent confidence limits for δ are $0.321 \pm 2.101(0.144) = 0.018$ and 0.624.

The values of the parameters γ and δ for the whole universe of 120,000 pairs of values of x and y in Table 6.24 have been computed; they are shown in Table 9.5. We can compare the values of the estimates c and d computed from our sample of 20 shirts with the true values of γ and δ in the universe. The amount of sampling error in each of the two estimates c and d is shown in the third column of the table. Incidentally, you may notice that the value 0.339 for δ is near the middle of the 95 per cent confidence interval 0.018 to 0.624 that we computed above.

TABLE 9.5

**Parameters in the Universe of 120,000 Shirts in Table 6.24
and the Estimates Computed from a Sample of 20 Shirts**

Parameter in the Universe	Estimate Computed From the Sample	Amount of Sampling Error in the Estimate
$\gamma = +4.051$	$c = +4.659$	$+0.608$
$\delta = +0.339$	$d = +0.321$	-0.018

The universe linear prediction equation $\mu'_{y \cdot x} = 4.051 + 0.339x$ and the sample linear prediction equation $m'_{y \cdot x} = 4.659 + 0.321x$ are plotted in Figure 9.3. Notice that the two straight lines are quite close to each other over the range of useful values of x from $x = 31$ to $x = 35$.

9.1.5 Standard error of estimation. By substituting any reasonable value of x into the equation $m'_{y \cdot x} = 4.659 + 0.321x$, we can obtain an estimate or prediction of the value of $\mu'_{y \cdot x}$ for that particular value of x. For example, if we let $x = 33$, we find that $m'_{y \cdot 33} = 15.252$. This can be interpreted as the best estimate of the average neck size of all the shirts in Table 6.24 for which the sleeve length is 33 inches. What are reasonable confidence limits to be attached to this estimate?

In order to determine confidence limits for estimates obtained from a linear prediction equation, we need a good estimate of the standard deviation of the errors of estimation based on the equation $\mu'_{y \cdot x} = \gamma + \delta x$ in the universe from which the sample was drawn. That is, we need an estimate of the standard deviation of the universe of values of $y - \mu'_{y \cdot x}$. Of course, if the universe satisfies perfectly the four requirements listed in section 9.1.3, then $\mu'_{y \cdot x}$ is the same as $\mu_{y \cdot x}$ and the standard deviation of the universe of values of $y - \mu'_{y \cdot x}$ is the same as the constant standard

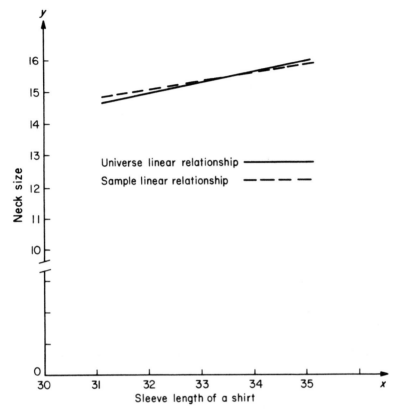

Fig. 9.3.   The Universe Linear Relationship and the Sample Linear Relationship
Between Neck Size and Sleeve Length

deviation $\sigma_{y \cdot x}$ mentioned in the fourth requirement. Our best available
estimate of the standard deviation of the universe of values of $y - \mu'_{y \cdot x}$
is the standard deviation of the values of $y - m'_{y \cdot x}$ for the sample. Let us
agree to use the symbol $s'_{y \cdot x}$ to stand for the standard deviation of the values
of $y - m'_{y \cdot x}$ in the sample.

Table 9.6 shows the 20 values of $m'_{y \cdot x}$ computed from the 20 values of
x in the sample. Subtract each value of $m'_{y \cdot x}$ from the corresponding value
of y in order to determine the amount of the discrepancy $y - m'_{y \cdot x}$ between
each of the 20 estimated values $m'_{y \cdot x}$ and the corresponding observed value
y. Some of these discrepancies are positive numbers and others are negative. This means merely that some of the dots in the scatter diagram in
Figure 9.2 are above the straight line and others are below it.

TABLE 9.6
The Discrepancies and Their Squares

Elementary Unit (Shirt)	Observed Value x	Observed Value y	Computed Value $m'_{y \cdot x}$	Discrepancy $y - m'_{y \cdot x}$	$(y - m'_{y \cdot x})^2$
A	31	15	14.610	0.390	0.152
B	32	14.5	14.931	−0.431	0.186
C	32	14.5	14.931	−0.431	0.186
D	32	15	14.931	0.069	0.005
E	32	16	14.931	1.069	1.143
F	33	14.5	15.252	−0.752	0.566
G	33	14.5	15.252	−0.752	0.566
H	33	14.5	15.252	−0.752	0.566
I	33	15	15.252	−0.252	0.064
J	33	15	15.252	−0.252	0.064
K	33	15.5	15.252	0.248	0.062
L	33	15.5	15.252	0.248	0.062
M	33	16.5	15.252	1.248	1.558
N	34	15	15.573	−0.573	0.328
O	34	15	15.573	−0.573	0.328
P	34	15.5	15.573	−0.073	0.005
Q	34	16	15.573	0.427	0.182
R	34	16	15.573	0.427	0.182
S	35	15.5	15.894	−0.394	0.155
T	35	17	15.894	1.106	1.223
Total	663	306.0	306.003	−0.003	7.583

For Table 9.6, notice that $\Sigma m'_{y \cdot x}$ is the same as Σy except for the small difference indicated by the digit 3 in the third decimal place of $\Sigma m'_{y \cdot x}$. Also, $\Sigma(y - m'_{y \cdot x})$, the sum of the discrepancies, is equal to zero except for a small difference. If we divide $\Sigma(y - m'_{y \cdot x}) = -0.003$ by 20, we see that the mean of the discrepancies is zero except for a small difference caused by the rounding off of decimals in the computation of the values of c, d, and $m'_{y \cdot x}$. In general, for all linear prediction problems of this type we have $\Sigma(y - m'_{y \cdot x}) = 0$ and therefore $m_{y - m'_{y \cdot x}} = 0$.

The squares of the discrepancies are given in the last column of Table 9.6. A number that is obtained by squaring of another number cannot be negative. Consequently, the sum of the squares of the discrepancies can never be negative and can never be zero except in the most unusual situation in which all the points of the scatter diagram lie exactly on the straight line. Because the mean of the discrepancies is zero, the standard deviation of the discrepancies is $s'_{y \cdot x} = \sqrt{7.583/20} = \sqrt{0.37915} = 0.616$.

$$s'_{y \cdot x} = \sqrt{\frac{\Sigma(y - m'_{y \cdot x})^2}{n}}$$

Statisticians usually call the value of $s'_{y \cdot x}$ that is obtained by this formula the *standard error of estimation* or the *standard error of a prediction*.

Obviously, it would require a very great amount of computational labor to calculate all the discrepancies and build a table similar to Table 9.6 if the number of pairs of values of x and y in the sample were some large number, such as 1000. Fortunately, elementary algebra enables us to devise a short cut that eliminates nearly all computational labor in determining the standard error of estimation. It is a good exercise in algebra to show that

$$s'_{y \cdot x} = s_y \sqrt{1 - r^2} \tag{9.3}$$

In our illustration, it is possible to find the value of $s'_{y \cdot x}$ by formula 9.3 without constructing Table 9.6. Using the totals in the bottom line of Table 9.3, we find that $s_y = 0.696$ and $r = +0.467$. Therefore, $s'_{y \cdot x} = 0.696\sqrt{1 - 0.467^2} = 0.696(0.884) = 0.615$.

The small difference between the value 0.615 obtained by using formula 9.3 and the value 0.616 obtained by the long method that required the construction of Table 9.6 is due to rounding off of decimals during the computations. Because the short method involves much less computation with approximate numbers, the value of $s'_{y \cdot x}$ obtained by formula 9.3 is the preferred value.

9.1.6 Confidence limits for an estimated average value of y. The value of d in our illustration was shown to be significant at the 5 per cent probability level. Let us find now 95 per cent confidence limits to be attached to estimates that are made by using the linear prediction equation determined by the sample.

The universe linear relationship equation here is $\mu'_{y \cdot x} = 4.051 + 0.339x$ and the sample linear relationship equation is $m'_{y \cdot x} = 4.659 + 0.321x$. The values of $\mu'_{y \cdot x}$ and $m'_{y \cdot x}$ for each of the specified values of x have been computed; the computed values are shown in Table 9.7. The differences given in the last column indicate the sampling errors that are present in the values of $m'_{y \cdot x}$ as estimates of the values of $\mu'_{y \cdot x}$.

If we repeated a very large number of times the process of drawing a sample of 20 values of y, always using the same 20 values of x, determining a new linear prediction equation for each new sample and predicting new values of $m'_{y : x}$, we would build up a normal frequency distribution of values of $m'_{y \cdot x}$ for each of the five specified values of x. The five means of these five distributions of $m'_{y \cdot x}$ would be the five values of $\mu'_{y \cdot x}$ shown in Table 9.7 that were determined by substituting the five specified values of x into the universe linear prediction equation $\mu'_{y \cdot x} = 4.051 + 0.339x$.

Mathematicians have shown that confidence limits for the unknown parameter $\mu'_{y \cdot x}$ can be determined by substituting the appropriate values of t, $s'_{y \cdot x}$, n, x, m_x, and s_x into

$$m'_{y \cdot x} \pm t s'_{y \cdot x} \sqrt{\frac{1}{n} + \frac{(x - m_x)^2}{n s_x^2}} \tag{9.4}$$

TABLE 9.7

**Errors in Estimating the Average Value of y in the Universe
by the Sample Linear Prediction Equation Instead of
the Universe Linear Prediction Equation**

Sleeve Length x	Average Neck Size in the Universe Computed from the Universe Linear Prediction Equation $\mu'_{y \cdot x}$	Average Neck Size in the Universe Estimated from the Sample Linear Prediction Equation $m'_{y \cdot x}$	Difference $\mu'_{y \cdot x} - m'_{y \cdot x}$
31 in.	14.57 in.	14.61 in.	−0.04 in.
32 in.	14.91 in.	14.93 in.	−0.02 in.
33 in.	15.25 in.	15.25 in.	0.00 in.
34 in.	15.59 in.	15.57 in.	0.02 in.
35 in.	15.92 in.	15.89 in.	0.03 in.

For our sample from Table 6.24, $n = 20$, $m_x = 33.15$, $s_x^2 = 1.0275$, $s'_{y \cdot x} = 0.615$, and $m'_{y \cdot x} = 14.61$ when $x = 31$. For $n' = n - 2 = 20 - 2 = 18$ and the 5 per cent probability level, Table 7.3 (p. 218) shows that $t = 2.101$. Consequently, 95 per cent confidence limits for $\mu'_{y \cdot x}$ when $x = 31$ are

$$14.61 \pm 2.101(0.615) \sqrt{\frac{1}{20} + \frac{(31 - 33.15)^2}{20(1.0275)}} = 14.61 \pm 0.68 = 13.93 \text{ and } 15.29.$$

This means that if we did not know the value of $\mu'_{y \cdot x} = \gamma + \delta x$ when $x = 31$, the sample of 20 shirts would enable us to state that the probability is about 0.95 that when $x = 31$ the value of $\mu'_{y \cdot 31}$ for the universe is between 13.93 inches and 15.29 inches. In this example, the value of $\mu'_{y \cdot 31}$ is not unknown. Table 9.7 shows that $\mu'_{y \cdot 31} = 14.57$. This value of $\mu'_{y \cdot 31}$ is between the confidence limits 13.93 and 15.29; in fact, it is almost at the middle of the confidence interval.

TABLE 9.8

**Ninety-five Per Cent Confidence Limits
for the Average Values of y**

Specified Sleeve Length x	Estimated Average Neck Size $m'_{y \cdot x}$	95 Per Cent Confidence Limits For Average Neck Size	
		Lower Limit	Upper Limit
31 in.	14.61 in.	13.93 in.	15.29 in.
32 in.	14.93 in.	14.49 in.	15.37 in.
33 in.	15.25 in.	14.96 in.	15.54 in.
34 in.	15.57 in.	15.19 in.	15.95 in.
35 in.	15.89 in.	15.29 in.	16.49 in.

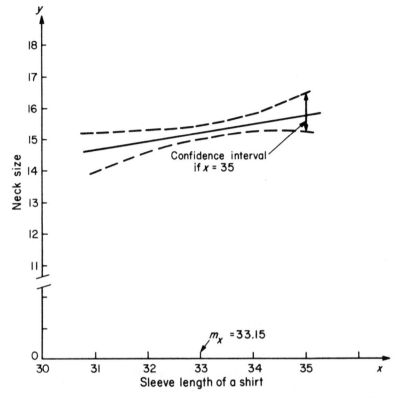

Fig. 9.4.   Ninety-Five Per Cent Confidence Limits for Average Neck Sizes

The 95 per cent confidence limits for $\mu'_{y \cdot x}$ at all the values of x are shown in Table 9.8. A graph of these 95 per cent confidence limits for $\mu'_{y \cdot x}$ is given in Figure 9.4. Notice that the upper and the lower boundaries of the confidence band are curved lines. The confidence band is narrowest at $x = m_x$ and it becomes wider and wider as the value of x that is substituted into formula 9.4 becomes farther and farther from m_x in either direction.

9.1.7 Confidence limits for an individual value of y. Sometimes it is either necessary, or at least interesting, to know something about how far it is probable that a value of $m'_{y \cdot x}$ might miss an individual value of y that is paired with a specific value of x in the universe from which the sample was drawn. It can be shown that confidence limits for this situation are

$$m'_{y \cdot x} \pm t s'_{y \cdot x} \sqrt{1 + \frac{1}{n} + \frac{(x - m_x)^2}{n s_x^2}} \qquad (9.5)$$

In our illustration, when $x = 31$ the 95 per cent confidence limits are

$$14.61 \pm 2.101(0.615)\sqrt{1 + \frac{1}{20} + \frac{(31 - 33.15)^2}{20(1.0275)}} = 14.61 \pm 1.46 =$$

$$13.15 \text{ and } 16.07$$

This means that if someone selected at random one shirt from the universe of 120,000 shirts, told us that the sleeve length of the shirt was 31 inches and asked us to estimate or predict the neck size of the selected shirt, the probability is about 0.95 that we would be correct if we stated that the neck size was between 13.15 inches and 16.07 inches.

TABLE 9.9

Ninety-five Per Cent Confidence Limits
for the Individual Values of y

Specified Sleeve Length x	Estimated Average Neck Size $m'_{y \cdot x}$	95 Per Cent Confidence Limits For Individual Neck Size	
		Lower Limit	Upper Limit
31 in.	14.61 in.	13.2 in.	16.1 in.
32 in.	14.93 in.	13.6 in.	16.3 in.
33 in.	15.25 in.	13.9 in.	16.6 in.
34 in.	15.57 in.	14.2 in.	17.0 in.
35 in.	15.89 in.	14.5 in.	17.3 in.

The 95 per cent confidence limits for an individual value of y that is paired with each of the five possible values of x are shown in Table 9.9. A graph of these confidence limits is given in Figure 9.5. Compare this graph with the graph in Figure 9.4. In both graphs, the upper and the lower boundaries are curved lines. In both, the confidence bands are narrowest at $x = m_x$ and they become wider and wider as x becomes farther and farther from m_x in either direction. As you would expect, the confidence band in Figure 9.5 is much wider than the confidence band in Figure 9.4; we expect the confidence interval for an average value of the variable in a distribution to be much narrower than the confidence interval for an individual value of the variable in the distribution.

If you examine the universe of 120,000 shirts, you will find that exactly 6 per cent of the shirts are outside of the 95 per cent confidence limits given in Table 9.9. However, only 1 per cent of the shirts are below the lower confidence limits and 5 per cent are above the upper confidence limits. This unequal division of the elementary units that are outside of the 95 per cent confidence limits probably is due mainly to the fact that the universe in Table 6.24 is not perfectly symmetrical with respect to the values of y, but may also be due in part to sampling error in the equation of the sample linear relationship.

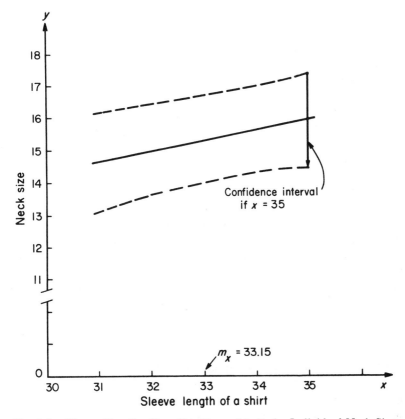

Fig. 9.5.   Ninety-Five Per Cent Confidence Limits for Individual Neck Sizes

9.1.8 Remarks about estimates from prediction equations. If you examine formulas 9.3, 9.4, and 9.5, you will notice some of the things that affect the reliability of estimates made by linear prediction equations. First, formula 9.3 shows that the reliability of an estimate depends upon the value of s_y in the sample. The smaller the value of s_y, the smaller the standard error of estimation will be. But if σ_y for the universe is large, then s_y for the sample likely will be large.

Second, the reliability of an estimate depends upon the value of r in the sample. In order to obtain an estimate that has high reliability, we need a sample in which the value of r is close to either $+1$ or -1. If the value of r for the sample is close to zero, then the linear prediction equation will be almost worthless as a basis for an estimate.

Third, the reliability of an estimate depends upon the value of n. In general, we need to have a large number of pairs of values of x and y in the sample in order to determine a linear prediction equation that will yield an estimate with high reliability.

Fourth, because s_x appears in the denominator of the last term in formulas 9.4 and 9.5, the larger the value of s_x, the smaller will be the size of the confidence interval for an estimate.

Fifth, the confidence band is narrowest when the value of x for which the estimate is to be made is at m_x.

Consequently, for high reliability in an estimate obtained from a linear prediction equation, it is desirable that s_y be small, that s_x be large, that r be close to either $+1$ or -1, that n be large, and that the value of x for which we estimate the value of y be reasonably close to the mean value of x in the sample.

Remember, too, that the value of $m'_{y \cdot x}$ for a particular value of x is likely to be a much better estimate of the average value of y for all the elementary units in the universe that have the particular value of x than it is of the value of y for one randomly selected elementary unit in the universe that has the particular value of x. People who forget this frequently fall into the trap of computing a linear prediction equation from a small sample or from a sample in which the value of r is not high. They are disappointed when the estimate they make for an individual elementary unit turns out to be practically useless because of the relatively large error of estimation to which the estimate is subject.

9.1.9 Interpolation and extrapolation. The values of x for which you will need to compute estimates $m'_{y \cdot x}$ may or may not be the same values of x that were specified for the sample and used in determining the linear prediction equation. Occasionally, you may desire to interpolate one or more additional values of x between the values of x in the sample and compute $m'_{y \cdot x}$.

It may be permissible also, if you have valid reason for desiring to do so, to substitute one or more values of x outside the range of the values of x in the sample and compute $m'_{y \cdot x}$. This type of estimation, usually called *extrapolation* and sometimes called *projection*, may be difficult to justify for a value of x that is far outside the range of the values of x that were used in determining the equation of the straight line. Besides, the confidence interval for such an estimate may be exceedingly large.

9.1.10 Caution about interpretation of r here. The value of r computed for the sample in this type of investigation should not be interpreted as a valid estimate of the linear correlation ρ between the values of x and y for all the elementary units in the universe from which the sample was drawn. The reason for this warning is that the values of x were not obtained by random sampling; they were specified deliberately in advance. Deliberately fixing the values of either variable that may be in a sample may influence the value of r produced by the sample. If a valid and reliable estimate r of the linear correlation ρ in the universe is needed, then one should use a sampling procedure such as that illustrated in Chapter 8 in which the values of both variables in the sample are determined by random chance.

9.1.11 Planning the investigation. In many practical applications of this method of estimation by using linear prediction equations, the procedure is as follows: Select n elementary units at random from the universe. Next, by a random method divide the set of n elementary units into the number of groups indicated by the number of different possible values of x in which you are interested. Then administer a treatment corresponding to the value of x to each elementary unit assigned to that value of x. The observed result of the treatment on a particular elementary unit provides the value of y to be paired with the value of x to which that particular elementary unit was assigned.

EXERCISES

1. Use the five values of $m'_{y \cdot x}$ that are given in Table 9.8 to compute 90 per cent confidence limits for the average neck size of all the shirts in Table 6.24 that have the sleeve length (a) 31 inches, (b) 32 inches, (c) 33 inches, (d) 34 inches, (e) 35 inches. Plot the boundaries of the 90 per cent confidence band on a graph. (*Ans.* (a) 14.05 and 15.17; (b) 14.57 and 15.29; (c) 15.01 and 15.49; (d) 15.26 and 15.88; (e) 15.39 and 16.39.)

2. Use the five values of $m'_{y \cdot x}$ that are given in Table 9.9 to compute 90 per cent confidence limits for the neck size of one shirt selected at random from all the shirts in Table 6.24 if the selected shirt has sleeve length (a) 31 inches, (b) 32 inches, (c) 33 inches, (d) 34 inches, and (e) 35 inches. Plot the boundaries of the 90 per cent confidence band on the same graph that was constructed in exercise 1. (*Ans.* (a) 13.41 and 15.81; (b) 13.80 and 16.06; (c) 14.16 and 16.34; (d) 14.46 and 16.68; (e) 14.71 and 17.06.)

3. Using the 20 values of x that are given in Table 9.2, draw a new sample of 20 values of y from Table 6.24 and carry out the analysis developed in sections 9.1.1 to 9.1.7.

4. Instead of using the 20 values of x that are given in Table 9.2, use each of the five different possible values of x four times and draw a random sample of four values of y from each of the five distributions of y in Table 6.24. Then carry out the analysis developed in sections 9.1.1 to 9.1.7.

5. In Table 6.24, there are 10 different neck sizes. Consider these neck sizes to be the values of x that are specified or fixed in advance. Draw two shirts at random from each of the ten distributions of y and carry out the analysis developed in sections 9.1.1 to 9.1.7.

6. Table 9.10 shows the relationship between speed and miles per gallon of gasoline for one automobile in a test. Construct a scatter diagram for these data. Is there any evidence to indicate that the relationship between speed and miles per gallon is not a linear relationship? (*Ans.* Yes.)

TABLE 9.10

**Relation Between Speed and Miles
per Gallon of Gasoline for a
Particular Automobile**

Speed in Miles Per Hour	Miles Per Gallon of Gasoline
x	y
10	17
20	20
30	21
40	20
50	18
60	17

7. Using the method developed in sections 9.1.1 to 9.1.7 as a model, determine the linear prediction equation for the data in Table 9.11. Determine 95 per cent confidence limits for the average amount of dextrose found by analysis for each of the actual amounts used in the mixtures. Plot the boundaries of the 95 per cent confidence band on a graph.

TABLE 9.11

**Determination of Dextrose
in Sugar Mixtures***

Actual Amount of Dextrose in Mixture (Milligrams)	Amount of Dextrose Found by Analysis (Milligrams)
0.5	0.46
0.5	0.51
0.8	0.79
0.8	0.78
0.8	0.79
1.0	1.03
1.0	1.04
1.2	1.22
1.2	1.20
1.2	1.20
1.2	1.21
1.2	1.22
1.5	1.50
1.5	1.50
1.6	1.57
1.6	1.58
1.6	1.58
2.0	2.07
2.0	1.99

Source: Emma J. McDonald, "Quantitative Chromatographic Procedure for Determining Dextrose in Sugar Mixtures," reprinted from *Analytical Chemistry* 29 (1957), pp. 32–34. Copyright 1957 by the American Chemical Society and reprinted by permission of the copyright owner.

8. With the method developed in sections 9.1.1 to 9.1.7 as a model, use the data given in Table 9.12 to compare the effectiveness of the alcohol method of analysis with the proposed method of analysis for the determination of the amount of calcium in compounds. For which of the two methods is the standard error of estimation the smaller? Graph the boundaries of the 99 per cent confidence bands for the two methods.

TABLE 9.12

Galvimetric Determination of Calcium in the Presence of Large Amounts of Magnesium*

Actual Amount of Calcium Present (Milligrams)	Amount of Calcium Found by Alcohol Method (Milligrams)	Amount of Calcium Found by Proposed Method (Milligrams)
4.0	3.7	3.9
8.0	7.8	8.1
12.5	12.1	12.4
16.0	15.6	16.0
20.0	19.8	19.8
25.0	24.5	25.0
31.0	31.1	31.1
36.0	35.5	35.8
40.0	39.4	40.1
40.0	39.5	40.1

Source: Wallace M. Hazel and Warren K. Eglof, "Determination of Calcium in Magnesite and Fused Magnesia," reprinted from *Analytical Chemistry* 18 (1946), pp. 759–760. Copyright 1946 by the American Chemical Society and reprinted by permission of the copyright owner.

9. By substituting the expression on the right-hand side of formula 9.2 for c in the equation $m'_{y \cdot x} = c + dx$ and then rearranging the terms, show that the linear prediction equation can be written in the form $m'_{y \cdot x} = m_y + d(x - m_x)$.

10. Prove that in the linear prediction equation $m'_{y \cdot x} = c + dx$ the expressions for the coefficients c and d may be written in the form

$$d = r \frac{s_y}{s_x} \quad \text{and} \quad c = m_y + r \frac{s_y}{s_x} m_x$$

and that, consequently, the linear prediction equation may be written in the form

$$m'_{y \cdot x} = m_y + r \frac{s_y}{s_x} (x - m_x)$$

11. Prove that $r = 0$ if $d = 0$ and that $d = 0$ if $r = 0$. (Hint: Use the expression given for d in exercise 10, and formulas 9.2.)

12. Prove that the sign of d always is the same as the sign of r. (Hint: Use the expression given for d in exercise 10 and the fact that, because the definition in formula 3.3 includes only the positive square root, all standard deviations must be positive quantities.)

13. Show that

$$s_d = \frac{s'_{y \cdot x}}{s_x \sqrt{n-2}}$$

14. Prove algebraically that the value of the coefficient d in the linear prediction equation $m'_{y \cdot x} = c + dx$ is not changed if a constant is added to (or subtracted from) every value of y in the sample.

15. Prove algebraically that $\sqrt{\Sigma(y - m'_{y \cdot x})^2 / n} = s_y \sqrt{1 - r^2}$.

9.2 RELATIVE EFFICIENCY OF TWO METHODS OF ESTIMATION

Suppose that instead of drawing the sample of 20 shirts so that there were some shirts in the sample with each of the five different possible sleeve lengths, we had restricted the investigation to one sleeve length, say, the 33-inch sleeve length. Then all of the 20 shirts in the sample would have had 33-inch sleeves.

Table 9.4 shows that the mean value of y, the neck size, for the sample would have been somewhere in the neighborhood of 15.19 inches and that the standard deviation of the values of y in the sample would have been somewhere in the neighborhood of 0.70 inches, varying somewhat from these values because of sampling error. By the method explained in section 7.3.1, the 95 per cent confidence limits for the mean $\mu_{y \cdot 33}$ would have been approximately 14.86 and 15.52.

The size of the confidence interval would have been $15.52 - 14.86 = 0.66$ inches. Using $m_y = 15.19$ and $s_y = 0.70$, we could construct a normal distribution of 48,600 shirts with 33-inch sleeves, by the method given in section 6.5.4 (p. 173). That is about all the practical information we could extract from a sample of 20 shirts with 33-inch sleeves.

Table 9.8 shows that for $x = 33$, the 95 per cent confidence limits determined by the sample of 20 shirts when some of the shirts in the sample had each of the five different possible sleeve lengths were 14.96 inches and 15.54 inches. The size of the confidence interval is $15.54 - 14.96 = 0.58$ inches, which is about 12 per cent smaller than the confidence interval obtained by the other experimental method. Consequently, the estimate of the mean value of y obtained by the linear prediction equation method has greater reliability than the estimate obtained by the other experimental method.

Another way to determine the relative efficiency of two experimental methods is to compare the standard deviations of the two estimates. If all 20 shirts in the sample had 33-inch sleeves, the standard deviation of the estimate of the mean value of y would be $0.70/\sqrt{20} = 0.16$. The standard deviation of the estimate of the mean value of y by the linear prediction equation method when $x = 33$ is

$$0.615 \sqrt{\frac{1}{20} + \frac{(33 - 33.15)^2}{20(1.0275)}} = 0.14$$

The difference between these two standard deviations is $0.16 - 0.14 = 0.02$, and $(0.02/0.16)(100) = 12.5$.

In other words, the standard deviation of the estimate by the linear equation method is about 12.5 per cent smaller than the standard deviation of the estimate by the other method. Now, we know that the standard deviation of the mean of a sample varies inversely as the square root of the size of the sample, as indicated by formula 7.2. Consequently, we can determine how large the sample would need to be in which all the shirts have 33-inch sleeves, in order to reduce the standard deviation of the mean to the same size as the standard deviation of the estimate obtained by the other method. This requires that $0.70/\sqrt{n} = 0.14$ and, consequently, $n = 25$. This is a 25 per cent increase in the size of the sample.

Although the sample containing 25 shirts with 33-inch sleeves would tell us nothing about the average neck size of shirts with other sleeve lengths, the sample of 20 shirts that we have been discussing in this chapter gave us estimates of the average neck sizes for all the other sleeve lengths with approximately the same degree of reliability as for the shirts with 33-inch sleeves. At the same time, the method gave us important information about the nature of the relationship between neck size and sleeve length of men's shirts and it produced an equation for estimating the average neck size of all the shirts that have any specified sleeve length.

How does it happen that so much more information, with higher reliability, is obtained from the sample of 20 shirts, containing shirts with five different sleeve lengths? The answer is that in this sample we are dealing with two correlated variables whereas the other sample had only one variable, neck size. The effect of the relationship between the two variables is to reduce the size of the standard deviation of the estimates, that is, to produce estimates with higher reliability for a given size of sample.

Collecting good data often is an expensive process, and increasing the size of the sample may increase the cost of the investigation considerably. Furthermore, for some reason it may not be feasible to control the experiment so that only one variable is involved. Frequently, it is possible to collect data about two related variables for each elementary unit in the sample almost as easily and economically as it is to collect data about only one variable for the same elementary units.

9.3 THREE TYPES OF DATA FOR FITTING PREDICTION LINES

It is necessary to distinguish between three different types of situation in which a statistician may need to find the "best fitting" prediction line for data. The three different types of situation arise out of the fact that three different types of data may serve as the basis for fitting and using a linear prediction equation.

1. For both variables x and y, the values that are to be used to determine the equation of the "best fitting" prediction line are merely fixed values that were not obtained by any random sampling procedure. That is, the values of neither of the two variables were determined by random chance. Then neither of the two variables can be considered to be a source of sampling error in connection with any estimate that is based on the linear prediction equation determined by the given data.

2. The values of one of the variables, say x, are arbitrarily specified or fixed in advance; then the values of the other variable, say y, to be paired with the arbitrarily specified or fixed values of x are determined by random chance. In such cases, the variable y must be considered to be a source of sampling error in connection with any estimate that is based on the linear prediction equation determined by the observed data, and the variable x must not be considered to be a source of sampling error in connection with such an estimate.

3. For the two variables x and y, the values of x as well as the values of y that are to be used to determine the equation of the prediction line are determined entirely by random chance operating on the elementary units in the universe. In these cases, both variables must be considered to be sources of sampling error in connection with any estimate that is based on the linear prediction equation determined by the observed data.

9.3.1 Neither variable a source of sampling error. In Table 9.13, neither the years nor the number of nurses was obtained by drawing a random sample. The years were selected deliberately because they are years for which published data were available. The numbers of nurses are merely the recorded totals corresponding to the deliberately chosen years. It may not seem reasonable to think of a distribution of values of y, the number of

TABLE 9.13

Number of Gainfully Employed Trained Nurses in the United States, 1870–1950*

Year	Number of Male Nurses	Number of Female Nurses
1870	50	1,154
1880	73	1,464
1890	383	4,206
1900	758	11,046
1910	5,819	76,508
1920	5,464	143,664
1930	5,452	288,737
1940	7,509	344,977
1950	11,185	463,495

**Source:* U.S. Census Reports.

nurses, corresponding to each value of x, the year.

The values of c and d in the equation $m'_{y \cdot x} = c + dx$ for data of this type may be computed by formulas 9.2. The linear prediction equation will be determined now for some of the data on female nurses. The data for the years 1870 to 1900 will not be used. It is commonly found in dealing with growth data that a straight line is not a good representation of the trend if all the data from the beginning are included. However, a straight line may be a satisfactory approximation to the trend over a limited period, provided, of course, that it is not the intention to project the straight line very far beyond either end of the period for which it is known to be a satisfactory representation of the trend of the relationship between the two variables.

Frequently, short cuts can spare you a great deal of arithmetic computation with large numbers. You can often use the idea illustrated in Table 9.14. Subtract 1930 from each value of x and then divide each of the results by 10. In this way, the simple numbers -2, -1, 0, 1, and 2 in column three are obtained.

TABLE 9.14

**Tabulation for Determining the Trend Line for the Data
on Female Nurses in Table 9.13 for 1910 to 1950**

Year x	x − 1930	$X = \dfrac{x-1930}{10}$	No. of Female Nurses y	Xy	X²
1910	−20	−2	76,508	−153,016	4
1920	−10	−1	143,664	−143,664	1
1930	0	0	288,737	0	0
1940	10	1	344,977	344,977	1
1950	20	2	463,495	926,990	4
		0	1,317,381	975,287	10

$$d = \frac{5(975287) - (0)(1317381)}{5(10) - (0)^2} = 97,528.7$$

$$c = \frac{1317381 - (97528.7)(0)}{5} = 263,476.2$$

$$m'_{y \cdot x} = 263,476.2 + 97,528.7X$$

$$m'_{y \cdot x} = 263,476.2 + 97,528.7\left(\frac{x - 1930}{10}\right)$$

$$m'_{y \cdot x} = -18,559,562.9 + 9,752.87x$$

If $x = 1910$, then $X = -2$, and $m'_{y \cdot x} = 68,419$. If $x = 1950$, then $X = +2$ and $m'_{y \cdot x} = 458,534$. These two pairs of values of x and $m'_{y \cdot x}$ were used to determine the position of the straight line in Figure 9.6.

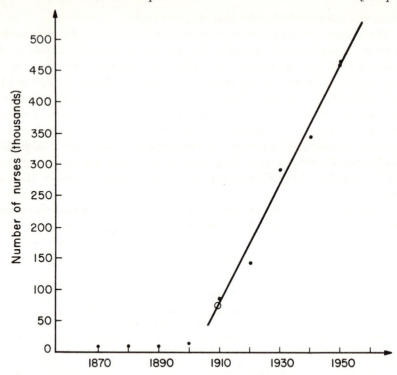

Fig. 9.6.   Number of Gainfully Employed Trained Female Nurses in the United
States, 1870–1950

We can estimate the number of gainfully employed trained female nurses
in the United States in 1956 in either of two ways. One way is to extend the
trend line on the graph until it crosses the vertical line for the year 1956 and
then read off the approximate value of $m'_{y \cdot x}$ from the vertical scale on the
graph. The resulting graphical estimate is approximately 515,000 nurses.
The other way is to substitute 1956 for x in the equation of the trend line.
The resulting estimate is $m'_{y \cdot x} = 517{,}051$ nurses.

If you attempt to use the equation here to estimate the number of nurses
for any year prior to 1900 you obtain a negative number, and that is non-
sensical as an estimate. The only circumstances under which it is safe to
make estimates $m'_{y \cdot x}$ for values of x that are far outside the range of the values
of x used to determine the equation are that you have other data or some
good reason for believing that the trend continues in the direction of the
straight line and that the value of x for which you are making the estimate
is reasonable under the conditions of the situation that you are studying.

Fitting a straight line for data of this type may be regarded as a problem
of geometry and algebra rather than a problem of statistical sampling and
statistical inference. Under that interpretation, it is not legitimate to speak

of a test of significance or of testing a hypothesis or of confidence limits for an estimate with data of this type. This should not be taken as implying that the important statistical ideas mentioned in the preceding sentence never can be used in connection with time series. Many eminent statisticians have given a great deal of thought to the development of statistical methods for dealing with time series, but their ideas and methods will not be discussed in this book.

9.3.2 One variable a source of sampling error. You have been given an illustration of this type of data already in sections 9.1 through 9.1.11. Experimental investigations in which the values of one of the two variables can be fixed in advance provide suitable circumstances for practical applications of linear prediction equations when it is desired to have confidence limits for the estimates or to apply tests of significance or tests of hypotheses.

9.3.3 Both variables as sources of sampling error. In Table 8.14, the values of neither of the two variables were specified in advance or controlled or restricted in any way while the elementary units for the sample were being selected. If a new random sample of 42 students were drawn from that graduating class, it is likely that a different set of values of x and a different set of values of y would be obtained and that both s_x and s_y would be changed in value.

Unless the scatter diagram indicates that a straight line is not a satisfactory representation of the trend for data of this type, the value of r computed from the sample data may be used as a valid estimate of the linear correlation ρ for the universe. Consequently, for data of this type, the test of significance for r given in section 8.4.7 may be used as a method of testing for the existence of a linear relationship of the form $\mu_{y \cdot x} = \gamma + \delta x$ between the two variables in the universe from which the sample was drawn.

If the value of r is significant at a satisfactory probability level, then formulas 9.2 may be used to compute the values of c and d by the "method of least squares" for data of this type. But it is conceivable that for data of this third type the formulas for c and d that would be determined by the "method of maximum likelihood" would be different from formulas 9.2 or that the formulas for computing confidence limits to be attached to estimates obtained from the linear prediction equation would be different from formulas 9.4 and 9.5. These possibilities should be borne in mind if you decide to use formulas 9.2 to determine the values of c and d in the equation $m'_{y \cdot x} = c + dx$ for sample data of this type and if you use formulas 9.4 and 9.5 to determine confidence limits.

It has been found frequently that if an equation determined by using formulas 9.2 is tested by applying the equation to new elementary units from the universe, the errors of estimation for these new units are larger than the errors of estimation for the elementary units in the sample. Consequently, if you ever find it necessary to determine a linear prediction equation from sample data, especially if both variables are sources of sampling error, then you should test the prediction equation on a second large random sample of elementary units from the universe.

This may seem to be and, to some extent it is, a rather discouraging outlook for useful applications of linear prediction equations, especially if both variables in the sample are sources of sampling error. The fact is that less is known by statisticians about the theory of estimation in situations in which both variables are subject to sampling error than is known about the theory of estimation in situations in which only one of the two variables is a source of sampling error. We still do not know some details of the mathematical theory of probability that must be the foundation for testing the significance of the coefficients in a linear prediction equation, for determining confidence limits to be attached to estimates, and for other forms of statistical inference about a universe when the values of both variables in the sample are sources of sampling error.[1]

A great deal of time and effort has been wasted in fruitless attempts to derive useful linear prediction equations. For example, one fine young man tried desperately for several years without success to develop a linear multiple prediction equation for predicting from pretest scores the ratings (Hooperatings and Nielsen ratings, for example) that radio and television programs would earn when they were broadcast for the public. In this way he hoped to be able to select for broadcasting only those programs that would be almost certain to earn at least specified minimum ratings. Three of the principal reasons why his efforts did not produce a practically useful equation are (1) it was necessary for him to work with very small samples, (2) most, if not all, of the variables in his samples were subject to sampling error, and (3) the ratio of the number of independent variables to the size of his samples was too great.

EXERCISES

1. Select a suitable universe of boys or girls and collect appropriate data from a sample to determine a linear prediction equation and a chart for predicting the average height of the boys or the girls at each year of age in the universe. Deter-

[1]Abraham Wald, "The Fitting of Straight Lines if Both Variables are Subject to Error," *Annals of Mathematical Statistics* XI (1940), pp. 284–300.

Paul Horst, ed., *The Prediction of Personal Adjustment*, Social Science Research Council, 230 Park Avenue, New York, N. Y., *Bulletin* 48 (1941).

M. S. Bartlett, "Fitting a Straight Line when Both Variables are Subject to Error," *Biometrics Bulletin* V (1949), pp. 207–212.

Albert Madansky, "The Fitting of Straight Lines When Both Variables are Subject to Error," *Journal of the American Statistical Association* 54 (1959), pp. 173–205.

mine 95 per cent confidence limits for the average height at each year of age and also for the height of an individual at each year of age.

2. Select a suitable universe of boys or of girls and collect appropriate data from a sample to determine a linear prediction equation and a chart for predicting the average weight of the boys or the girls at each year of age in the universe. Determine 95 per cent confidence limits for the average weight at each year of age and also for the weight of an individual at each year of age.

3. Select a suitable universe of boys or girls and collect appropriate data from a sample for the purpose of estimating the strength of the correlation between height and weight for all the boys or all the girls in the universe.

4. The Committee on Careers in Nursing has stated that, according to a Gallup Poll, nursing is rated the most popular profession for young women. Project the trend line in Figure 9.6 and obtain from the graph a prediction of the number of gainfully employed female trained nurses for 1960. Obtain a prediction for 1960 by using the equation determined in section 9.3.1 for female nurses. (*Ans.* 542,677)

5. Table 9.15 shows the annual death rates due to accidents involving motor vehicles for two age-groups for certain years from 1913 to 1958. Construct a graph containing two broken lines to represent the data. If you were asked to describe for the class the information contained in the table and in your graph, what are the most important features to which you would invite their attention? Determine the equation of the straight trend line for the death rate in the 15–24 age group during the period from 1913 to 1958. Plot the trend line. Use the equation to forecast the death rates for the 15–24 age-group for the years 1962 and 1963. Determine the equation for the 25–44 age group and make similar forecasts for that group.

TABLE 9.15

**Death Rates Due to Motor Vehicle
Accidents, 1913–1958***

Per 100,000 Population

	Death Rate	
Year	People of Age 15 to 24	People of Age 25 to 44
1913	3.1	4.1
1918	6.9	8.3
1923	13.8	14.6
1928	22.0	20.5
1933	24.7	23.4
1938	25.4	22.5
1943	20.6	16.1
1948	32.5	19.8
1953	38.9	24.5
1958	36.6	22.3

Source: Accident Facts, various editions,
National Safety Council, Chicago, Illinois.
Reprinted by permission of the Council.

6. Plot the data on gas-heated homes and oil-heated homes given in Table 9.16. Is there any evidence in the graph to indicate that smooth curves might be better indicators of the trends here than straight lines?

TABLE 9.16

**Gas Heating Units and Oil Heating Units
in U.S. Homes, 1949–1955***

Year	No. of Homes with Gas Units	No. of Homes with Oil Units	Oil Lead over Gas
1949	3,120,000	4,491,000	1,371,000
1950	4,087,000	5,172,000	1,085,000
1951	4,700,000	5,705,000	1,005,000
1952	5,371,000	6,346,000	975,000
1953	6,077,000	7,018,000	941,000
1954	6,959,000	7,606,000	647,000
1955	8,000,000	8,208,000	208,000

*Source: The New York Times, April 1, 1956. Copyright by *The New York Times*. Reprinted by permission.

7. Table 9.17 presents an illustration of the improvements that have been made in medical care during this century. Find the equation of the best linear trend. Plot the trend line. Make a graphical estimate of the maternal death rate due to childbirth in 1962. Use the equation to obtain an estimate for 1962. Can you suggest any reason why it would be unwise to attempt to use your graph or your equation to make estimates for 1970 or 1975?

TABLE 9.17

**Death Rate of Mothers
Due to Childbirth***

Year	Maternal Death Rate Per 10,000 Live Births
1920	79.9
1925	64.7
1930	67.3
1935	58.2
1940	37.6
1945	20.7
1950	8.3
1955	4.7
1960	3.2

*Source: Statistical Abstract of the United States.

8. Find the linear prediction equation for the 12 pairs of values of intelligence quotient and grade point average in Table 8.16. You may use the summations that are shown in Table 8.16. (*Ans.* $m'_y._x = -2.74 + 0.0414x$). What is the predicted average G.P.A. for students in that college whose IQ is 100? (*Ans.* 1.40.) What is the predicted average G.P.A. for students whose IQ is 115? (*Ans.* 2.02.) Plot the straight line on a scatter diagram. What is the value of the standard error of estimation here? (*Ans.* 0.367 by formula 9.3.) Compute 95 per cent confidence limits for the G.P.A. of an individual whose IQ is 100 and for an individual whose IQ is 115. How do you interpret each pair of these confidence limits in terms of letter grades? Are these confidence intervals small enough to be of practical usefulness in advising a high school student about the degree of success that he or she might expect in that college? (*Ans.* No.)

    9. Repeat exercise 8 using the data for 42 students in Table 8.14, and compare the two sets of results. You may use the summations that you computed in exercise 6 on page 275 if you have done that problem. (*Ans.* $m'_y._x = -2.68 + 0.0391x$.)

    10. Try to obtain more recent data for Tables 9.13, 9.15, 9.16, and 9.17.

9.4 FOUR METHODS FOR ESTIMATING A UNIVERSE TOTAL

Frequently one of the objectives of an investigation is to estimate the total with respect to some characteristic of the elementary units in a universe. It was shown in section 7.5.3 that one way to obtain this estimate is to multiply the mean m of the sample by the number N of elementary units in the universe. This method usually is called the blow-up method.

If in addition to data with respect to the characteristic for a sample, suitable supplementary information is available, there may be better methods for estimating the total for the universe. Often it is easy to obtain suitable supplementary information. Sometimes the information is in the internal records of the organization that needs the estimate. In many other situations, some of the information may have been collected and published by government agencies, by trade associations, or by others. For example, the censuses of business and the censuses of population and housing made by the United States Government are common sources of such supplementary information. Often, however, it is necessary to include in the plans for the investigation the collection of at least some of the needed supplementary information.

In order to manage their businesses efficiently, many businessmen need to have monthly estimates of their own sales and their competitors' sales. Sample surveys sometimes can be used for this purpose. For example, suppose that at the end of July of the current year you wish to estimate the total sales during July of all the supermarkets in a city in which there are 125 supermarkets, and that from a list of the 125 supermarkets you plan to select 10 stores for the sample.

It probably would be desirable to stratify the universe of 125 supermarkets by annual dollar volume of sales before drawing the random sample of 10 supermarkets. However, it likely would be impossible to obtain the annual dollar volume of sales for each of the 125 supermarkets. Often a bit of ingenuity on the part of the researcher will produce a way to circumvent such a difficulty. For example, one research man has found that often it is sufficient for his purposes if the supermarkets in a universe are stratified by the number of checkout counters. Table 9.18 shows the kind and the amount of information that might be collected from the 10 super-markets in the sample.

Let us use the small Greek letter ω (omega) to stand for a true total for the universe, and let us use w to stand for an estimated total for the universe.

TABLE 9.18

Dollar Volume of Sales of 10 Supermarkets in a City

Store No.	*Sales in July of Current Year (in Thousands of Dollars)* y	*Sales in Preceding Calendar Year (in Thousands of Dollars)* x
1	59.0	736.2
2	53.8	687.1
3	35.7	531.3
4	48.2	625.8
5	59.6	770.4
6	43.7	547.9
7	24.2	390.6
8	63.6	822.0
9	37.5	534.5
10	29.3	409.2
Total 454.6		6055.0

9.4.1 The blow-up method. The only data needed for this method are the data in the second column of Table 9.18 and the fact that the total number of supermarkets in the universe is $N = 125$.

$$w = 125\left(\frac{454.6}{10}\right) = 5682.5 \text{ thousands of dollars}$$

In general,

$$w = \frac{N\Sigma y}{n}$$

For the other three methods, we need the supplementary information consisting of (1) the data in the third column of Table 9.18 and (2) the fact that the true total sales for the universe of 125 supermarkets during the preceding calendar year was $\omega_0 = \$84,218,700$, that is, $\omega_0 = 84,218.7$ thousands of dollars.

9.4.2 The ratio method.

$$w = \frac{454.6}{6055.0}(84218.7) = 6323.1 \text{ thousands of dollars}$$

$$= \frac{\text{total for sample in July}}{\begin{array}{c}\text{total for sample in}\\\text{preceding year}\end{array}} \text{(true total for universe in preceding year)}$$

In general,

$$w = \frac{\Sigma y}{\Sigma x}(\omega_0)$$

9.4.3 The difference method.

$$w = 125\left(\frac{454.6}{10}\right) + \frac{1}{12}\left[84218.7 - 125\left(\frac{6055.0}{10}\right)\right] = 5{,}682.5 + 710.9$$
$$= 6393.4 \text{ thousands of dollars}$$
$$w = \frac{\text{blow-up estimate}}{\text{for July}} + \frac{1}{12}\left[\begin{array}{c}\text{true universe total}\\\text{in preceding year}\end{array} - \begin{array}{c}\text{blow-up estimate for}\\\text{preceding year}\end{array}\right]$$

In general,

$$w = \frac{N\Sigma y}{n} + c\left(\omega_0 - \frac{N\Sigma x}{n}\right)$$

where c is a constant that may be needed to change the supplementary information into the same unit of measurement as the sample data. For example, the fraction $\frac{1}{12}$ is needed in the example in order to change the annual data to an average monthly basis.

9.4.4 The linear equation method.
This method begins by plotting the pairs of values of x and y for the sample in the scatter diagram shown in Figure 9.7. There is no indication in Figure 9.7 that a straight line is not a satisfactory representation of the trend. Next, use formulas 9.2 to find c and d for the ten pairs of values of x and y. The equation is $m'_{y \cdot x} = -9.604 + 0.09094x$.

To estimate the total sales of the 125 supermarkets in July, we must find the total of the 125 values of $m'_{y \cdot x}$ that would be obtained by substituting the sales of each of the 125 supermarkets during the preceding year for x in the equation. Of course, we cannot do this because we know the individual sales of only ten of the stores for the preceding year. An easy way around this difficulty involves a bit of algebra employing the sigma notation for summation. Consider the equation

$$m'_{y \cdot x_i} = c + dx_i \qquad \text{for} \qquad i = 1, 2, 3, \ldots, N$$

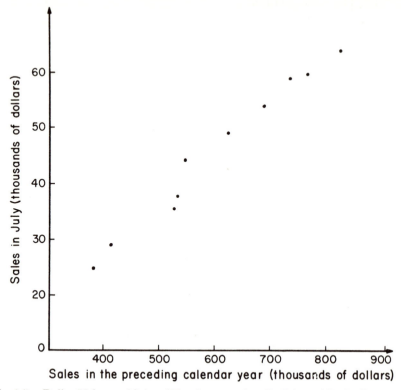

Fig. 9.7.   Dollar Volume of Sales of Ten Supermarkets In July and In the Preceding
Calendar Year

Then

$$w = \sum_{i=1}^{N} m'_{y \cdot x_i} = \sum_{i=1}^{N} c + \sum_{i=1}^{N} dx_i$$

$$= Nc + d \sum_{i=1}^{N} x_i$$

$$= Nc + dw_0$$

In our example,

$$w = 125(-9.604) + 0.09094(84218.7)$$
$$= 6458.3 \text{ thousands of dollars}$$

We now have four estimates of the total sales of the 125 supermarkets
in the universe during July. Some of these estimates may be less than the

true total sales during July and some of them may be greater, or all of them may be greater than the true total or less than the true total. In general, the linear equation method is the most satisfactory of the four estimates because it is based on more of the available information than any of the other three.

In order to make our illustration more concrete, let us consider the ten supermarkets in Table 9.18 to be a universe. Three random numbers were drawn in the range from 1 to 10 and the three stores chosen as a sample of the ten stores turned out to be numbers 1, 7, and 9. Table 9.19 is similar to Table 9.18 but contains only the data for the three stores in the new sample. Table 9.20 shows a comparison of the four estimates of the total sales in July for the ten supermarkets, with the true total sales of 454.6 thousands of dollars. The fact that the linear equation estimate happens to be exactly equal to the true total is, of course, merely a lucky accident that was not expected when this illustration was being prepared.

TABLE 9.19

Dollar Volume of Sales of Three Supermarkets in a City

Store No.	Sales in July of Current Year (in Thousands of Dollars) y	Sales in Preceding Calendar Year (in Thousands of Dollars) x
1	59.0	736.2
7	24.2	390.6
9	37.5	534.5
Total	120.7	1661.3

TABLE 9.20

Comparison of Estimates Obtained by Four Methods with the True Value

Method of Estimation	Estimate w	True Value ω	Error in Estimate	Percentage Error in Estimate
Blow-up Method	402.3	454.6	−49.3	10.8
Ratio Method	439.9	454.6	−14.7	3.2
Difference Method	445.4	454.6	−9.2	2.0
Linear Equation Method	454.6	454.6	0.0	0.0

The information about sales in the preceding year is not the only kind of supplementary information that could have been used in the last three methods of estimation described above. Almost any kind of relevant and meaningful information that was highly correlated with the sales data for July could have been used as suitable supplementary information.

These four methods of estimation can be used frequently for estimating parameters other than the total value for all the elementary units in a universe. For example, although the proportion of the elementary units in a random sample that possess an attribute may be used as an estimate of the proportion of the elementary units in the universe that possess the attribute, a linear equation estimate of the proportion (or of the total) of the elementary units in the universe that possess the attribute might be a better estimate than the observed proportion in the sample.

EXERCISES

1. Draw a new random sample of three supermarkets from the universe consisting of the ten stores listed in Table 9.18 and compute four types of estimate of the total July sales of the ten stores in the universe. Compare your estimates with those in Table 9.20.

2. Show that, because of the way in which c is defined by formula 9.2, the last formula in section 9.4.4 can be written in the form

$$ w = \frac{N\Sigma y}{n} + d\left(\omega_o - \frac{N\Sigma x}{n} \right) $$

Compare this form with the formula for the difference method.

9.5 CHOOSING APPROPRIATE METHOD OF CORRELATION ANALYSIS

An important difference between the linear correlation coefficient method described in Chapter 8 and the linear prediction equation method described in Chapter 9 should be clearly understood. In using the linear prediction equation method, the values of one variable usually are considered to be fixed values chosen in advance and not containing any sampling errors. The values of the other variable are considered to be values obtained by random sampling and, consequently, are subject to sampling errors.

The linear prediction equation method is especially appropriate for application in many experimental situations in which it is desirable to try certain specific values of one variable in order to study the effects. For example, if a manufacturer of an antibiotic desired to study the effects of different amounts of the antibiotic added to the feed of hogs on the length of time required to bring the hogs to marketing condition, it is likely that he would try certain specific amounts of the antibiotic added to the feed mixture; it is very unlikely that he would determine the amount of the antibiotic to be added for each trial by a random chance method such as choosing a random number from Table A.15.

Computation of the linear correlation coefficient is the appropriate method if the sample values of both variables were obtained by random sampling and if we are interested mainly in estimating the strength of the linear relationship between the two variables in the universe from which the sample was drawn.

When we use the linear prediction equation method, we usually are interested in the way in which one variable, called the independent variable, can be used to predict or estimate the values of another variable, called the dependent variable. Neither of the two variables has greater primary or prior importance than the other insofar as the linear correlation coefficient method is concerned. In other words, when we use the linear correlation coefficient method, it is not necessary to think of one of the variables as being the dependent variable and the other the independent variable.

Statisticians do not always need to know that one variable is the cause and the other variable is the effect. All that they need to know for many purposes is that the variables are correlated and the strength of that correlation. It is quite common for two variables to be highly correlated and very useful to a statistician even though neither is the cause or the effect of the other; both variables may be effects of a third variable about which it is not feasible for the statistician to obtain the kind of data that he might desire to have.

For example, it is well known that there is a relationship between the heights of members of a family; if the boys in a family are tall, there is a tendency for the girls in that family to be tall; if the boys in a family are short, there is a tendency for the girls in that family to be short. If we are interested mainly in the strength of the association or correlation between the heights of boys and the heights of their sisters, it is neither necessary nor reasonable to consider a boy's height as the independent variable and his sister's height as the dependent variable. There is an association between the heights of brother and sister, but the height of neither one is dependent upon the height of the other; both have the tendency to similarity in height because of heredity, that is, because of the traits that have been passed on to both of them by the generations of their relatives who preceded them.

Under circumstances such that it is not reasonable to consider one of the two variables as the independent variable and the other variable as the dependent variable, but rather to think of both variables as being associated, the appropriate way to describe the association is to compute the correlation coefficient.

9.6 REMARKS

Businesses hire statisticians to try to discover relationships between past or present conditions and future events, so that they may forecast sales of their products. If they can make good forecasts, then they can plan production wisely and maximize their profits and the earnings of their employees. If they cannot make as good forecasts as their competitors, they may become bankrupt in a short time.

If you apply for work with a large industrial firm or governmental organization, it is likely that they will give you tests to discover the kinds of work you can do best. If the personnel director had not established the existence of a relationship between a person's score on the tests and success in certain kinds of work, there would be no use in giving the tests.

If school administrators did not know that there are relationships between your high school marks, your interests, your aptitudes, your attitudes, other aspects of your personality, and success in the trades, in college, and in the professions, they would not need to have any of this information about you when you apply for entrance to a trade school, a college, a professional school.

In business forecasting, in psychological research, in personnel selection, and in other fields of application, statisticians find that some variables correlate more highly than others. Consequently, a great deal of research goes on all the time in these fields, to find variables that correlate more and more highly so that estimates or predictions with higher and higher reliability may be obtained.

For example, the Bureau of Home Economics in the United States Department of Agriculture made a study of 35 body measurements of nearly 150,000 boys and girls, age four to seventeen years, throughout the United States. The purpose of the investigation was to devise a better system of sizing for the children's clothing industry.

Of the 20 different predictors of body measurements studied, age was the poorest of all. The Bureau stated:

> Age is now used as a basis for sizing practically all children's garments and patterns, and many persons have surmised that this is the underlying cause of size difficulties. A great many children of exactly the same age have entirely different dimensions and body proportions.
>
> Eighteen important body measurements in addition to age and weight were selected for study. The correlation coefficient was calculated for the 190 possible pairs of the 20 items. In general terms the correlation coefficient is an index which measures the closeness with which one measurement can be predicted from another . . . . Thus, of the entire set, age is the least highly correlated with all the other measurements.[2]

The Bureau concluded that age alone is the poorest possible basis for sizing any kind of garments for children, and that the size of children can be estimated best from a combination of two measurements; a vertical length, such as height; and a girth, such as hip girth. This idea is similar to the long-accepted idea that in order to select a shirt to fit himself a man specifies two measurements, sleeve length and neck size.

[2]U. S. Department of Agriculture Miscellaneous Publication No. 365, *Children's Body Measurements for Sizing Garments and Patterns*. Washington, D. C.: U. S. Government Printing Office, 1939, p. 2.

An example of the ways in which statistical thinking can play a worthwhile part in the informal affairs of life, such as social conversation and hobbies, was published in the *Scientific Monthly* and in the *Literary Digest* some years ago.[3]

Bert E. Holmes found that the method of least squares and linear prediction equations helped him to spend his leisure time on summer evenings in what was for him an interesting manner. At the same time, his keen, observing, and inquiring mind learned more about the secrets of nature around him than he could have known without these statistical ideas.

Holmes observed that the crickets chirped at different rates on cool evenings and warm evenings. By counting and recording the number of chirps per minute during the late summer and early fall, and then applying "the method of least squares," Holmes was able to develop a linear prediction equation that could be used to estimate the prevailing temperature. The formula was so reliable that usually the error in the temperature estimate was less than one degree Fahrenheit. His equation is $t = 36.9 + 0.258c$ where t is the estimated temperature and c is the number of chirps per minute made by a cricket.

At the beginning of the summer in 1959, the U. S. Weather Bureau began issuing reports several times daily of what it calls a temperature–humidity index (THI). The purpose of the index is to describe numerically the average human discomfort resulting from the combined effects of temperature and relative humidity. The linear prediction equation used for this purpose is $THI = 0.4t + 15$ where t is the sum of the dry-bulb temperature and the wet-bulb temperature at the time for which the THI is computed. An ordinary thermometer indicates the dry-bulb temperature. The wet-bulb temperature is read on an aspirating psychrometer, an instrument for measuring humidity accurately by means of evaporation. A Weather Bureau scientist says that a few people are uncomfortable even before the THI reaches 70, over half are uncomfortable when it passes 75, and at 80 or above almost everyone is uncomfortable and most of us are downright miserable.

REFERENCES

George W. Snedecor, *Statistical Methods*. Ames, Iowa: Iowa State College Press, 1956.

R. Clay Sprowls, *Elementary Statistics*, (Chapters 11 to 13 on analysis of time series). New York: McGraw-Hill Book Co., Inc., 1955.

[3]Bert E. Holmes, "Vocal Thermometers," *Scientific Monthly* 25 (September 1927), pp. 261–264.

Chapter 10

INDEX NUMBERS

10.1 INTRODUCTION

The Bureau of Labor Statistics in the U. S. Department of Labor computes and publishes an index of certain important components of the cost of living every month. It is called the Consumer Price Index. Table 10.1 contains values of this index for the years 1913 to 1962.

TABLE 10.1
Consumer Price Index*
(1957–1959 = 100)

Year	Index	Year	Index	Year	Index	Year	Index
1913	34.5	1927	60.5	1941	51.3	1955	93.3
1914	35.0	1928	59.7	1942	56.8	1956	94.7
1915	35.4	1929	59.7	1943	60.3	1957	98.0
1916	38.0	1930	58.2	1944	61.3	1958	100.7
1917	44.7	1931	53.0	1945	62.7	1959	101.5
1918	52.4	1932	47.6	1946	68.0	1960	103.1
1919	60.3	1933	45.1	1947	77.8	1961	104.2
1920	69.8	1934	46.6	1948	83.8	1962	105.4
1921	62.3	1935	47.8	1949	83.0	1963	
1922	58.4	1936	48.3	1950	83.8	1964	
1923	59.4	1937	50.0	1951	90.5	1965	
1924	59.6	1938	49.1	1952	92.5	1966	
1925	61.1	1939	48.4	1953	93.2	1967	
1926	61.6	1940	48.8	1954	93.6	1968	

**Source:* U. S. Department of Labor, Bureau of Labor Statistics.

Whenever the Consumer Price Index shows that the cost of living for employees has risen, some employers raise wages the proper amount to take care of the increased living expense. In this way, the employees' wages always will purchase at least as much food, clothing, recreation, and so on

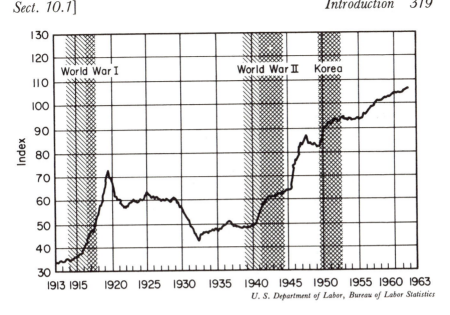

Fig. 10.1.   Consumer Price Index

for their families as they were able to buy in previous months. Here are some excerpts from a recent U. S. Government news release.

> Living costs declined by two-tenths of one percentage point in December because of lower food prices, the Labor Department reported Friday. It was the largest decline for any month in four years.
>
> Despite the December drop, wage adjustments under labor contracts based on the government index readings over a period of months will bring pay increases of 4 cents an hour for about 270,000 truck drivers in most parts of the nation. About 3,500 New York Harbor seamen will get a 1.2 per cent pay boost. An estimated 18,000 other workers in various small industries will receive a one-cent hourly increase.
>
> On the other hand, nearly 30,000 workers will get a pay cut of one cent an hour, based on different living cost adjustment plans.

Look at the figures for the Consumer Price Index in Table 10.1 and as they are represented in the graph in Figure 10.1. The annual average of the index for 1962 is 105.4. This means that the prices paid by city wage-earner and clerical-worker families in the United States were 5.4 per cent higher in 1962 than they were during the period 1957–1959, which is the base period or reference period for the Consumer Price Index. Similarly, the annual average for the year 1937 is 50.0, which means that the prices of these commodities in 1937 were only 50.0 per cent as high as the prices of the same commodities during the 1957–1959 period.

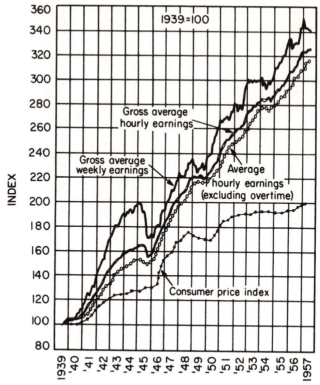

U. S. Department of Labor, Bureau of Labor Statistics

Fig. 10.2.   Trends of Earnings in Manufacturing and Consumer Price Index

There are a great many practical situations in which index numbers are useful. By using index numbers, people often are able to organize and present data so that the interpretation needed for certain purposes can be made most easily. For example, see Figure 10.2.

Imagine all the price and sales data that would need to be collected every month and all the calculations that a worker or his union or his employer would be required to do in order to figure out the cost of living every month! The United States Government statisticians collect all the necessary information, do all the calculating, and publish the results monthly for everyone to use. The interpretation and application is so simple and easy that every worker can know by glancing at the index numbers how the cost of living stands for him and his family.

Of course, there are many other kinds of index numbers besides cost-of-living indexes. Illustrations can be found in the financial section of many newspapers, and in government publications such as the *Monthly Labor Review* and the *Survey of Current Business*.

Why do people need all these index numbers? We need many kinds of
index numbers because the social and economic world in which we live is
complex and dynamic; it is changing continually. If we wish to be able
to exert the controls that are necessary for protection in the midst of all these,
we must have some way to measure and indicate the relationship between
our present standing—social, economic, and so on—and that of an earlier
date. And, of course, we want to know what the immediate future will
hold for us. Several important relationships are illustrated in Figure 10.2.

Have employment opportunities in the steel industry increased during
recent years? How do the current average wages of store clerks and of
carpenters compare with the average wages of workers in those occupations
in 1939, that is, before World War II? What has been the trend in earnings

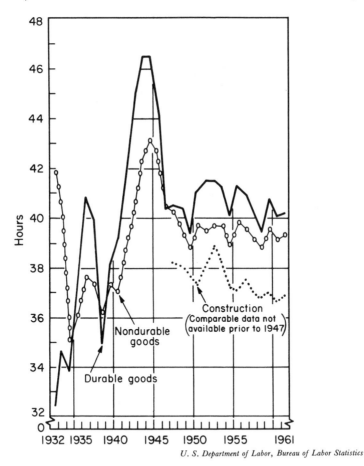

U. S. Department of Labor, Bureau of Labor Statistics

Fig. 10.3.   Average Number of Hours Worked Weekly in Three Industrial
Classifications

of store clerks and of carpenters during the past twenty-five years? In which industries are jobs increasing? In which occupations are opportunities for work decreasing? Is crime in our community increasing or decreasing? What are the conditions of health in our city? What is happening in the stock market these days? What is the trend in the prices of farm products? To provide easily understood answers to questions such as these is the purpose of index numbers.

10.2 CONSTRUCTION OF A SIMPLE INDEX

The data in Table 10.2 give some information about recent changes in working conditions in the manufacturing industries. For example, the effect of World War II in counteracting the modern trend toward a shorter work week can be seen. Similar information is contained in Figure 10.3.

TABLE 10.2

Average Number of Hours Worked Per Week by Production Workers in Manufacturing Industries*

Year	Average Number of Hours Worked Per Week
1939	37.7
1940	38.1
1941	40.6
1942	42.9
1943	44.9
1944	45.2
1945	43.4
1946	40.4
1947	40.4
1948	40.1
1949	39.2
1950	40.5
1951	40.7
1952	40.7
1953	40.5
1954	39.7
1955	40.7

**Source:* U. S. Department of Labor, Bureau of Labor Statistics.

Let us find the ratio of the average number of hours worked per week in 1940 to the average number of hours worked per week in 1939. To do this, we divide the figure for 1940 by the figure for 1939. The result is $38.1/37.7 = 1.01$, which tells us that, on the average, the employees worked 1 per cent more hours per week in 1940 than they did in 1939. Similarly, the ratio 40.6/37.7 = 1.08 tells us that the average work week in 1941 was 8 per cent longer than it was in 1939 in these industries.

Index numbers usually are results of finding ratios. Frequently, for reasons of simplicity, the ratios are multiplied by 100 in order that they may be expressed as percentages. For some purposes, it is sufficient to write the index numbers in the form of the nearest whole numbers. For other purposes, it is often desirable to retain one decimal place in the index numbers. If the two ratios that we worked out above are multiplied by 100, we obtain the index numbers 101 and 108.

Select a certain period, for example, a certain day or week or month or year or group of years, and compare the data for all other periods with the data for the selected period. The particular period that is selected for this purpose is called the *base period*.

The index numbers are computed by dividing the figure for the base period into the figures for all the other periods, and then multiplying by 100 in each case. Obviously, the index number for the base period itself will be 100, because dividing a number by itself gives one, and then multiplying by 100 gives 100.

The index numbers for the average number of hours worked per week in the manufacturing industries during the years 1939 to 1955 are tabulated in Table 10.3. The base period is the year 1939.

TABLE 10.3

**Index of Average Number of Hours Worked
Per Week by Production Workers
in Manufacturing Industries
(1939 = Base Period)**

Year	*Index Number*
1939	100
1940	101
1941	108
1942	114
1943	119
1944	120
1945	115
1946	107
1947	107
1948	106
1949	104
1950	107
1951	108
1952	108
1953	107
1954	105
1955	108

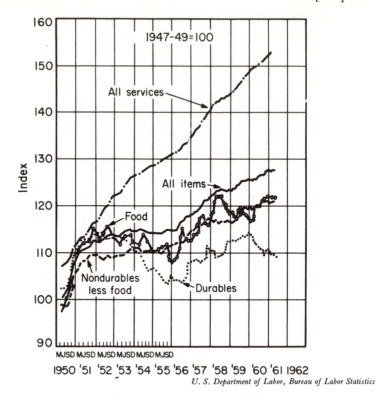

U. S. Department of Labor, Bureau of Labor Statistics

Fig. 10.4.   Consumer Price Index (Commodities and Services)

10.3 CONSTRUCTION OF A COMPLEX INDEX

The Consumer Price Index involves the following components: food; housing; clothing; transportation; medical care; personal care; reading and recreation; and other goods and services, such as tobacco products and alcoholic beverages. (See Figure 10.4.) Altogether, there are approximately 300 different items in the Consumer Price Index. The food component alone contains about 90 different items. (See Tables 10.4 and 10.5.)

The Consumer Price Index will be used to illustrate the principles by which several different items may be combined to produce a composite index number. Most of the statements that will be made here about the Consumer Price Index have been taken from publications of the Bureau of Labor Statistics.

10.3.1 Determination of the purpose of the index. Without a precise formulation of the purpose for which the index is to be used, it may be impossible to know how to carry out properly the steps involved in construction of the index.

For example, if the purpose of the index is to show the changes that take place in the cost of living as a result exclusively of changes in the prices of the various items that families customarily purchase, then the *quantities of the items* that are used in computing the index should remain fixed from period to period. The Consumer Price Index is an index of this type. If the quantities used in computing the index numbers, as well as the prices, were permitted to change from period to period, changes in the index would be due partly to changes in prices and partly to changes in the quantities used in computing the index numbers.

On the other hand, if the purpose of the index is to show the changes that take place in industrial production as a result exclusively of changes in the quantities of the various materials produced, then the *prices of the items* that are used in computing the index numbers should remain fixed from period to period. The important Index of Industrial Production that is computed and published monthly by the Federal Reserve Board is an index of this type.

The Consumer Price Index sometimes is used as a measure of changes in the purchasing power of the consumer's dollar. (See Figure 10.5.) Insofar as the Index may be interpreted as a measure of changes in the purchasing power of a consumer's dollar, the method for determining the purchasing power of a dollar at a given date as compared with its purchasing power during the base period of the index requires that $1.00 be divided by the index number for the given date and multiplied by the index number for the base period, namely, 100. In other words,

$$\text{purchasing power of a consumer's dollar in period } k \text{ as compared with the base period} = \frac{\$1.00}{I_k}(100)$$

For example, from the data in Table 10.1, we may determine that the purchasing power of a consumer's dollar in 1962 as compared with the purchasing power of a consumer's dollar in the period 1957–1959 was

$$\frac{\$1.00}{105.4}(100) = \frac{\$100}{105.4} = \$0.95$$

Similarly, the purchasing power of a consumer's dollar in 1940 as compared with the purchasing power of a consumer's dollar in the period 1957–1959 was

$$\frac{\$1.00}{48.8}(100) = \frac{\$100}{48.8} = \$2.05$$

Thus, in 1940, a consumer's dollar purchased about as much in goods and services as could be bought for $2.05 during the period from 1957–1959.

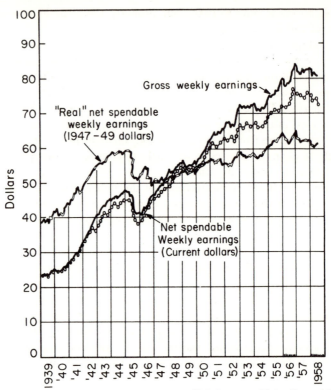

Fig. 10.5. Gross and Net Spendable Weekly Earnings Compared with Real Net Spendable Earnings Expressed in 1947–49 Dollars

If a man earned $125.00 per week in 1962, these earnings would buy for his family only approximately as much in consumer goods and services as $125.00(0.95) = $118.75 bought during the period 1957–1959. It is sometimes stated that the man's "real wages" during 1962 were only $118.75 per week in relation to the purchasing power of a consumer's dollar during the period from 1957 to 1959.

10.3.2 Determination of the field to be represented by the index. For many years the Consumer Price Index was popularly called a cost-of-living index. The Bureau of Labor Statistics now advises those who use the Consumer Price Index not to consider it a cost-of-living index because the Bureau claims that the Index may not represent properly the whole field of the cost of living. The field that the Bureau tries to represent by the Consumer Price Index is the field of retail prices of goods and services customarily purchased by city wage-earner and clerical-worker families to maintain their level of living.

Wage-earner and clerical-worker families are moderate-income families. These families represent about 64 per cent of all the persons living in urban places and about 40 per cent of the total population of the United States. The average size of the families included in the Index was estimated to be about 3.3 persons, and their average family income after taxes in 1952 was estimated to be about $4,160. Families with incomes of $10,000 or more after taxes are excluded. Consequently, the Consumer Price Index does not reflect purchases of families at the lowest and the highest extremes of the income scale, who live at different levels.

The Consumer Price Index measures only changes in prices; it tells us nothing about changes in the kinds and amounts of goods and services families buy, or about the total amount families spend for living, or about the differences in living costs in different places.

10.3.3 Selection of a representative sample of items. There are so many different consumer items available today that it would be impossible to use all of them in a statistical survey. Furthermore, new consumer products and services are being introduced all the time. But, as indicated earlier, it is not desirable to change the items used in the index from month to month. How may a person select a representative sample of items for an index?

Selection of the goods and services on which the revised Consumer Price Index is based depended on the answers to the following questions: (1) How important is each item in the purchases of the families of city wage-earners and lower salaried workers? (2) How similar or dissimilar are the price changes over time for the various goods and services purchased by these families?

The consumer expenditure survey that was conducted by the Bureau of Labor Statistics in 1950 provided the information required to answer the first question, by obtaining data on the kinds of things families bought, the amounts they bought, and the amounts they paid for the things they bought. While these records were being obtained from families, the Bureau also secured prices of hundreds of items in retail stores. These special collections of data, combined with other available price data, helped to identify "price families," that is, groups of items that had shown similar price movements in the past and thus might be expected to show similar price changes in the immediate future.

The representative sample of cities used in the survey included the 12 largest urban areas with populations of more than 1,000,000 people and 85 of the large, medium-size, and small cities. The sample was selected so as to account for the characteristics of different types of city that affect

TABLE 10.4
Units and Average Retail Prices of the Food Items in the Consumer Price Index (August 1962)

Food Item	Unit	Price (Cents)	Food Item	Unit	Price (Cents)
Cereals and bakery products:			Oranges, size 200	doz.	79.0
Flour, wheat	5 lb.	57.3	Lemons	lb.	19.5
Biscuit mix	40 oz.	47.8	Grapefruit	each	15.5
Macaroni	lb.	24.8	Peaches	lb.	16.8
Corn meal	lb.	14.1	Grapes, seedless	lb.	25.9
Rice, short grain	lb.	19.3	Watermelons	lb.	4.0
Rice, long grain	lb.	21.6	Potatoes	10 lb.	68.5
Rolled oats	18 oz.	23.8	Sweet potatoes	lb.	17.3
Corn flakes	12 oz.	27.5	Corn-on-cob	doz.	68.8
Bread	lb.	21.2	Onions	lb.	11.7
Soda crackers	lb.	31.0	Carrots	lb.	15.8
Cookies, cream sandwich	lb.	52.4	Lettuce	head	16.5
Meats, poultry, and fish:			Celery	lb.	15.7
Round steak	lb.	106.5	Cabbage	lb.	8.4
Sirloin steak	lb.	110.7	Tomatoes	lb.	21.4
Chuck roast	lb.	61.3	Beans, green	lb.	21.9
Rib roast	lb.	83.5	Canned fruits and vegetables:		
Hamburger	lb.	51.5	Orange juice	46-oz. can	41.1
Veal cutlets	lb.	148.3	Pineapple juice	46-oz. can	32.7
Pork chops, center cut	lb.	94.3	Peaches	# 2½ can	33.1
Roast pork, loin half	lb.	66.4	Pineapple	# 2 can	39.1
Bacon, sliced	lb.	74.0	Fruit cocktail	# 303 can	25.9
Ham, whole	lb.	62.4	Corn, cream style	# 303 can	20.0
Lamb, leg	lb.	73.3	Peas, green	# 303 can	22.6
Frankfurters	lb.	63.5	Tomatoes	# 303 can	15.6
Luncheon meat	12-oz. can	50.4	Tomato juice	46-oz. can	32.4
Frying chickens, ready-to-cook	lb.	40.4	Baby foods	4½-5 oz.	10.7
Ocean perch, fillet, froz.	lb.	50.2	Dried prunes	lb.	41.1
Haddock, fillet, frozen	lb.	55.1	Dried beans	lb.	17.3
Salmon, pink	16-oz. can	77.5	Soup, tomato	11-oz. can	12.4
Tuna fish, chunk	6-6½-oz. can	35.1	Beans with pork	16-oz. can	15.0
Dairy products:			Pickles, sliced	15 oz.	26.6
Milk, fresh, (grocery)	qt.	24.3	Catsup, tomato	14 oz.	23.1
Milk, fresh, (delivered)	qt.	26.1	Coffee	1-lb. can	71.3
Ice cream	½ gal.	85.5	Coffee, instant	6 oz.	90.3
Butter	lb.	74.6	Tea bags	pkg. of 16	24.4
Cheese, American process	lb.	36.1	Cola drink	carton, 72-oz.	50.1
Milk, evaporated	14½-oz. can	15.4	Shortening, hydrogenated	3-lb. can	88.1
Frozen fruits and vegetables:			Margarine, colored	lb.	28.1
Strawberries	10 oz.	27.1	Lard	lb.	20.0
Orange juice concentrate	6 oz.	20.0	Salad dressing	pt.	38.3
Lemonade concentrate	6 oz.	13.2	Peanut butter	lb.	57.8
Peas, green	10 oz.	20.8	Sugar	5 lb.	58.6
Beans, green	9 oz.	22.9	Corn syrup	24 oz.	27.4
Potatoes, french fried	9 oz.	18.8	Grape jelly	12 oz.	29.6
Fresh fruits and vegetables:			Chocolate bar	1 oz.	4.5
Apples	lb.	21.1	Eggs, Grade A, large	doz.	51.3
Bananas	lb.	15.9	Gelatin, flavored	3-4 oz.	9.9
			Potato chips	4 oz.	27.9

the ways in which families spend their money. The most important characteristics were size, climate, density of the population, and level of income in the community. In this way, 97 cities were selected to represent all types of cities in the United States.

In each city, the Bureau selected a representative sample of families from the entire population, including all family types and income classes. Interviewers visited and interviewed each family and obtained a complete record of the kinds, qualities, and amounts of foods, clothing, furniture, and all other goods and services the family bought in 1950, together with the amount spent for each item. Then the family records of all wage-earner and clerical-worker families of two or more persons were averaged together for each city, to form the basis for index-weight determination.

10.3.4 Determination of weights for the items. How much of each item should be included? The most common way to answer this question is to say that the quantity of each item should be determined so that all the items will be included in quantities that are proportional to the average quantities of those items used by the families in the universe to which the index is to be applied.

Table 10.4 shows the food items used in the Consumer Price Index and the quantity or unit of each item included in computing the index of the cost of food at home for August 1962. For example, the quantity or unit of flour is five pounds. The quantity or unit of fresh eggs is one dozen, Grade A, large.

If the prices of the items were averaged without weights representing their relative importance in family spending, the price of a low-priced item such as a loaf of bread would be dwarfed by the price of a high-priced item such as a new automobile. For the Consumer Price Index, the price of each item is multiplied by a factor that represents the average amount of the item purchased by wage-earner and clerical-worker families during a specified period, usually a year.

The Bureau of Labor Statistics publishes annually a list of the items included in the Consumer Price Index as of the preceding December. The list shows also the relative importance or weight of each item in the index calculation. Table 10.5 shows part of the list for December 1952 and December 1961.

The figures of relative importance for any specific period are the percentage distribution of the value weights that enter into the calculation of the Index for that period. The relative importance of an item does not remain constant. Although the quantities and the qualities of items priced for the Index are held constant, prices do change, and these changes in prices affect the value weights. Because prices for all items do not change in exactly the same degree, the relative importance of items varies from time to time.

TABLE 10.5

**Examples of Items Included in the Consumer Price Index
and Their Relative Importance in the Index**

Item	Per Cent of Each Item to Total		Item	Per Cent of Each Item to Total	
	Dec. 1952	Dec. 1961		Dec. 1952	Dec. 1961
Transportation	11.3	11.7	Personal care	2.0	2.3
Private	10.0	10.0	Men's haircuts	0.6	0.8
Automobiles, new	2.9	2.8	Permanent wave	0.1	0.1
Automobiles, used	2.0	1.7	Shampoo and wave set	0.2	0.2
Auto repairs	1.1	1.2	Toilet soap	0.2	0.3
Tires	0.3	0.3	Cleansing tissues	0.1	0.1
Gasoline	2.2	2.4	Toothpaste	0.2	0.2
Motor oil	0.2	0.2	Shampoo	0.1	0.1
Auto insurance	1.0	1.1	Shaving cream	0.1	0.1
Auto registration	0.3	0.3	Face cream	0.1	0.1
Public	1.3	1.7	Face powder	0.1	0.1
Transit fares	1.0	1.4	Razor blades	0.1	0.1
Railroad fares	0.3	0.3	Sanitary napkins	0.1	0.1

The relative importance of items in the Index shows only how families would be spending their money if they had continued to buy the same kinds and quantities of goods and services that they purchased in the period on which the Index value weights are based. Actually, consumers vary the amounts and kinds of things they buy as prices change, so that the relative importance figures may, as time passes, no longer reflect actual expenditures. When the difference becomes so great as to render the current relative importance patterns unrealistic, revision of the Index weights becomes necessary.

10.3.5 Collecting the data for an index. For a price index, it is necessary to find the price of the proper quantity of each item. As a matter of fact, it is necessary to know the price of the assigned quantity of each item at each of several specific dates, if we wish to compare the price today with the price at other dates in recent years, and if we wish to know what the trend in the price has been.

The Bureau of Labor Statistics collects prices in 46 cities selected from the 97 cities that were used in the 1950 survey of family expenditures. The 46 cities include the 12 largest cities in the United States, 9 other large cities, 9 medium-size cities, and 16 small cities. In each city, the Bureau selected a list of stores and all other types of establishments where families of wage and salary workers buy goods and services. This list includes representative chain stores, independent stores, and department and specialty stores. Prices reported by these stores are averaged together for each city to determine average price changes.

Samples of independent food stores were selected from listings of all outlets in each city, stratified by type of store in terms of foods sold (meat markets, supermarkets, etc.), size of store as measured by annual sales volume, and geographic location within the city. All important chain-store systems are included in the sample for each city. Prices obtained from chain and independent stores are averaged separately and combined with weights based on sales volume data.

Samples of rental units are selected, by probability sampling procedures, from block listings of the total housing rental market in each city, stratified according to block density and location within the area. Rental data are collected from about 30,000 tenants each month.

For small cities, where mail-order buying is important, prices are obtained from catalogs and are included in the calculation of price changes for each city.

Food prices are collected at approximately the middle of each month by local, part-time agents. These often are housewives who meet United States Civil Service requirements for work of this kind and who are specially trained to follow standard pricing procedures and specifications. Commodities and services other than food are priced by the Bureau's full-time field agents, who are carefully chosen through the Civil Service system, and who are intensively trained for this work. Rent information is collected by mail every month from each city.

The Bureau states that approximations of the sampling errors for the average prices indicate that the sampling errors are relatively small— about 5 per cent in most cases but somewhat larger for articles and services that are difficult to describe and to price.

10.3.6 Selecting a base period for the index. People usually prefer to compare present conditions with conditions in a base period that is not too far back in time. The base period should be closely related also to the time period when the pattern of spending habits of consumers was determined. A base period of a year is preferable to any shorter period, because seasonal price fluctuations that might be predominant in a month or other short period are averaged out if the base period is a year. Frequently, a two-year period or a five-year period is used as base period for an index number.

Another important purpose of a series of index numbers is to show the trend in conditions. When this is the primary purpose, it is frequently most convenient to select as the base period the first period for which the data are known or the first period in which we are interested—but this is not always so.

Whenever an index number has been constructed on the basis of a certain base period, it is usually possible to change the base period later on. That is, the previously computed index numbers often can be converted easily to new index numbers in terms of the new base period. The Consumer Price Index is an example of an index for which the base period was changed—from 1925–1929 to 1935–1939 after World War II. On January 1, 1953, the base period was changed to 1947–1949. And on January 1, 1962, the base period for the Consumer Price Index was shifted to 1957–1959.

If an attempt were made to use a new list of items and new weights for those items in order to compute the historical series of Consumer Price Indexes all the way back to the year 1913, certain technical and practical difficulties would be encountered. It would be necessary to find the prices of the items for each year back to 1913. However, no price information exists for the early years for some of the items in the current Consumer Price Index. For example, television sets, nylon goods, and frozen foods were not available to consumers before World War II.

Usually, when the base period is shifted, the revised index numbers for the years prior to the new base period are computed by using the old list of items and the old weights. In other words, the index numbers for years prior to the new base period are revised by merely shifting the base period without changing the items or their weights. This is accomplished by multiplying the old index numbers by 100, the index for the old base period, and dividing by the old index for the new base period.

10.3.7 Computing the index number. After we have collected the data for all the items, we need to find the weighted average of the prices of the items in the sample. The type of formula that is usually employed in the calculation of price index numbers is

$$I_k = \frac{\Sigma p_k q_a}{\Sigma p_o q_a}(100)$$

In other words, multiply the price p_k of each item in the given period, (that is, in the kth period) by the quantity weight q_a determined for the item in a certain period (namely, period a), add these products for all the items, then divide by the similar sum of products for the base period and multiply the quotient by 100 in order to express the result in the form of a percentage.

A similar formula may be written for the calculation of quantity index numbers

$$I_k = \frac{\Sigma p_a q_k}{\Sigma p_a q_o}(100)$$

10.3.8 Limitations in use. The Consumer Price Index is specifically designed to measure the average change in prices of goods and services bought by city wage-earner and clerical-worker families. Consequently, the Index must be applied with caution to other situations.

The Index represents all wage-earner and clerical-worker families, but not necessarily any one family or small group of families. There are limitations on the application of the Index to very low or very high income groups, to elderly couples, to single workers, or to other groups whose level or manner of living and spending are different from the average of all worker families. To the extent that these special groups spend their income differently and are therefore differently affected by price changes, the Index is not exactly applicable. On the other hand, when the Index is applied to all city families or to the total city population, the limitations are not considered by the Bureau to be serious, because the wage-earner and clerical-worker family group represents such a large proportion (nearly two-thirds) of these populations.

Comparison of city indexes shows only that prices in one city may have changed more or less than in another city. City indexes do not measure differences in price levels between cities.

10.4 REMARKS

Have you noticed how important good sampling is in the construction of index numbers? Sampling is involved in several ways in the Consumer Price Index. It enters into the selection of items to represent the field to be described by the Index; the items were determined by interviewing a sample of the city wage-earner and clerical-worker families to determine what items they buy, and then a sample of these items was chosen for the Index. Sampling enters into the determination of the weights for the items. And it enters again in the collection of the price data for each item; price changes are based on prices collected in about 2000 food stores and 4000 other retail stores and establishments; rents are obtained from about 30,000 tenants. Thus the Consumer Price Index involves samples of families, samples of items, samples of stores, and samples of cities. There is even a "sample" of time, because information is collected only at certain periods.

Some people probably would tell you that it is too difficult and too expensive to follow scientific sampling procedures in constructing index numbers. If that is their argument, they should be asked if they would approve of the use of unscientifically constructed index numbers to determine their wages and other matters of importance.

You ought to observe how the idea of correlation can be of great usefulness in the very practical problem of constructing index numbers. For example, even though American families buy thousands of different kinds and qualities of variously priced goods and services, prices of some groups of items may change at about the same rate. For instance, prices of different articles of men's cotton clothing may increase or decrease at about the same time and at about the same rate, so that a change in the price of one item—say, a business shirt—may be representative of the changes in all items. Similarly, the price of fresh milk may be used to typify changes in the prices of buttermilk and various other milk products. Price changes of round steak, chuck roast, rib roast, and hamburger may be used to represent the price changes of all beef products. Changes in doctor's fees for appendix and tonsil operations may be used to represent changes in doctors' fees for surgery in general.

In March 1959, the Labor Department asked Congress to appropriate funds to pay for a revision of the Consumer Price Index. The reason given by the Department for recommending the revision is that American spending habits have changed a great deal since 1950. For example, more money is spent now on television, other home appliances, automobiles, travel, education, and medical care. Smaller proportions of income are now spent on food and dress clothes. The Department estimated that the work of the revision could be completed before January 1964, and that the cost of the revision would be $4,600,000.

EXERCISES

1. Bring Table 10.1 for the Consumer Price Index up to date.

2. Bring Figure 10.1 for the Consumer Price Index up to date.

3. Which components, if any, of the Consumer Price Index have been increasing in price in recent months? Which components, if any, of the Consumer Price Index have been decreasing in price in recent months?

4. Assuming that the Consumer Price Index is a measure of the change in the purchasing power of a consumer's dollar, use the latest available value of the Consumer Price Index to find the current purchasing power of a consumer's dollar as compared with the amount of consumer goods and services that a dollar would have purchased during the period 1957–1959. Relative to 1957–1959 consumer price conditions, what are the "real wages" now of a man whose weekly earnings at present are $105.00 per week?

5. Has industrial production been increasing or decreasing in the United States during recent months? Use the Federal Reserve Board Index of Industrial Production as a basis for your answer to this question.

6. Does the condition indicated by the Index of Industrial Production apply necessarily to the condition of industrial production in a specific city, such as Detroit, Michigan? (*Ans.* No.)

7. Obtain some series of index numbers for the occupation or occupations in which you are most interested and see what you can learn from them.

8. Obtain a series of index numbers for employment in a manufacturing industry, for example, electrical machinery manufacturing, which is growing in New England. Make a graph to illustrate the situation. Make a similar study for a manufacturing industry, such as textile-mill products, which is declining in employment possibilities in New England.

REFERENCES

Philip M. Hauser and William R. Leonard, *Government Statistics for Business Use.* New York: John Wiley & Sons, Inc., 1956.

The Federal Reserve Bulletin, issued monthly by the Board of Governors of the Federal Reserve System, Washington, D. C.

Survey of Current Business, issued monthly by the U. S. Department of Commerce, Office of Business Economics, Washington, D. C.

Statistical Abstract of the United States, published annually by the U. S. Department of Commerce, Bureau of the Census, Washington, D. C.

Chapter 11

REPORTS

11.1 INTRODUCTION

In this course it is assumed that every student in the class will participate in at least one statistical investigation of a problem that is important and interesting to him or her. No statistical investigation is complete until a report of the study has been prepared and delivered. In fact, in most practical situations, the report is the main thing in which your employer is interested.

There is very valuable educational experience to be gained by telling others about your investigations and answering their questions. It is a well-known psychological fact that a person understands better and remembers longer the details of experiences if he has discussed them with someone at the time.

Businessmen and government officials are appalled sometimes at the incompetence of high school, and even college, graduates when they are required to report to their employers about projects on which they have been engaged. Time after time, supervisors must send back material to be rewritten in order that the reports may be understandable and useful.

One of the reasons for the difficulty these employees have in preparing reports is that many of the projects on which they work are statistical—in part, at least. These young men and women never had an opportunity to learn anything about the statistical method of thinking and about methods for the presentation of statistical data when they were in school.

Preparing and delivering reports on your investigations will give you excellent opportunities to develop clear, concise, accurate, and interesting expression. Even correct ideas do not satisfy people if the ideas are concealed by vague language.

Interpreting the results of your investigations, describing them to others, and answering their questions will give you opportunities for clear thinking, logical consistency, and lucid exposition. You will learn the value of forming habits of orderly, systematic thinking and writing. This means that you will have learned how to talk and write your way to higher success in school and in your jobs after you finish school.

By means of your report, you will share knowledge of methods and results with other students who did not participate in your investigation. This will help all members of the class to gain an appreciation of the contributions that mathematical thinking makes to the study of life situations. Report after report will stress the usefulness of samples, probabilities, averages, percentages, standard deviations, correlations, index numbers, tables, graphs, formulas, and the like.

At the same time, you will increase your vocabulary of useful technical terms in the easiest and best way. Technical terms that you learn by memorizing are forgotten quickly. But technical words and phrases that you learn gradually through daily experience with them in your investigations and your discussions and your reports will become permanent parts of your useful, practical knowledge.

11.2 THE PROGRESS REPORT

We are interested in two kinds of reports. The first kind is what is called a *progress report*. It is just what its name implies. It describes (1) the assignment undertaken, (2) the procedures followed since the investigation began or since the last progress report, (3) the ways in which mathematical thinking was useful, (4) any special difficulties encountered and how they were treated, (5) the results obtained, (6) any considerations affecting the reliability of the results or conclusions, (7) any ideas as to how the method could have been improved, and (8) any relation to the work of other committees or subcommittees. Sometimes a progress report may suggest the logical steps to be taken next.

The chairman of the committee should be responsible for arranging for the progress reports, by consultations with subcommittees or individual members of the class. Individuals have the duty of being efficient and conscientious about their assignments and of being punctual in making their reports, so that they will not inconvenience the committee and the class. Punctuality and efficiency are as important in the work of a committee as they are in a football team. Each chairman should try to develop a group spirit, so that his committee members will try to see that their committee is the most capable and effective in the class.

The progress reports ordinarily will be made in writing, and they will be handed to the chairman. Sometimes, however, it will be quite proper for the chairman to arrange for oral progress reports in class or outside class.

11.3 THE FINAL REPORT

The second kind of report is called the *final report*. The final report may be made by the chairman of the committee or by a member of the committee to whom that privilege has been delegated by the chairman. It must be an integration, organization, and summary of all the important features of the investigation.

The final report usually is written first and then delivered orally. It should contain (1) a precise statement of the problem as formulated by the committee; (2) a brief account of all the processes and techniques used during the investigation, including the aspects of mathematical thinking that contributed to the success of the study; (3) the source and amount of data obtained; (4) the method by which the data were obtained; (5) any shortcomings of the data; (6) any insurmountable difficulties encountered in analyzing the data; (7) the type of solution obtained; (8) conclusions about facts and relationships; (9) any assumptions that might modify one's confidence in the conclusions; (10) an evaluation of the whole investigation, suggestions for improving and expediting the work in case another similar investigation should be needed, and questions that remain unanswered; (11) some of the important ways in which people could use the results or the techniques to make life more interesting and more profitable; (12) credits to those who participated for the contributions that they made; and (13) a copy of any forms and/or instructions used in the investigation (often these are placed in an appendix of the report).

If possible, the final report, or at least a summary of its most important parts, should be duplicated so that each member of the class may have a copy. After the report has been delivered to the class, the students who did not participate in the investigation should be given an opportunity to ask pertinent questions and to have their problems with respect to the report clarified. All members of the class should be held responsible for a knowledge of the general nature of the study reported, the most important aspects of mathematical thinking employed, and the main conclusions reached.

Practical applications of the statistical concepts and techniques that you learn can be made quite meaningful to the class. For example, the student who made the final report on a reading preference survey (see Table 5.5, p. 156) mentioned that a person operating a book store or a lending library could keep records of book requests for a while and then make similar analyses as an aid in keeping in stock those books for which the demand is greatest.

In the class discussion after the report had been made, several students who had not participated in the study of reading preferences suggested other practical applications of this type of statistical investigation and of the statistical techniques of analysis and interpretation that were used. Thus, many members of the class benefited from that report.

11.4 PRESENTATION OF DATA

The value of a report is directly dependent upon the quality of the presentation. The form in which complex data are presented may make big differences in what a person can learn from the data, the speed and ease with which he can learn it, and his ability to recall the information at a later date.

There are three ways to organize and present data: the *textual* or narrative form, the *tabular* form, and the *graphical* form. A hundred years ago, data usually were presented in the textual or narrative form; tables and graphs were seldom used. Nowadays, data usually are presented in tabular and graphical form. These two forms tend to be much easier for most people to read, especially if a large amount of related information is involved. Also, tables and graphs present information compactly.

Here is an illustration of the textual or narrative form of presentation.

The average American city family in 1952 spent 30.1 per cent of its total cost-of-living expenditures for food; 32.0 per cent for housing; 9.7 per cent for clothing; 11.0 per cent for transportation; 4.7 per cent for medical care; 2.1 per cent for personal care, such as toilet articles and beauty shop and barber services; 5.4 per cent for reading and recreation, including radio and television sets; and 5.0 per cent for miscellaneous goods and services, such as tobacco, alcoholic beverages, legal services, banking fees, etc. These figures were determined by the Bureau of Labor Statistics by interviewing a representative sample of about 12,000 families out of the more than 18 million wage-earner and clerical-worker families living in more than 3000 cities and towns in the United States.

One of the most common and helpful uses nowadays of the textual or narrative form of presentation of data is in the comments that often accompany a table or chart. These comments help to draw the attention of the reader to the highlights, the most important facts and relationships in the table or chart. For example, sometimes the report on a statistical research project is presented in book form, with tables on the right-hand pages and comments about the most important features of each table on the page opposite. Also, many good statistical reports begin with a few paragraphs in which the highlights of the tables and charts that make up the main part of the report are described in textual or narrative form.

Here is a rule: When you are presenting numerical information in the textual or narrative form, use written numbers (for example, twelve) for making comparisons, and use numerals (for example, 12) for listing specific amounts.

Tables are a form of mathematical symbolism, and they are intended to be reading aids and computational aids. If a table is too long or too complicated it ceases to be helpful in one or both of these ways, and then it is not a good presentation of data.

Tables, graphs, charts, and other visual aids should be used whenever they are helpful in the presentation. Often when facts and relationships are not clearly visible in a table, they may become more obvious and meaningful if the data are presented in graphical form.

It has been said that one picture is worth a thousand words. A good graph functions as a pictorial telegram. It attracts attention, arouses interest, and delivers a message. This leads to recognition or understanding of significant facts and relationships. Because graphs permit a person to obtain a concrete and comprehensive image of the subject matter being presented, they are among the most effective ways for presenting statistical data for certain purposes. Graphs have been used frequently throughout this book in order to help you to visualize the ideas and understand them more clearly.

Some types of graphical presentation make striking displays but are of little use for purposes of computation and estimation. Other types of graphical presentation are very useful for analyzing data but are not likely to arrest the attention of the casual reader. And, of course, propagandists often use graphs deliberately to deceive and mislead the public.

Every good statistical organization has a set of rules that its staff uses as a guide in preparing data for presentation. Consequently, if you should go to work for a statistical organization, inquire immediately about such rules. Procure a copy of the organization's manual on presentation of data, if the organization has one, and study it thoroughly before you waste your time and cause yourself embarrassment by having work rejected because the form in which you presented it to your supervisor is unacceptable. Ignorance of or disregard of rules for presentation of data is one of the most common reasons why some young statistical workers start off on the wrong foot in their jobs.

Furthermore, even if you have worked successfully for one statistical organization, do not take it for granted when you move to another organization that the same rules will apply. Every group has some special rules of its own and the sooner you find out about them and adopt them, the better it will be for you. Consequently, it would be fruitless to attempt here to give you hard-and-fast rules for the presentation of data.

There is another reason why it is not feasible to give one set of rules for all situations. The method by which the report is to be reproduced controls a great many aspects of the presentation of data. For example, if a report is to be printed in regular book form either by linotype or by monotype process, a great many things are possible that are impossible if the report is to be produced on an office typewriter and then reproduced by mimeograph or some photo-offset process. For reports that are to be printed from linotype or monotype, the most widely accepted rules are those in the *Style Manual* which can be purchased from the United States Government Printing Office, Washington, D. C.

For the guidance of students in preparing reports and for people who do not work with a statistical organization but occasionally need to present statistical material in tabular form, here are some suggestions that, if followed, are almost certain to result in tables as attractive and effective as it is possible to produce on a typewriter. Table 11.1 will serve to illustrate most of these rules.

TABLE 11.1

**TYPES OF MOTION PICTURES PREFERRED
BY YOUNG PEOPLE IN MARYLAND***

Preferred Type of Movie	All		Boys		Girls	
	Number	Per Cent	Number	Per Cent	Number	Per Cent
Musical comedy	1,939	21.4	954	19.9	985	23.1
Historical	1,907	21.0	931	19.4	976	22.9
Action and western	1,468	16.2	1,118	23.3	350	8.2
Love story	1,223	13.5	230	4.8	993	23.3
Mystery	834	9.2	484	10.1	350	8.2
Gangster and G-men	514	5.7	446	9.3	68	1.6
Comedy of manners	459	5.1	225	4.7	234	5.5
News and education	422	4.7	269	5.6	153	3.6
Other types	292	3.2	139	2.9	153	3.6
Total	9,058	100.0	4,796	100.0	4,262	100.0

*Source: Howard M. Bell, <u>Youth Tell Their Story</u>. Washington, D. C.: American Council on Education, 1938, Table 71, p. 172. Reproduced here with the permission of the Council.

First of all, there are seven main parts of a simple table. At the top of the table is the table number. Next comes the title or caption. Then there are the column headings. The left-hand column is called the stub. The part in which the data appear is called the body of the table. Then there usually is a row of totals. Finally, footnotes often are needed to complete a tabulation.

1. The table number should be centered above the table. Use capital letters for the word "table." Do not put a period after the numeral. Number

all tables in a report consecutively.

2. Skip a line space. That is, turn the typewriter roller as for a double space. The title or caption should be centered at the top of the table. Do not permit the title to extend beyond the edges of the table. If the title is too long for one line, split it so that the second line is shorter than the first line, and if a third line is necessary it should be shorter than the second line; in this way a long title forms an inverted pyramid. Single space the lines in a title. Use capital letters in the title. Do not put a period at the end of the title.

3. Turn the typewriter roller as for a single space. If the table has more than two columns, make a double ruling with the underlining key. Tables that have more than two columns should be ruled both horizontally and vertically, but they should be left open at both sides. If a table has only two columns, there should be no ruling.

4. Center each column heading in the space provided for it. Capitalize the first letter in every noun, pronoun, adjective, adverb, and verb in the column headings.

5. Capitalize only the first letter of the first word in each item in the stub (capitalize proper nouns and proper adjectives, of course). If an item in the stub runs over into a second line, indent the second line a couple of spaces. Subheadings in the stub should be indented a couple of spaces. The word "total" should be indented a couple of spaces in the stub.

6. Notice that the ruling above the line of totals in the body of the table should not be extended through the stub. Make a single ruling at the bottom of the table.

7. In a long column of decimal fractions, the zero preceding the decimal point may be omitted from all entries except the first and the last. A dollar sign usually is needed only at the first entry in the column and at the total of the column.

8. Ordinarily, the numbers in a column should be aligned vertically by the decimal points. Plus and minus signs in a column should be aligned vertically.

9. Within a table, you may use single spacing or double spacing. If there are only a few lines in the table, use double spacing. If there are a great many lines, use single spacing, except for a double space after every fifth or every tenth line.

10. A footnote may be used to indicate the source of borrowed data. Footnotes may be used also to explain special aspects of the data. If there is only one footnote for a table, use an asterisk to indicate where the footnote applies to the table. If there are two or more footnotes, use the small letters a, b, etc., to relate the footnotes to the table. Place the asterisk or letter a half space above the line of the number or expression to which it refers.

11. If a table is so wide that it reaches across two pages facing each

other, the only caption that appears on the second page is, for example, "TABLE 1 — Continued." The horizontal rulings and all typing must run directly across both pages in perfect alignment.

Each graph or chart in a report usually is assigned a number. A common practice is to label the graphs and charts as Fig. 1, Fig. 2, etc., the label being placed a short distance below the horizontal scale of the graph or chart. For typewritten reports, it is suggested that you arrange the title in a way similar to that suggested above for the title of a table, but put the title of a graph or a chart at the bottom of the graph or the chart.

Judicious use of cross-hatching and colors often adds to the effectiveness of graphs and charts. Kits of adhesive tape containing a variety of cross-hatchings and colors in many widths are available, and they help to simplify the construction of effective graphs and charts.

11.5 REMARKS

Reports should be brief and to the point. In general, they will tend to be too long rather than too short. Lucid brevity is a characteristic of good reports. They should also be impersonal; the statistician cannot afford the luxury of emotion or personal animosity when he is preparing or presenting a report.

Frequently, students are nervous, self-conscious, or embarrassed when they are called upon to deliver a memorized speech on a subject in which they have had no real part or about which they have no genuine interest. On the other hand, it usually happens that a student who has been an active, interested participant in the investigation of a significant situation is both able and eager to tell his classmates, teachers, and others what he has learned. You have a right to be permitted to bring your experience to an appropriate conclusion.

REFERENCES

Herbert Arkin and Raymond R. Colton, *Graphs, How to Make and Use Them*. New York: Harper and Brothers, 1940.

Lyndon O. Brown, *Marketing and Distribution Research*. New York: The Ronald Press Company, 1949, Chap. 25.

Bruce L. Jenkinson, *Bureau of the Census Manual of Tabular Presentation*. Washington: U. S. Government Printing Office, 1949.

Rudolf Modley and Dyno Lowenstein, *Pictographs and Graphs*. New York: Harper and Brothers, 1952.

Mildred B. Parten, *Surveys, Polls and Samples*. New York: Harper and Brothers, 1950, Chap. 17.

Mary Eleanor Spear, *Charting Statistics*, New York: McGraw-Hill Book Co., 1952.

A Manual of Style. Chicago: University of Chicago Press, 1956.

Kate L. Turabian, *A Manual for Writers of Term Papers, Theses, and Dissertations*. Chicago: University of Chicago Press, 1955.

APPENDIX

A.1 GOOD SYMBOLS ARE USEFUL TOOLS

In statistics you must work with two different kinds of quantities: (1) a universe parameter, and (2) an estimate of this parameter obtained from a sample drawn out of the universe. Most people do not like to permit one symbol to stand for two different kinds of thing at the same time. It is easier to keep track of what is going on if a separate symbol is used for each kind of thing.

For this reason, many statisticians have adopted the following rule: Use small, that is, "lower case," Greek letters to stand for universe parameters; use small Roman (Latin, English) letters to stand for estimates obtained from samples. A few exceptions are necessary, however. For example, there is no letter in the Greek alphabet that corresponds to the Roman letter q.

Roman Letter For Estimate from a Sample	*Greek Letter* For Parameter in a Universe	
a	α	alpha
b	β	beta
c	γ	gamma
d	δ	delta
m	μ	mu
p	π	pi
r	ρ	rho
s	σ	sigma
w	ω	omega

The capital, that is, the "upper case," letter N is reserved for the number of elementary units in a universe, and the "lower case" letter n is reserved for the number of elementary units in a sample.

Frequently Used Mathematical Symbols

Symbol	Meaning
$=$	equals, is equal to
$\neq$	is not equal to
$\doteq$	is approximately equal to, approximately equals
$>$	is greater than. For example, $3 > 2$, $-1 > -7$.
$<$	is less than. For example, $8 < 11$, $-14 < +5$.
$\geq$	is greater than or equal to, is not less than
$\leq$	is less than or equal to, is not greater than
$+$	plus
$-$	minus
$\pm$	plus and minus, plus or minus
$\div$	division. For example, $6 \div 3 = 2$.
$/$	division. For example, $1/2$ and $(x+2)/(x-3)$.
$\lvert x \rvert$	absolute value of x, that is, the value of x ignoring the sign. For example, $\lvert +5 \rvert = 5$ and $\lvert -5 \rvert = 5$.
Σ	summation. Greek capital letter sigma.
$!$	factorial. For example, $4! = 4(3)(2)(1) = 24$.
$\binom{n}{r}$	number of combinations of n things r at a time
$(\ )$	parentheses
$\{\ \}$	brace brackets
$[\ ]$	square brackets
$\sqrt{\ }$	square root. For example, $\sqrt{25} = 5$.
∞	infinity. For example, $+\infty$ and $-\infty$, that is, plus infinity and minus infinity.

A.2 THE ACCURACY OF COMPUTATIONS INVOLVING APPROXIMATE NUMBERS

Carrying out a statistical investigation usually involves a considerable amount of arithmetical computation. Frequently, many of the numbers that are involved in statistical computations are approximations. If one or more of the numbers involved in a computation is an approximation, the result of the computation is also an approximation. Consequently, it is important to know the rules indicating the degree of accuracy to be expected in the results of a computation that involves one or more approximate numbers. Before stating these rules it is necessary to define what we mean by the significant digits of an approximate number.

In general, the significant digits of an approximate number are the digits to the left of which and to the right of which are no figures other than zeros. For example, the approximate numbers 37.503, .0037503, and 37503000 have the same five significant digits, namely, 3, 7, 5, 0, and 3. The two zeros at the left end of .0037503 and the three zeros at the right end of 37503000 are not significant digits; they serve only to force the decimal point to the desired place. That is the general rule for determining the number of significant digits in an approximate number. There are, however, a few exceptions. In an approximate number such as 2.370 the zero at the right end of the number is not needed to fix the position of the decimal point; therefore, if a zero is written in that position, it should be interpreted as meaning that the correct figure in that place is zero and then the zero should be counted as a significant digit. Consequently, do not write a zero at the right end of an approximate number such as 2.370 unless you intend that the zero shall be counted as a significant digit. Another exception arises occasionally because some mathematicians have adopted the practice of writing a decimal point at the right end of a number such as 260. to indicate that the zero is to be counted as a significant digit. In other words, they interpret the approximate number 260 as having two significant digits and the approximate number 260. as having three significant digits.

A.2.1 Multiplication and division. If an approximate number that has n_1 significant digits is multiplied by or divided by an approximate number that has n_2 significant digits, the number of significant digits in the result in either case is the smaller of the numbers n_1 and n_2. For example: $123000 \div 2.369 = 51900$ and $(23.76)(0.034) = 0.81$.

A.2.2 Roots and powers. If an approximate number has n significant digits, then both a root and a power of the number have n significant digits. For example: $\sqrt{1470000} = 1210$ and $(0.023)^2 = 0.00053$

A.2.3 Addition and subtraction. In addition and subtraction of approximate numbers, the accuracy of the answer does not only depend upon the number of significant digits in the individual numbers added or subtracted. It also depends upon the positions of the decimal points in the individual numbers. The approximate number that has its right-hand significant digit farthest to the left determines the number of significant digits in the result of addition or subtraction. For example:

23.506	23.506
15700	0.00267
0.00267	23.503 (difference)
15700 (sum)	

A.2.4 Remarks. Although the rules stated above may be proved to be correct statements of the average degree of accuracy to be expected in computations involving approximate numbers, it is not always possible to follow them strictly in statistical work. Frequently, statistical analysis involves a long series of computations involving additions, subtractions, multiplications, divisions, and square roots, and it may not be possible to foresee at the very beginning of the calculations how many significant digits may be lost during the course of the long series of computations. Consequently, when a statistician knows or has some reason to expect that the number he obtains as a result of one computation may be involved in several other computations before the analysis is completed, he usually retains a couple of digits beyond those that are significant in accordance with the rules stated above. Unless this is done, there will often be no significant digits in the final result. Nevertheless, it is important that the statistician be aware of what he is doing.

In connection with the use of random samples to obtain estimates of the parameters in universes, sampling error is involved in determining the numbers of digits that should be considered significant or retained in the final expressions for the estimates. The sampling error of an estimate usually is indicated by the standard deviation of the estimate.

In general, consideration of the sampling error justifies the retention of one more significant digit in the final expression for the mean of a sample than were in the original observed data, especially if n, the size of the sample, is as great as 100. For formula 7.2 shows that the standard deviation of the mean of a sample in which $n = 100$ is 10 times smaller than the standard deviation of the values of the variable.

Similar considerations suggest that in the final expression for the standard deviation of the variable in a sample we may retain one more significant digit than were in the original observed data.

It is customary in statistical work to write the values of proportions to two or three decimal places, depending upon the size of the sample; unless the value of n is at least as great as 100, not more than two decimal places should be retained in the final expression for the proportion possessing an attribute in a sample. In the final expression for the standard deviation of a proportion it is usually desirable to retain one more decimal place than the number of decimal places retained in the proportion itself.

Two decimal places usually are retained in the final expression for the value of a correlation coefficient obtained from a sample. Occasionally, it may be desirable to retain three or even four decimal places in the value of a correlation coefficient, especially if the size of the sample is extremely large or if the value of the correlation coefficient is extremely close to $+1$ or -1.

Under most circumstances only two decimal places will be retained in the values of z and t obtained from samples. However, you should be able to recognize exceptions to this statement when they arise in your work.

A.2.5 Computation of a square root. Every time you need to find the standard deviation of a sample, you will find it necessary to obtain a square root of the variance of the sample. A square root of a given number is another number that when multiplied by itself produces the given number. Tables of square roots are available in nearly all schools and libraries; your teacher can show you how to use a table to find a square root.

Several different methods may be used for computation of a square root of a number. One method will be demonstrated for you here. Let us suppose that we need to find a square root of 213.61 and that the number 213.61 is an approximate number correct to five significant digits.

The method begins with inspection of the number 213.61 to find some number that when multiplied by itself will produce a number not far from 213 or 214. For example, we might notice that $(14)(14) = 196$ and $(15)(15) = 225$. Consequently, we may say that the square root we are seeking is between 14 and 15, and we may use 14.5 as a first approximation to the square root of 213.61. The next step consists in dividing 213.61 by 14.5, obtaining the quotient 14.73, retaining at least one more significant digit than is in the divisor, 14.5. Now we must find the average of the divisor and the quotient, that is, $(14.50 + 14.73)/2 = 14.615$. This number, 14.615, will be used as a second approximation of the desired square root. When we divide 213.61 by 14.615, we obtain the quotient 14.6158, retaining at least one more significant digit than there is in the divisor 14.615. Next, we find the average of the divisor and the quotient, that is, $(14.6150 + 14.6158)/2 = 14.6154$.

Because the original number 213.61 is correct to only five significant digits, we are justified in retaining five and only five significant digits in the square root. Consequently, we round off the last digit in 14.6154 and take 14.615 as a square root of 213.61.

It can be proved that each new average we obtain by this method is a better approximation of the square root than the preceding average.

If you need to find a square root of a number such as 0.00021361, you can use as your first approximation the number 0.0145 and proceed by division and averaging to show that the desired square root is 0.014615 correct to five significant digits.

A.3 COMBINING ESTIMATES FROM DISPROPORTIONATE SUBSAMPLES

When a total sample is made up of two or more subsamples, the estimate of the total-universe parameter furnished by the total sample should be computed as a "weighted average" instead of a "simple average" of the estimates provided by the subsamples. Otherwise, an incorrect over-all estimate is likely to be obtained, unless you have the relatively rare situation in which all the strata in the universe contain exactly the same number of elementary units and there is one subsample from each stratum.

For example, suppose that you draw a random sample of the total student body to make an attitude survey on a proposal among a universe of 1000 men and women in a school where 30 per cent of the students are women and 70 per cent are men. Suppose that the survey shows that 65 per cent of the women are in favor of the proposal and only 48 per cent of the men are in favor of the proposal. What is the correct estimate of the per cent of this universe of men and women who are in favor of the proposal? The correct estimate is not $(65 + 48)/2 = 113/2 = 56.5$ per cent. The correct answer is

$$\frac{30(65) + 70(48)}{30 + 70} = \frac{1950 + 3360}{100} = \frac{5310}{100} = 53.1 \text{ per cent}$$

You can verify this estimate very easily simply by computing the total number of men and women who are in favor of the proposal and dividing this total by 1000. In general, the weighted average p of the two proportions p_1 and p_2 obtained from two subsamples drawn from two strata of a universe is

$$p = \frac{w_1 p_1 + w_2 p_2}{w_1 + w_2}$$

where the weights w_1 and w_2 are proportional to the sizes of the two strata in the universe. You can extend this idea very easily to the weighted average of three, four, or more estimates of p from subsamples.

The general formula for a weighted average of any set of constants $c_1, c_2, c_3, \ldots, c_k$ is

$$c = \frac{w_1 c_1 + w_2 c_2 + w_3 c_3 + \ldots + w_k c_k}{w_1 + w_2 + w_3 + \ldots + w_k}$$

where $w_1, w_2, w_3, \ldots, w_k$ are the weights.

A.4 PROBABILITY OF BEING IN A SAMPLE DRAWN WITHOUT REPLACEMENT

Let us borrow from algebra a formula that tells us the number of different combinations containing v things that it is possible to make out of n different things. The symbol and the formula for this are

$$\binom{n}{v} = \frac{n!}{v!(n-v)!}, \quad n \geq v$$

where, by definition,

$$n! = n(n-1)(n-2) \ldots (2)(1)$$
$$v! = v(v-1)(v-2) \ldots (2)(1)$$
$$(n-v)! = (n-v)(n-v-1)(n-v-2) \ldots (2)(1)$$

These last expressions are called "factorials." For example, we read

$n!$ as "*n* factorial." For our purposes in this book, we may define a factorial as the product of a positive integer and all the lesser integers on down to 1. We need an additional definition, namely, that $0! = 1$.

In connection with the formation of combinations of *n* things taken *v* at a time, it is important to remember that merely changing the order in which the items appear in a combination does not produce a different combination. For example, the combination ABC of the three letters A, B, and C is not a combination different from ACB or BAC or CAB or BCA or CBA.

The number of different samples containing *n* elementary units that it is possible to obtain from a universe of *N* different elementary units by random sampling without replacement is, then, $\binom{N}{n}$. Now, set aside the specified elementary unit, say, A, and then see how many different combinations of $n-1$ elementary units we can make out of the $N-1$ elementary units left in the universe; obviously, the number of combinations is $\binom{N-1}{n-1}$. If the elementary unit A is added to each of these combinations, we will have $\binom{N-1}{n-1}$ samples of size *n* each of which includes the specified elementary unit A.

Therefore, the probability that any specified elementary unit will be included in a random sample of size *n* drawn without replacement from a universe containing *N* elementary units is

$$P = \frac{\binom{N-1}{n-1}}{\binom{N}{n}}$$

which, when simplified by dividing out all the factors common to the numerator and the denominator, reduces to

$$P = \frac{n}{N}$$

A.5 SOME TECHNIQUES IN THE ANALYSIS OF SERIAL DISTRIBUTIONS

A.5.1 Short cut to reduce difficulty of computation. If a person does statistics the hard way, he will have a very large amount of arithmetical computation, and no one is fond of that kind of work. The good statistician uses mathematical principles to devise short cuts that will eliminate a great deal of the arithmetical computation. Let us consider the following illustration of one of the best ways to reduce the amount and difficulty of the arithmetical computation in statistical analysis.

Let us look again at the data in Table 3.2. Suppose that we do not know the mean of the pay rates of the 12 truck drivers and that someone asks us to guess approximately what the mean of the pay rates is. By merely looking at the 12 numbers and without doing any computation we would probably guess that the mean is approximately $1.80. Now, build a new table by subtracting our guessed mean from each of the values of the variable. The values obtained by subtraction are written in the third column of Table A.1.

TABLE A.1

Driver's Initials	Hourly Rate of Pay (Dollars) x	x − 1.80
E. K.	1.35	−0.45
H. B.	1.50	−0.30
M. C.	1.55	−0.25
H. V.	1.55	−0.25
L. W.	1.65	−0.15
T. A.	1.80	0.00
A. J.	1.80	0.00
J. S.	1.80	0.00
R. Q.	1.95	0.15
B. M.	2.05	0.25
F. T.	2.15	0.35
V. P.	2.33	0.53
Total	21.48	−0.12

TABLE A.2

Driver's Initials	Hourly Rate of Pay (Dollars) x	x − 1.75
E. K.	1.35	−0.40
H. B.	1.50	−0.25
M. C.	1.55	−0.20
H. V.	1.55	−0.20
L. W.	1.65	−0.10
T. A.	1.80	0.05
A. J.	1.80	0.05
J. S.	1.80	0.05
R. Q.	1.95	0.20
B. M.	2.05	0.30
F. T.	2.15	0.40
V. P.	2.33	0.58
Total	21.48	+0.48

Add all the numbers in the third column and write the total at the bottom of the column. This total is −0.12. Divide the total by 12, that is, divide the total by the number of values of the variable in the sample. The result is −0.12/12 = −0.01. Add −0.01 to $1.80, the guess that we made at the beginning. The result is $1.79. Compare this result with the value that we found for the mean of these data in section 3.3.4. The two values are exactly the same.

Suppose we guess that the mean is $1.75 rather than $1.80. If we subtract 1.75 from each of the values of the variable in the second column of Table A.2, we obtain the numbers in the third column of the Table. The total of the third column of Table A.2 is +0.48. Divide this total by 12. The result is +0.48/12 = +0.04. Add +0.04 to the guess, namely, $1.75. The result is $1.79—exactly the same answer we obtained before.

No matter what your guess happens to be, you will always obtain the exact value of the mean of the sample if you follow the steps that have been used in the two examples given above. A proof of this statement will be given now.

Let the n values of the variable in a sample be the n values of x that are given in the first column of Table A.3. Let us represent our guessed value of the mean of the values of x by the constant c. Now, subtract the constant c from each value of the variable and write the results of these subtractions in the second column. The best label for the heading of the second column is $x - c$, because that symbol indicates exactly how the numbers in the second column were obtained; they were obtained by subtracting c from each value of x.

TABLE A.3

Variable	
x	x − c
x_1	$x_1 - c$
x_2	$x_2 - c$
x_3	$x_3 - c$
.	.
.	.
.	.
x_n	$x_n - c$
Σx	$\Sigma x - nc$
	$= \Sigma(x - c)$

We wish to prove that the mean m of the values of the variable x in a sample may be obtained by the formula

$$m = \frac{\Sigma(x - c)}{n} + c$$

Proof: We know by definition that $m = \Sigma x/n$. The total of the second column in Table A.3 shows that $\Sigma(x - c) = \Sigma x - nc$. Consequently,

$$\frac{\Sigma(x - c)}{n} + c = \frac{\Sigma x - nc}{n} + c = \frac{\Sigma x}{n} - \frac{nc}{n} + c = m - c + c = m$$

We shall make use of this privilege of subtracting a constant from the original data in many of the tables in the remainder of this book. Remember that the purpose of subtracting a constant from every value of the variable in a sample is to shorten and simplify the arithmetical computation that must be done in the statistical analysis of data. One of the ways in which this principle shortens and simplifies the computations is by giving us smaller numbers to work with. Everyone prefers to add or subtract or multiply small numbers rather than large ones.

Although we have proved that the principle holds true no matter what constant we subtract, nevertheless some constants simplify the work more than others. Consequently, if you wish to take full advantage of the privilege of subtracting a constant in analyzing data, you should use good judgment in selecting the constant to be subtracted. Here are some suggestions to guide you.

First, the nearer you guess to the mean, the smaller will be the absolute values of the numbers in the $(x - c)$ column of your table. Try to make a guess that is not far from the mean of the sample, but do not waste time trying to guess extremely close to the mean; any constant in the general neighborhood of the mean may be used.

Second, choose for your constant a number that is as similar to the original values of the variable as possible. For example, if the original values of the variable are all whole numbers, choose a whole number as the constant, so that the subtractions will result in whole numbers rather than decimal fractions. If the original values of the variable all end in .5, then choose a constant that ends in .5, so that the results of the subtractions will be whole numbers. If all of the values of the variable happen to be even numbers, then choose an even number as the constant. In other words, choose as the constant a number that will produce the simplest set of numbers when you subtract it from each of the values of the variable in the sample.

A.5.2 Short method for computing the standard deviation. The long method for finding the value of the standard deviation of a sample requires that you first find the mean of the sample. In the short method, you do not need to know the value of the mean in order to find the standard deviation of the sample. Table A.4 is an example of the kind of table that may be used for finding the value of the standard deviation by the short method. The original values of the variable are written in the first column.

Next you choose some convenient constant to subtract from every value of x that appears in the first column of the table. In Table A.4, the number 1.80 was chosen as the constant because it would produce some zeros in the second column and because all the numbers in the second column would be easy to square for the third column. The easiest numbers to add or multiply are zeros. The next easiest numbers to add or multiply are ones.

As soon as the convenient constant has been chosen, it is subtracted from the original values of the variable and the results are written in the second column of the table. The symbol that best describes the numbers in the second column is $x - 1.80$. Then each of the numbers in the second column is squared and the result is written in the third column. The symbol that best describes the way in which the numbers in the third column were obtained is $(x - 1.80)^2$. $\Sigma(x - 1.80) = -0.12$ and $\Sigma(x - 1.80)^2 = 0.9284$.

The standard deviation of the sample is obtained by the following easy computation:

$$s = \frac{1}{12}\sqrt{12(0.9284) - (-0.12)^2} = \frac{1}{12}\sqrt{11.1408 - 0.0144} = \frac{1}{12}\sqrt{11.1264}$$

$$= \frac{3.336}{12} = 0.278 = \$0.28$$

TABLE A.4

Hourly Rate of Pay x	x − 1.80	(x − 1.80)²
$1.35	−0.45	0.2025
1.50	−0.30	0.0900
1.55	−0.25	0.0625
1.55	−0.25	0.0625
1.65	−0.15	0.0225
1.80	0.00	0.0000
1.80	0.00	0.0000
1.80	0.00	0.0000
1.95	0.15	0.0225
2.05	0.25	0.0625
2.15	0.35	0.1225
2.33	0.53	0.2809
	−0.12	0.9284

If you review the steps performed in finding the standard deviation for the sample in Table A.4, you will see that the formula for the short method of finding the standard deviation for any serial distribution of values of a variable x in a sample is

$$s_x = \frac{1}{n}\sqrt{n\Sigma(x - c)^2 - [\Sigma(x - c)]^2}$$

where n is the number of values of the variable in the sample and c is the arbitrary constant that you choose to subtract from the values of x.

You may wish to construct tables similar to Table A.4 by using values other than 1.80 for the constant. For example, let $c = 1.75$ or 1.85 or 1.82. In every instance, you should obtain the answer $s = \$0.28$. However, the arithmetical computation may be slightly more difficult than it was when 1.80 was chosen as the constant for this sample.

Because it is permissible to use any value for c in any problem of finding the standard deviation of a serial distribution in a sample, let us see what happens to the last formula if we choose the value zero for c. Substituting zero for c in the formula gives us the new formula

$$s_x = \frac{1}{n}\sqrt{n\Sigma x^2 - (\Sigma x)^2} \tag{3.4}$$

Formula 3.4 is one of the best known of all formulas for the standard deviation. It is used frequently by statisticians who have automatic electric calculating machines, because those machines multiply and add large numbers almost as quickly as they multiply and add small numbers. Table A.5 shows the tabulation that may be used if you intend to find the value of the standard deviation of a serial distribution by formula 3.4.

TABLE A.5

x	x^2
1.35	1.8225
1.50	2.2500
1.55	2.4025
1.55	2.4025
1.65	2.7225
1.80	3.2400
1.80	3.2400
1.80	3.2400
1.95	3.8025
2.05	4.2025
2.15	4.6225
2.33	5.4289
21.48	39.3764

By substituting 21.48 for Σx and 39.3764 for Σx^2 into formula 3.4, we find that

$$s = \frac{1}{12}\sqrt{12(39.3764) - (21.48)^2} = \frac{1}{12}\sqrt{11.1264} = \frac{3.336}{12} = \$0.28$$

We have, then, three methods for computing the standard deviation of a serial distribution in a sample. Formula 3.3 usually is considered to be the definition of the standard deviation of a serial distribution in a sample, but it generally is the poorest of the three formulas for computation purposes. Formula 3.4 is sometimes suitable for use when automatic electric calculating machines are available; in those situations it is not necessary to construct a table such as Table A.5, because the values of the variable may be squared and accumulated in the machine to show the only quantities needed for formula 3.4, namely, the totals Σx and Σx^2 and the number n. Formula 3.4 is useful also for some theoretical purposes in advanced courses in statistics.

A.5.3 Finding m and s at the same time. Have you noticed that the short method for finding the standard deviation of a serial distribution in a sample enables us to find the mean of the sample at the same time without additional tabulations? Table A.4 contains all the numbers that are in Table A.1 which was used in computing the mean by the short method. Consequently, both the mean and the standard deviation may be obtained from Table A.4. You have been shown already how to compute the standard deviation from Table A.4. To find the mean from Table A.4, compute

$$m = \frac{\Sigma(x - 1.80)}{12} + 1.80 = \frac{-0.12}{12} + 1.80 = -0.01 + 1.80 = \$1.79$$

Form the habit of computing the mean and the standard deviation of a sample at the same time by the short method. This habit will eliminate a lot of useless computation. Furthermore, as suggested earlier in this book, it is a good idea to think of the mean and the standard deviation together as a powerful team of basic estimates about the two most important parameters of the universe from which the sample was drawn. Remember, also, that the mean and the standard deviation always are expressed in the same unit of measurement as the original data in the sample.

A.5.4 Remarks about the sigma notation. In the previous pages, we have made use of some important rules for working with the sigma notation for summation. The most important of those rules are

$$\Sigma(x + y) = \Sigma x + \Sigma y$$
$$\Sigma(x - y) = \Sigma x - \Sigma y$$
$$\Sigma kx = k\Sigma x \text{ if } k \text{ is a constant}$$
$$\Sigma k = nk \text{ if } k \text{ is constant}$$

In each of the four rules stated above, the right-hand side of the equation is equal in value to the left-hand side. This means that you are free to employ whichever side of the equation is easier to compute or serves your purpose better at any particular moment. In general, you probably will find computations easier if you use the expressions on the right-hand side of the equations for most of the situations described in this book.

Here is a word of warning about two mistakes frequently made by students when they are learning to work with the sigma notation for summation. If you make either of these mistakes in analyzing the data in your samples, your results are likely to be very seriously in error. The first mistake that is often made is substituting the value of Σx^2 for $(\Sigma x)^2$. The second serious mistake is substituting $(\Sigma x)(\Sigma y)$ for Σxy. If you think clearly about the operations that are indicated by the symbols in each of these situations, you will see why the substitutions cannot be made without danger of error.

For example, Σx^2 indicates that each of the values of x is squared and, after that, the squared numbers are summed. On the other hand, $(\Sigma x)^2$ indicates that the values of x are summed and, after that, the sum is squared. Practice with four or five values of x and you will soon understand the principles that are involved. The mistakes are caused by performing the operations of addition and multiplication in the wrong order.

EXERCISES

1. How do you decide what to use as the column heading when you are inserting a new column in a table for computation purposes?

2. What is the sigma notation for the summation $a_1 + a_2 + a_3 + a_4 + a_5$?

$$\left(Ans. \; \sum_{i=1}^{5} a_i.\right)$$

3. Given $x_1 = 7$, $x_2 = 6$, $x_3 = 2$, and $x_4 = 1$, write out the individual terms in the summations Σx and Σx^2. Show that Σx^2 is not equal to $(\Sigma x)^2$. (*Ans.* $\Sigma x = 7 + 6 + 2 + 1 = 16$, $(\Sigma x)^2 = (16)^2 = 256$, $\Sigma x^2 = 7^2 + 6^2 + 2^2 + 1^2 = 49 + 36 + 4 + 1 = 90$.)

4. Given $x_1 = 3$, $x_2 = 1$, $x_3 = 5$, $x_4 = 0$, and $x_5 = 4$, find the values of Σx and Σx^2. Is Σx^2 equal to $(\Sigma x)^2$ here? (*Ans.* $\Sigma x = 13$, $(\Sigma x)^2 = (13)^2 = 169$, and $\Sigma x^2 = 51$).

5. Can you give a logical reason why Σx^2 is not equal to $(\Sigma x)^2$ for the n values of x in a sample?

6. Using the values of x given in exercise 3, write out the individual terms in $\Sigma 3x$ and $3\Sigma x$ and show that each expression is equal to 48. Evaluate similar examples, if necessary, to convince yourself that

$$\sum_{i=1}^{n} kx_i = k \sum_{i=1}^{n} x_i \; , \; \text{if } k \text{ is a constant}$$

7. If $x_1 = 7$, $x_2 = 7$, $x_3 = 7$, $x_4 = 7$, $x_5 = 7$, and $x_6 = 7$, prove that here $\Sigma x = \Sigma 7 = 6(7) = 42$. Evaluate similar examples, if necessary, to convince yourself that

$$\sum_{i=1}^{n} k = nk, \; \text{if } k \text{ is a constant}$$

8. What is the sigma notation for the summation $x_1y_1 + x_2y_2 + x_3y_3 + x_4y_4$?

$$\left(Ans. \; \sum_{i=1}^{4} x_i y_i\right)$$

9. Given $x_1 = 2$, $x_2 = 3$, $x_3 = 1$, $y_1 = 1$, $y_2 = 5$, $y_3 = 4$, write out the individual terms in the summations Σx, Σy, and Σxy. (*Ans.* $\Sigma x = 6$, $\Sigma y = 10$, $\Sigma xy = (2)(1) + (3)(5) + (1)(4) = 2 + 15 + 4 = 21$.) Is Σxy equal to $\Sigma x\Sigma y$ here? (*Ans.* No, $\Sigma xy = 21$ and $\Sigma x\Sigma y = (6)(10) = 60$.) Evaluate similar examples, if necessary, to convince yourself that, in general,

$$\sum_{i=1}^{n} x_i y_i \neq \left(\sum_{i=1}^{n} x_i \right)\left(\sum_{i=1}^{n} y_i \right)$$

10. Can you give a logical reason why, in general, Σxy is not equal to $\Sigma x\Sigma y$ for the n paired values of x and y in a sample?

11. Show that the quantity $n\Sigma x^2 - (\Sigma x)^2$ in formula 3.4 cannot be negative for real values of x. In other words, prove that for every sample, $n\Sigma x^2 - (\Sigma x)^2 \geq 0$. (Hint: Begin with $\Sigma(x - m)^2 \geq 0$.)

12. Prove that if every value of the variable in a distribution is multiplied by the same positive constant, the values of the mean and the standard deviation are multiplied by the constant. First, convince yourself that this is true by multiplying every value of *x* in Table 3.2 by the same positive constant, such as 2 or 3, and computing the new mean and the new standard deviation; then generalize the idea. What happens to the variance of a distribution when every value of the variable is multiplied by the same positive constant?

13. What happens to the value of the mean of a distribution if the same constant is subtracted from every value of the variable in the distribution? (*Ans.* The new mean can be found by subtracting the constant from the old mean.) What happens to the variance and the standard deviation of a distribution if the same constant is subtracted from every value of the variable in the distribution? (*Ans.* The variance and the standard deviation remain unchanged in value.) Apply these principles to the distributions in Tables 1.4 and 1.5.

14. Prove algebraically that $\dfrac{1}{n}\sqrt{n\Sigma x^2 - (\Sigma x)^2} = \sqrt{\dfrac{\Sigma(x-m)^2}{n}}$.

15. Prove algebraically that $\dfrac{1}{n}\sqrt{n\Sigma(x-c)^2 - [\Sigma(x-c)]^2} = \sqrt{\dfrac{\Sigma(x-m)^2}{n}}$.

A.5.5 Rank of an elementary unit. The order in which things stand or occur in a distribution is often of considerable interest. For instance, a man likes to be one of the most skillful in his group both at work and at play. A woman may not like to be the tallest in a large group of people. In other words, people seem to like to rank things, that is, to label them according to their relative magnitudes or importance.

TABLE A.6

Hourly Rates of Pay of Twelve Light-Truck Drivers in Boston

Driver's Initials	Hourly Rate of Pay	Driver's Rank
V. P.	$2.33	1
F. T.	2.15	2
B. M.	2.05	3
R. Q.	1.95	4
J. S.	1.80	6
A. J.	1.80	6
T. A.	1.80	6
L. W.	1.65	8
H. V.	1.55	9.5
M. C.	1.55	9.5
H. B.	1.50	11
E. K.	1.35	12

Mathematics provides a simple method for this kind of ordering and labeling. For example, let us attach ranks to each of the 12 truck drivers in Table 3.2. We do this by giving the rank 1 to the driver with the highest pay rate, V. P., and entering this rank in the first row and third column of Table A.6. Similarly, we assign rank 2 to F. T. because he had the second-highest rate of pay; we assign ranks 3 and 4 to B. M. and R. Q., and so on. It is easy to see how these first four ranks were determined.

But it would not be correct to assign to the next three drivers, each of whom received $1.80 per hour, the ranks 5, 6, and 7, because that would be the way we would rank them if no two of them received the same rate of pay. Probably you would not consider it fair to give three different ranks to three students who received exactly the same mark on an examination. Statisticians resolve this difficulty by giving to each of the three individuals the average of the ranks that they would have received if they had all been different. Therefore, we give to each of the three drivers who earned $1.80 per hour the average of the ranks 5, 6, and 7, namely, $(5 + 6 + 7)/3 = 18/3 = 6$.

Next in order is L. W., and he is given rank 8. No one is given rank 5 or rank 7 in this table because 5 and 7 were averaged with 6, and three sixes were assigned instead of 5, 6, and 7.

If M. C. had received less pay per hour than H. V., they would be ranked 9 and 10. But, because both of them received $1.55 per hour, we assign to each of them the average of 9 and 10, that is, $(9 + 10)/2 = 19/2 = 9.5$. Then we assign rank 11 to H. B. and rank 12 to E. K.

If four people should have the same value of the variable, write down the four ranks that they would be given if all their values were different. Then find the average of those four numbers. Use this average rank as the rank of each of the four individuals. Do not assign any of the four averaged numbers to any other individual in the table.

A little experimentation will show you that whenever an even number of individuals are identical in value they will be given a rank ending in .5. Whenever an odd number of individuals have the same standing in a distribution, they receive a whole number as their rank.

Ranking is a simple and often useful method for describing the order of importance of elementary units in a distribution. But it is not a completely satisfactory method; it has shortcomings, especially for small groups. In the first place, ranking individuals in order of merit does not take into account the amount of difference between two individuals. That is, it does not tell us the importance of the difference between two ranks. The first person may be slightly or tremendously superior to the second person. For example, in an examination there might be ten test points' difference between the first and second students, but only one test point between the second and third students. Consequently, a difference of one unit in rank indicates that one individual is superior to the other, but it does not indicate the amount of superiority. In other words, ranking the elementary units in a sample produces a partly quantitative classification; it does not produce a truly quantitative classification.

A small amount of thinking will convince you that there is another serious disadvantage in describing a person merely by giving his rank in a distribution. For example, suppose that you are told that a student has rank 3 in some group. That does not tell you very much of value about the person. Suppose that there were only three in the group; then the person is the lowest in the group. However, if there were 100, the person is very near the top of the group. You can see that rank by itself does not tell us much about a person unless it is combined with the number in the group.

A.5.6 Percentile rank. Statisticians have figured out how to calculate one number that has some of the combined effect of the rank and the number in the group. This number is called the percentile rank, usually abbreviated to P.R.

Percentile ranks are useful for comparing the relative standings of the members of a particular group. They are useful also for comparing a particular person's standing in two or more groups. Colleges and employment offices usually ask the principals of high schools to give them the percentile ranks of high school graduates. Use the formula that we are going to develop now to compare your own standing in several of the courses that you have taken.

Percentile rank is not difficult to understand. It simply means assigning each person a position on a scale that begins with 0 and ends with 100. Each person is given an equal amount of the scale and the middle value of his part of the scale is his percentile rank.

For example, there are 12 individuals in the distribution in Table A.6. Let us draw a scale on a straight line and divide the distance from 0 to 100 into 12 equal parts. Each part will be $8\frac{1}{3}$ units. Now, what is the percentile rank of V. P., the top individual in the table? In other words, what is the middle value of the section of the scale occupied by V. P.?

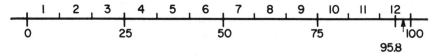

There are 11 individuals below V. P. and, therefore, there are 11 parts of the scale below the beginning of V. P.'s part of the scale. The middle of V. P.'s part of the scale is at the point that marks 11.5 parts of the scale. Consequently, the percentile rank for V. P. is 11.5 (100/12) = 95.8 (96, if rounded off to the nearest whole number).

It is customary to write the percentile rank to the nearest whole number if n, the number of individuals in the distribution, is less than 100. If n is between 100 and 1000, the percentile rank usually is given to one decimal place.

Here is a simple way to determine the proper percentile rank of any individual in a distribution. First, determine the proper ranks for the individuals. Let R be the rank of a person in a distribution and let n be the number of individuals in the distribution. Then the number of spaces in the scale below the person is $n - R$. To move up to the middle of the person's portion of the scale, add half a section to $n - R$ obtaining $(n - R + 0.5)$ sections of the scale. But each section of the scale contains $100/n$ units, because the scale distance from 0 to 100 is divided equally among the n individuals. Consequently, the percentile rank of the person whose rank is R is

$$\text{P.R.} = (n - R + 0.5)\left(\frac{100}{n}\right) \qquad (\text{A.1})$$

Show yourself how to develop this formula by actually determining the middle value in each of the sections of the scale for the 11 individuals other than V. P. in Table A.6. The correct answers are given in the last column of Table A.7.

The three people who earned $1.80 per hour have the same rank and the same percentile rank. In this way, the formula for percentile rank is fair to all of them. The same is true for the two drivers who earned $1.55 per hour. These groups of equal individuals are assigned the percentile rank corresponding to the middle point of the part of the scale that belongs to the group.

TABLE A.7

Hourly Rates of Pay of Twelve Light-Truck Drivers in Boston

Driver's Initials	Hourly Rate of Pay	Driver's Rank R	Driver's Percentile Rank P.R.
V. P.	$2.33	1	96
F. T.	2.15	2	88
B. M.	2.05	3	78
R. Q.	1.95	4	71
J. S.	1.80	6	54
A. J.	1.80	6	54
T. A.	1.80	6	54
L. W.	1.65	8	38
H. V.	1.55	9.5	25
M. C.	1.55	9.5	25
H. B.	1.50	11	12
E. K.	1.35	12	4

Suppose that you stand seventh in a class of twenty-two students in English composition and ninth in a class of thirty students in history. In which class do you have the higher relative standing? An easy way to answer this question is to compute your percentile rank for each of the two subjects and compare the two percentile ranks. Your percentile rank in English composition is $(22 - 7 + 0.5)(100/22) = 70$. Your percentile rank in history is $(30 - 9 + 0.5)(100/30) = 72$. Consequently, your relative standing in history is higher than your relative standing in English composition.

A person is said to be at the 99th percentile or to have percentile rank of 99 if only 1 per cent of the people in the distribution are superior to him. Similarly, a person is said to be at the 63rd percentile or to have percentile rank of 63 if 37 per cent of the people in the distribution are better than he.

Percentile ranks have one serious shortcoming. In most distributions, there is a tendency for the individuals or the items to cluster heavily around the middle of the range of the variable. And there usually are only a few widely separated individuals or items near the extremes of the range of the variable. Consequently, great care must be taken in interpreting differences in percentile ranks.

For example, because of the heavy clustering of test scores near the middle of the range, the student whose percentile rank is 60 may be only two or three test points superior to the student whose percentile rank is 40. In other words, a difference of 20 in percentile ranks near the middle of the distribution may mean only two or three additional correct answers on a test. On the other hand, because of the wide separation of test scores near the extremes of the range, the student whose percentile rank is 99 may have earned ten test points more than the student whose percentile rank is 98. But only one percentile separates them.

Therefore, when you are interpreting percentile ranks, remember that a difference of one percentile around the middle of the distribution, that is, near P.R. = 50, is much less important or significant than a difference of one percentile near the ends of the range, that is, near P.R. = 99 and P.R. = 1.

When we are dealing with truly quantitative data, the method of making comparisons by reducing the data to standard units often is preferred over the percentile-rank method. One reason for this preference is the role that the mean and the standard deviation have in the normal probability law. The percentile rank has no such direct relationship to the normal probability law or to any other fundamental law of statistics.

EXERCISES

1. Assign the proper rank and percentile rank to each student in Table 3.3 (p. 67).

2. Assign the proper rank and percentile rank to each student in the distribution that you obtained in exercise 1 on page 79.

3. How do you interpret a person's percentile rank? In other words, how would you explain to someone what his or her percentile rank in a distribution means?

4. The amount of difference between the elementary units rated is not taken into account by which of the following methods for making evaluative ratings? (a) the ordered array method ☐ (b) the ranking method ☐ (c) the percentile rank method ☐ (d) all of the above ☐.

5. Assume that the following pairs of numbers represent the percentile ranks of two individuals among several hundred students who took the same achievement test. Which pair of percentile ranks ordinarily would represent the least amount of difference in achievement between the two individuals? (a) 20 and 30 ☐ (b) 45 and 55 ☐ (c) 70 and 80 ☐ (d) 85 and 95 ☐.

6. The main weakness of the percentile rank method of classifying elementary units is the fact that (a) a person's age has a great deal to do with his percentile rank ☐ (b) there is a tendency for randomly selected elementary units to cluster around low values of the variable ☐ (c) percentile ranks are not expressed in convenient and common units that are applicable to large distributions ☐ (d) any given percentile interval, for example, five percentile points, does not represent the same amount of difference between two elementary units at different places on the percentile scale ☐.

7. Which of the following methods of classifying elementary units may have a fundamental relation to the normal distribution? (a) rank ☐ (b) percentile rank ☐ (c) standard score ☐.

8. An important advantage in using the standard score instead of the percentile rank is that the standard score (a) is based on a scale from 1 to 100, is convenient to use, and is easily understood ☐ (b) is easily adaptable to special kinds of distributions ☐ (c) is adaptable to all age groups and all universes ☐ (d) uses units that reflect equal amounts of difference between elementary units in all parts of the scale ☐.

A.6 SOME TECHNIQUES IN THE ANALYSIS OF FREQUENCY DISTRIBUTIONS

A.6.1 How to construct a frequency distribution. When a person has finished collecting the data in a statistical investigation, he usually has either a series of numerical measurements, a series of check marks, or a stack of questionnaires. For example, Table 1.3 contains the numerical counts of the red beads in each of 200 samples. You have already seen two types of frequency distribution, Tables 3.10 and 3.11, that can be constructed for the serial distribution in Table 1.3. Here is how Table 3.10 was constructed.

First, Table 1.3 was examined carefully for the smallest and the largest values of the variable. The smallest number of red beads in any of the 200 samples was 69. The largest number of red beads in any of the samples was 111. Then, in the first column of Table 3.10, all the integers from 69 to 111 were written. Next, it was noticed that the number of red beads in the first sample was 95 and a stroke or tally mark was placed in the second column of Table 3.10 opposite the number 95. The next value of the variable in Table 1.3 is 83, and so a stroke or tally mark was placed opposite the number 83 in Table 3.10. The next value of the variable in Table 1.3 is 93, and so a tally mark was placed opposite the number 93 in Table 3.10. This procedure was continued until 200 tally marks had been placed in Table 3.10 to correspond to the 200 values of the variable in Table 1.3.

A simple and convenient way to do such a job is for one student to read aloud slowly the values of the variable while another student places the tally marks in the frequency table. The process should be repeated in order to minimize the chance of mistakes in the frequency table.

Notice that the fifth tally mark in each row of Table 3.10 is a horizontal stroke across the preceding four strokes. This is called the "cross-fives" method of tallying. It makes the counting of the tally marks easy.

Next, the tally marks in each row were counted and the numbers of tally marks were recorded in the third column. If a possible value of the variable did not occur in the serial distribution, there were, of course, no tally marks opposite that value. In such situations, a dash or a zero should be placed opposite that value of the variable; for example, the values 70, 71, 108, and 110 did not occur in Table 1.3. The numbers in the third column of Table 3.10 were then added and the total was written at the bottom of the column. If the total had turned out to be different from 200, it would have been a signal that a mistake had been made in the work. This is an important check.

In Table 3.10, a blank line was left after every tenth row. This adds to the attractive appearance of the table and helps the reader to locate quickly the information contained in any specified row of the table.

A.6.2 Choosing appropriate class boundaries. Some preliminary decisions must be made before you can construct a grouped frequency table such as Table 3.11. For example, it is necessary to decide how wide or coarse the grouping will be. And it is necessary, too, to decide whether the range will be divided into equal or unequal parts.

There are no hard-and-fast rules for making these preliminary decisions. However, it often is desirable and convenient to have about twelve to fifteen groups or classes in the frequency table. Occasionally, it may be advisable to use either a larger or a smaller number of groups. Notice that in Table 3.11 there are no dashes or zeros in the frequency column. If you find after you have constructed a grouped frequency table that there are vacant classes, usually that is an indication that you made a poor choice in determining the classes, especially if the vacant classes are not near the ends of the range of the variable. The mathematical analysis may become very complicated if you divide the range into unequal intervals for the grouping process.

This is how it was decided to form the classes or groups in Table 3.11. First, it was noticed that the range of the variable in Table 1.3 was from 69 to 111. That is, there were 43 units in the range. If three units of the range were to be assigned to each class, there would be fifteen classes. As indicated above, fifteen often is a suitable number of classes for a frequency distribution. If two units of the range had been assigned to each class there would have been twenty-two classes, and that probably would have been too many, especially with a total frequency as small as 200. If four units of the range had been assigned to each class there would have been eleven classes. But, as will be shown in the following paragraphs, there would have been disadvantages in assigning an even number of units of the range to each class in this example.

Now, how was the first class interval actually determined? Instead of beginning with the smallest value of the variable, 69, and taking 69–71 as the first class interval, work was begun at the middle of the range. The middle of the range is 90 in this example. Because it had already been decided to assign three units of the range to each class, 89–91 was taken as a class interval that would include 90.

Then, working in both directions from the middle of the range, the other classes were determined, until both ends of the range were reached. You will notice that in order to avoid having only two units of the range as the first and the last classes, we were forced to go one unit beyond each end of the observed range. That does no harm, and, as indicated earlier, it prevents the necessity for the complicated arithmetic that would be involved in analysis of the data if some of the exhibited class intervals were unequal to others.

The midpoint in each class is a very important number. In the 89–91 class, the midpoint is 90, and that is a typical value of the variable in the distribution. Similarly, the midpoint in the 74–76 class is 75. That also is a typical value of the variable in the distribution. The same is true for the midpoint in every one of the other class intervals.

Now suppose that only two units of the range had been assigned to each class. Then, beginning with 90 at the middle of the range, we might have taken as a class interval the values 89–90. The midpoint in this class interval would be 89.5, and that is not a typical value of the variable in the distribution. It is impossible to have 89.5 red beads in a sample; the number of red beads in a sample must be a whole number. All the other midpoints in the classes would also have been fractions. Consequently, none of the midpoints of the class intervals would have been typical values of the variable in the distribution. The same kind of situation would have resulted if we had assigned four units of the range to each class interval. However, this is not a serious matter.

You can see that it is usually desirable, but not necessary, to take care in setting up groups or classes in order that the midpoint in each class interval will be a meaningful, typical value of the variable in the distribution.

Sometimes the values of the variable in a serial distribution tend to cluster or bunch around certain numbers. For such distributions, it usually is advisable to form the classes in such a way that as many as possible of the clusters or bunches of values of the variable will fall at or near the midpoints of the class intervals, or so that the class marks will be approximately the averages of the clustering points in the intervals. For example, weekly salaries often cluster at amounts such as $110.00, $112.50, and $115.00.

Confusion sometimes arises in the constructing of frequency distributions when the upper limit of one class is exactly the same as the lower limit of the next class. For example, suppose that you were constructing the frequency distribution in Table 3.16. That is, suppose that you had decided to use the class limits and class intervals shown in column one of Table 3.16 and that, as another person read aloud the hourly earnings of the truck drivers, you were placing the tally marks beside the appropriate classes. If the other person read to you the value "two dollars," where would you place the tally mark? In other words, does $2.00 belong in the $1.90–$2.00 class or in the $2.00–$2.10 class?

There are several methods for resolving such confusing situations. The method that seems to be most widely used and that is used in this book is to adopt the rule that when the upper limit of one class is exactly the same as the lower limit of the next class, a value of the variable that falls exactly on the overlapping boundary between two classes always will be placed in the higher of the two classes. In the illustration given in the preceding paragraph, the value $2.00 would belong in the $2.00–$2.10 class rather than in the $1.90–$2.00 class.

Several textbooks in statistics recommend another method that will prevent confusion of the kind mentioned above. They recommend that class boundaries be established to one more significant digit or one more decimal place or a half unit beyond the actual recorded data. If this method is followed in the construction of every grouped frequency distribution, the upper limit of each class will be the same as the lower limit of the next higher class, but none of the recorded values of the variable will fall exactly at the class boundaries. Consequently, there can be no confusion about the class in which a recorded value of the variable should be placed. If this plan were followed, the boundaries of the first class in Table 3.16 would be $1.195 and $1.295. The boundaries of the second class would be $1.295 and $1.395. The boundaries of the third class would be $1.395 and $1.495. It is clear, then, that the value $2.00 would belong in the class that had the boundaries $1.995 and $2.095. In spite of the obvious merits of this method, it does not seem to be widely used in practical situations.

A.6.3 The tally marks. As soon as you have prepared a tabular form in which the classes that you have chosen are listed in the first column, you are ready to go ahead with the construction of the grouped frequency distribution. Here again, one student might read aloud the values of the variable in Table 1.3 and another student might make a tally mark in column two of the tabular form opposite the class interval in which each value of the variable lies. The result is shown in Table 3.11.

The frequencies in the third column of Table 3.11 were determined by counting the tally marks opposite each class interval. Then the frequencies were added, giving 200 as the total frequency, which is correct, because there were 200 values of the variable in Table 1.3.

Notice that the tally marks in Table 3.11 form a graphic representation of the frequency distribution. Turn the page on which Table 3.11 is printed so that the left-hand edge of the page is at the bottom and note the bell-shaped figure formed by the tally marks.

A.6.4 True class limits. A person ordinarily says that he is eighteen years old from the first moment of his eighteenth birthday until the last moment before his nineteenth birthday. Consequently, the ages in years and decimal fractions of a year that could be in the first class in Table 3.8 (p. 81) are represented by all the numbers from $18.000\ldots$ to $20.999\ldots$, with indefinitely long strings of zeros and nines, respectively. However, because we cannot deal easily with such a cumbersome quantity as an indefinitely long string of nines, we say that the true class limits for the first class are 18 and 21 and we interpret them as including exactly 18 and all other ages up to but not including exactly 21 years.

Similarly, the 21–23 age class includes all the men in the sample who married at any time from the first moment of their twenty-first birthday up to but not including the first moment of their twenty-fourth birthday. The true class limits for the last class are 57 and 60, that is, including exactly fifty-seven and all other ages up to but not including exactly sixty years.

A.6.5 The class interval. Now, suppose that you are given a grouped frequency distribution that someone else has constructed. Do you know how to determine the value of i, the class interval, that was used by the other person? It is easy to make a mistake in determining the class interval in such a situation. If you do make a mistake in determining the value of i in such a table, it is likely that you will obtain wrong answers for the median, the mean, and the standard deviation of the distribution.

Here is one way in which you can always find the correct value of the class interval for a grouped frequency distribution. Find the true lower limit of the variable in the first class and the true lower limit of the variable in the second class. Then the difference between these two lower limits is the value of i. In Table 3.8, $i = 21 - 18 = 3$ years.

Of course, you would obtain the same result if you found the difference between the true upper limit of the variable in the first class and the true upper limit of the variable in the second class. You may use any two adjacent classes instead of the first two classes if you prefer to do so.

Here are some additional illustrations that may help you to interpret properly the grouped frequency distributions with which you come into contact in your reading or in your work. All of these types of grouping are in common usage.

(a) Weights of students, the original measurements having been made to the nearest pound.

Recorded Measurement	No. of Students	True Class Limits	Class Midpoint
90–99	2	89.5–99.5	94.5
100–109	7	99.5–109.5	104.5
etc.	etc.	etc.	etc.

The class interval is $i = 99.5 - 89.5 = 10$ pounds.

(b) Weights of students, the original measurements having been made to the nearest half pound.

Recorded Measurement	No. of Students	True Class Limits	Class Midpoint
90–99.5	3	89.75–99.75	94.75
100–109.5	6	99.75–109.75	104.75
etc.	etc.	etc.	etc.

The class interval is $i = 99.75 - 89.75 = 10$ pounds.

(c) Weights of students, the original measurements having been recorded to the lower pound.

Recorded Measurement	No. of Students	True Class Limits	Class Midpoint
90–99	1	90–100	95
100–109	4	100–110	105
etc.	etc.	etc.	etc.

The class interval is $i = 100 - 90 = 10$ pounds.

(d) The number of children in the families of unemployed fathers in a city.

Recorded No. of Children	No. of Families	True Class Limits	Class Midpoint
0–2	46	0–2	1
3–5	87	3–5	4
6–8	73	6–8	7

The class interval is $i = 3 - 0 = 3$ children.

EXERCISES

1. Construct frequency distributions similar to Tables 3.10, 3.11, 3.12, and 3.13 for the data showing the numbers of red beads in the large number of samples that the class drew from the universe of 2000 red beads and 3000 white beads.

2. What is the range of the score in Table 3.7? Construct an ungrouped frequency distribution for the data in Table 3.7.

3. Construct a grouped frequency distribution, with eleven classes or groups, for the data in Table 3.7. What is the size of the class intervals? (*Ans. i* = 3.)

4. What is the value of i, the class interval, in each of the distributions in Tables 3.11, 3.12, 3.13, 3.15, 6.3, and 6.4? (*Ans.* 3, 0.012, 0.012, \$5, 10, 5.)

5. What are the true class limits for the second class in Table 3.17? (*Ans.* 17 and 18.)

6. What are the true class limits for the third class in Table 6.3? (*Ans.* 40 and 49.)

7. What are the true class limits for the last class in Table 6.4? (*Ans.* 60 and 64.)

8. What are the true class limits for the fourth class in Table A.14? (*Ans.* 100 and 124.)

9. Find the true class limits, the class midpoints, and the length of the interval in each class for each of the following four frequency distributions:

(a) Weekly earnings of employees in an office, recorded to the nearest dollar.

Recorded Earnings	No. of Employees	True Class Limits	Class Midpoint
\$90–\$99	3		
80–89	8		
etc.	etc.		

The length of the interval in each class is $i =$

(b) Weekly earnings of employees in an office, recorded to nearest half dollar.

Recorded Earnings	No. of Employees	True Class Limits	Class Midpoint
\$90.00–\$99.50	5		
80.00–89.50	7		
etc.	etc.		

The length of the interval in each class is $i =$

(c) Weekly earnings of employees in an office, recorded to the lower dollar, that is, disregarding all fractional parts of a dollar.

Recorded Earnings	No. of Employees	True Class Limits	Class Midpoint
\$90–\$99	4		
80–89	9		
etc.	etc.		

The length of the interval in each class is $i =$

(d) Number of employees in retail stores in a city.

No. of Employees	No. of Retail Stores	True Class Limits	Class Midpoint
1–5	17		
6–10	39		
etc.	etc.		

The length of the interval in each class is $i =$

A.6.6 The mode. The mode or modal value of the variable in a universe is the value of the variable that occurs more frequently than any other value of the variable in the universe. We often use the most frequently occurring value of the variable in a sample as an estimate of the mode in the universe from which the sample was drawn. Geometrically, the mode is the value of the variable corresponding to the highest point on the graph for the distribution in the universe.

A.6.7 Mode estimated from ungrouped frequency distribution. The estimated mode that we obtain from an ungrouped frequency distribution is the value of the variable for which the frequency is the largest in the sample. For example, using Table 6.22, we may estimate that the mode or modal size of shoe for the women in the universe from which the sample was drawn is 7½. Similarly, using Table 6.20, we may take the neckband size 15 as an estimate of the most common or most popular neckband size in the universe of shirts from which the sample was drawn.

A.6.8 Mode estimated from grouped frequency distribution. It is difficult to make a definite estimate of the mode in a universe if all we know about the universe is the information contained in a grouped frequency distribution, such as that in Table 3.8. One way to obtain such an estimate would be to fit a smooth frequency curve to the distribution very carefully and then use the value of the variable corresponding to the highest point of the smooth curve as the estimate of the mode in the universe from which the sample was drawn. Ordinarily, such a process requires more effort than the obtained estimate of the mode is worth in practice.

Statisticians usually are content to indicate the modal class in a grouped frequency distribution, that is, the class or group corresponding to the largest frequency in the distribution. For example, in Table 3.8, the modal class is the age 24–26 class, because that class has the largest number, 73 men, of any of the classes in the table. On this basis, we may estimate that the most common age at time of first marriage for the universe of men from which the sample was drawn is somewhere between the beginning of the twenty-fourth year of age and the end of the twenty-sixth year of age.

An important shortcoming of the modal class is that it may be seriously affected by the choice of grouping, as, for example, fine grouping or coarse grouping. Below is an illustration of this idea. In the distribution on the left, the class interval is 10 years and the modal class is the 20–29-year age group. In the distribution on the right, the same data have been thrown into coarser groups with a class interval of 20 years, and the modal class now is the 30–49-year age group.

Age	f		Age	f
10–19	5		10–29	20
20–29	15		30–49	22
30–39	13		50–69	9
40–49	9		Total	51
50–59	5			
60–69	4			
Total	51			

Of course, the idea of using a coarsely grouped frequency distribution to estimate the mode or the modal class is a rather artificial procedure in most practical statistical investigations. If you have the actual data in the form in which the individual values of the variable were collected, you can determine a modal value or modal class of the variable in the sample by examining the original data.

For example, it is not necessary for us to use Table 3.8 to find a modal class for the sample of ages of 242 men at time of first marriage. While the sample was being drawn, a record was made of each of the 242 ages, and by going through that list we find that the modal year of age at time of first marriage for these 242 men is twenty-five years; 29 of the men married for the first time when they were twenty-five years of age, and smaller numbers of the men married at each of the other ages from eighteen to fifty-nine years. (See the data in exercise 7 on page 385.)

Usually, it is much easier to find good estimates of the median and the mean from a grouped frequency distribution than it is to find a good estimate of the mode from such a distribution.

A.6.9 The median. The purpose in finding the median of a sample is to obtain an estimate of the value of the variable in the universe that divides the distribution in the universe into two parts such that 50 per cent of the elementary units in the universe have values of the variable that are less than or equal to the median and the other 50 per cent have values of the variable greater than or equal to the median. That is why the median of a sample is considered to be an estimate of the central value of the variable in the universe from which the sample was drawn.

A.6.10 Median estimated from ungrouped frequency distribution.

An ungrouped frequency distribution is somewhat similar to a serial distribution that has been arranged in the form of an array, that is, in the order of magnitude of the values of the variable. The numbers in the frequency column merely save us the trouble of repeating the same value of the variable many times.

Consequently, the median of an ungrouped frequency distribution is the value of the variable for the elementary unit at the middle of the distribution. If the total number of elementary units, n or Σf, is an odd number, then the middle elementary unit is the $(n+1)/2$th elementary unit in the distribution. Therefore, if n is odd, count down (or up) through the frequency column until you accumulate $(n+1)/2$ elementary units. The value of the variable opposite this point is the median of the distribution.

If the total number of elementary units, that is, the total frequency, in the distribution is an even number, then the median of the distribution is the value of the variable midway between the values of the variable corresponding to the two elementary units nearest to the middle of the distribution. In other words, if n is an even number in an ungrouped frequency distribution, the median is the value of the variable midway between the values of the variable for the $(n/2)$th elementary unit and the $(n/2 + 1)$th elementary unit. Therefore, if n is an even number in an ungrouped frequency distribution, count down (or up) through the frequency column until you have accumulated $n/2$ elementary units and write down the value of the variable for this elementary unit. Also write down the value of the variable for the $(n/2 + 1)$th elementary unit. Take the midpoint between these two values of the variable as the median of the distribution.

For example, if you construct an ungrouped frequency distribution from the data in Table 3.7, as was suggested in exercise 2 on page 369, you will have $n = 65$, an odd number. Consequently, you need to locate the $(65 + 1)/2 = 33$rd elementary unit. Counting down through the frequency column until you accumulate 33 individuals you arrive at the frequency 6, which is on the line for which the value of the variable is 54. Therefore, the median of the ungrouped frequency distribution obtained from Table 3.7 is a score of 54. This is the same value of the median that you found if you solved exercise 2 on p. 80. You can easily prove to yourself that you obtain the same value of the median by counting upward through the frequency column until you accumulate 33 students.

In Table 3.10, the value of n is 200, an even number. To find the median of the ungrouped frequency distribution, we need to locate the $200/2 = 100$th elementary unit and the $200/2 + 1 = 101$st elementary unit. Counting down through the frequency column until we accumulate 100 samples, we arrive at the frequency 11, which is on the line for which the value of the variable is 90 red beads. Counting down through the frequency column until we accumulate 101 samples, we arrive at the frequency 11 again, on the line for which the value of the variable is 90 red beads. Therefore, the median of the distribution in Table 3.10 is 90 red beads. Prove to yourself that you obtain the same result if you count upward through the frequency column.

A.6.11 Median estimated from grouped frequency distribution. The median value of the variable in a grouped frequency distribution is the value of the variable that may be considered as splitting the distribution into two parts so that 50 per cent of the total number of elementary units are on one side and 50 per cent are on the other side. For example, the median for Table 3.8 is the age before which 50 per cent of the men married and after which the other 50 per cent of the men married.

The total number of elementary units (men) in the sample is 242. Fifty per cent of 242 is 121. We need to find the age before which 121 of the men married. The other 121 men will be considered as having married after reaching this age. We count down through the frequencies in the second column, beginning with the frequency corresponding to the youngest age interval, until we have accumulated 121 men. We find $5 + 38 + 73 = 116$, which is less than 121. But, if we add 56 to this number we obtain 172 men, which is greater than 121 men.

We conclude that the median age is somewhere in the 27–29 years interval, because we need 5 men out of this class to give us 121. There are 56 men in this class. Some of them probably married before they reached the median age, and the others married after they reached the median age. How can we estimate the point in the 27–29-year interval at which the 121st man may be considered to have married?

We need to count off the first 5 men of the 56 men in the fourth class. Perhaps the easiest way to learn how to do this is to use a picture of a scale along a straight line.

Each of the class intervals covers a three-year period. Let us represent the three-year period corresponding to the 27–29 years class by a straight line segment 3 inches long. Mark age 27 at one end of the line segment and age 30 at the other end of the three-inch segment. Divide the segment into 56 small, equal parts, so that one small part may be assigned to each of the 56 men who married during that three-year age interval. Each small part will be 3/56 of an inch in length.

Now count off the first five of these small parts, beginning at the point marked age 27 on the scale. Mark the end of the fifth small step. That point represents the median of the distribution. What is its value? Well, it is 27 plus 5 small parts, each of which is 3/56 of an inch in length. Consequently, the median is at the point

$$27 + 5(3/56) = 27 + 15/56 = 27 + 0.27 = 27.27 \text{ or } 27.3 \text{ years of age}$$

Usually, the graphical method just illustrated is the easiest way to compute the median for a grouped frequency distribution. However, the same idea can be represented by the formula

$$\text{median} = L + \left(\frac{n}{2} - \Sigma f_{BL} \right) \frac{i}{f_c}$$

In general, the value of the variable such that p per cent of the distribution may be considered to have lower values of the variable is called the pth percentile and may be symbolized by P_p. The median is the 50th percentile, namely, P_{50}. Other frequently used percentiles are the 25th and the 75th, which are called the first quartile Q_1 and third quartile Q_3, respectively. Any percentile can be computed graphically or by the following formula, which is merely a generalization of the formula for the median.

$$P_p = L + \left(\frac{pn}{100} - \Sigma f_{BL} \right) \frac{i}{f_c} \tag{A.2}$$

where

$L =$ the value midway between the true lower limit of the class in which the pth percentile falls and the true upper limit of the next lower class

$n =$ the number of elementary units in the distribution

$\Sigma f_{BL} =$ the sum of the frequencies in the classes below L

$i =$ the size of the class interval

$f_c =$ the frequency in the class in which the pth percentile falls

For example, the median of the distribution in Table 3.8 is

$$P_{50} = 27.0 + \left(\frac{242}{2} - 116 \right) \frac{3}{56} = 27.0 + (121 - 116) \frac{3}{56} = 27.0 + \frac{15}{56} = 27.3 \text{ years}$$

because here $L = 27$ years exactly, $n = 242$ men, $\Sigma f_{BL} = 5 + 38 + 73 = 116$ men, $i = 3$ years, and $f_c = 56$ men.

The quantity defined by the following formula is used sometimes as a crude indication of the variability of a distribution.

$$\text{semi-interquartile range} = \frac{Q_3 - Q_1}{2}$$

EXERCISES

1. What is the value of the mode in each of Tables 6.20, 6.21, and 6.22? (*Ans.* size 15, size 33, and size 7½.)

2. What is the modal class in Table 3.15 (p. 98)? (*Ans.* $75 and under $80.)

3. What is the modal class in Table A.14 (p. 384)? (*Ans.* 125–149 patients.)

4. What is the modal value in Table 6.1 (p. 160)? (*Ans.* 9.)

5. What is the median neck size of the shirts in Table 6.20? (*Ans.* size 15.)

6. What is the median sleeve length of the shirts in Table 6.21? (*Ans.* size 33.)

7. What is the median shoe size (length) in Table 6.22? (*Ans.* size 7½.)

8. What is the median weekly earnings of the stenographers in Table 3.15? (*Ans.* $80.02.)

9. What is the median number of patients seen during the week by a doctor in Table A.14? (*Ans.* 137.5.)

10. What is the median age at marriage of the skilled men in Table 3.17? (*Ans.* 27.3 years.)

11. What is the median age at marriage of the skilled women in Table 3.17? (*Ans.* 25.7 years.)

12. What is the median age at marriage of the unskilled men in Table 3.17? (*Ans.* 25.6 years.)

13. What is the median age at marriage of the unskilled women in Table 3.17? (*Ans.* 23.0 years.)

14. What are the medians for the distributions in Tables 3.11, 3.12, and 3.13? (*Ans.* 90.0, 0.360, and 0.000.)

15. What are the medians for the distributions in Tables 6.1, 6.3, and 6.4? (*Ans.* 9, 72.3, and 32.0.)

16. Compute the median rank for each type of book in Table 5.5.

17. Compute Q_1, Q_3, and the semi-interquartile range for each type of book in Table 5.5. Can you think of any reason why it is more appropriate to use the median and the semi-interquartile range rather than the mean and the standard deviation as estimates of the central value and the variability in the universe from which the sample in Table 5.5 was drawn (p. 156)? (*Ans.* The classification in Table 5.5 is only partly quantitative.)

A.6.12 The mean and the standard deviation. In section A.5, we saw that considerable computational labor is saved if we compute the mean and the standard deviation at the same time. Furthermore, just as there were long methods and short methods for computing the mean and the standard deviation of a serial distribution, so also there are long methods and short methods for computing the mean and the standard deviation of a frequency distribution. The long method may be better for explaining the theory, but the short methods save us a lot of tedious arithmetical computation.

A.6.13 Long method for determining *m* and *s* in ungrouped frequency distribution. Table A.8 shows all the tabulations that are needed in a long method for computing the value of the mean *m* and the standard deviation *s* for the sample in the ungrouped frequency distribution in Table 6.22.

TABLE A.8

Tabulation for Long Method of Finding Mean and Standard Deviation for the Ungrouped Frequency Distribution in Table 6.22

Shoe Size (Length) x	No. of Women f	xf	x²f
4	31	124	496
4½	100	450	2,025
5	258	1,290	6,450
5½	494	2,717	14,943.50
6	820	4,920	29,520
6½	1,154	7,501	48,756.50
7	1,354	9,478	66,346
7½	1,449	10,867.5	81,506.25
8	1,297	10,376	83,008
8½	1,054	8,959	76,151.50
9	849	7,641	68,769
9½	491	4,664.5	44,312.75
10	454	4,540	45,400
10½	92	966	10,143
11	103	1,133	12,463
Total	10,000	75,627.0	590,290.50

The first two columns of Table A.8 are copied directly from Table 6.22. The third column is needed for finding the value of m for the sample. Make sure that you understand clearly the meaning of the numbers in column three. Remember that to find the value of m for a sample, we first must add all the values of the variable in the sample. The values of the variable are represented by the symbol x.

Perhaps the simplest way to explain this situation is to ask you to imagine how this sample containing 10,000 shoe sizes would appear if we arranged it in the form of an ordered-array serial distribution. First, we would write the number 4 thirty-one times. Next, we would write the number 4½ or 4.5 one hundred times. Then, we would write the number 5 two hundred fifty-eight times. And so on, until we had written 10,000 shoe sizes altogether. Next, we would add the 10,000 numbers (shoe sizes). What would be the easiest way to do this addition? No doubt, you will agree that instead of adding thirty-one 4's, it is easier to multiply 4 by 31. That operation gives us 124, which is the first number in the third column of Table A.8. Similarly, we would avoid unnecessary labor by using multiplication instead of addition to find each of the other partial sums in the total of the 10,000 values of x in the array.

Consequently, the number 75,627.0, which is the total of the third column, is exactly the same total we would obtain if we added the 10,000 values of x in the ordered-array serial distribution of shoe sizes. In order to find the mean, all we need do is to divide the total of column three by the number of elementary units in the sample, $n = 10,000$. Therefore,

$$m = \frac{75627.0}{10000} = 7.5627 = 7.56, \text{ if we retain only two decimal places.}$$

We can generalize this process for finding the mean m of any sample that is tabulated in the form of an ungrouped frequency distribution by

$$m_x = \frac{\Sigma xf}{n} \quad \text{or} \quad \frac{\Sigma xf}{\Sigma f}, \quad \Sigma f = n \tag{3.7}$$

The fourth column of Table A.8 is used for finding the standard deviation for the sample by a method that follows exactly the same principle that was the basis of formula 3.4 for the standard deviation of a serial distribution. If you have difficulty understanding column four, think again of the ordered-array serial distribution of 10,000 shoe sizes and the operations you would perform with those 10,000 values of x if you used formula 3.4 to find the value of s. The total of the fourth column of Table A.8 is exactly the same as the total of the 10,000 squares of the values of x in the ordered-array serial distribution. Therefore,

$$s = \frac{1}{10000}\sqrt{10000(590290.50) - (75627.0)^2}$$

$$= \frac{1}{10000}\sqrt{183461871} = \frac{13540}{10000} = 1.354 = 1.35, \text{ if we retain two decimal places}$$

In general, the formula for this method of finding the standard deviation of any sample that is tabulated in the form of an ungrouped frequency distribution is

$$s_x = \frac{1}{n}\sqrt{n\Sigma x^2f - (\Sigma xf)^2} \tag{3.8}$$

A.6.14 Short method for determining m and s in ungrouped frequency distribution. Table A.9 shows the tabulations that are needed in a somewhat shorter and easier method for finding the mean and the standard deviation of the ungrouped frequency distribution in Table 6.22. The short method involves two ideas: the subtraction of a suitable constant c from every value of the variable x, and the division of each of these differences by the constant step k of the distribution. These two operations have been carried out in columns three and four of Table A.9. Here $c = 7.5$ or $7\frac{1}{2}$. Usually, it is best to choose one of the values of x as the constant c. If the largest frequency is near the middle of the distribution and if the frequencies in the upper and lower halves of the frequency column are fairly well balanced, choose a value of x near the middle of the range as the constant c. If the larger frequencies happen to be toward one end of the range of values of x, then choose a value of x somewhat toward that end of the range as the constant c.

TABLE A.9

Tabulation for Short Method of Finding Mean and Standard Deviation for the Ungrouped Frequency Distribution in Table 6.22

Shoe Size (Length) x	No. of Women f	x − 7½ = d	d ÷ ½ = h	hf	h²f
4	31	−3.5	−7	−217	1,519
4½	100	−3	−6	−600	3,600
5	258	−2.5	−5	−1,290	6,450
5½	494	−2	−4	−1,976	7,904
6	820	−1.5	−3	−2,460	7,380
6½	1,154	−1	−2	−2,308	4,616
7	1,354	−0.5	−1	−1,354	1,354
7½	1,449	0	0	0	0
8	1,297	0.5	1	1,297	1,297
8½	1,054	1	2	2,108	4,216
9	849	1.5	3	2,547	7,641
9½	491	2	4	1,964	7,856
10	454	2.5	5	2,270	11,350
10½	92	3	6	552	3,312
11	103	3.5	7	721	5,047
Total	10,000			+1,254	73,542

The constant step in this distribution is a half size, that is, $k = \frac{1}{2}$. Column four is formed by dividing each of the differences in column three by $\frac{1}{2}$. Dividing by $\frac{1}{2}$ is equivalent to multiplying by 2. The letter h is used here to stand for the quotients in column four.

$$m = \left(\frac{1}{2}\right)\left(\frac{1{,}254}{10{,}000}\right) + 7.5 = 0.0627 + 7.5 = 7.5627 = 7.56$$

$$s = \frac{(\frac{1}{2})}{10{,}000}\sqrt{10{,}000(73{,}542) - (1{,}254)^2} = \frac{1}{20{,}000}\sqrt{733{,}847{,}484} =$$

$$= \frac{27{,}089.6}{20{,}000} = 1.354 = 1.35$$

These are the same values that we obtained for m and s by the long method.

If k is the constant step in the values of the variable, c is an arbitrary constant, and $h = (x - c)/k$, the general formulas for the short method of finding the mean and the standard deviation of a sample that is tabulated in the form of an ungrouped frequency distribution are

$$m_x = k\left(\frac{\Sigma hf}{\Sigma f}\right) + c$$

$$s_x = \frac{k}{n}\sqrt{n\Sigma h^2 f - (\Sigma hf)^2}$$

A.6.15 Long method for finding *m* and *s* in grouped frequency distribution. Table A.10 shows the tabulations needed in one method for finding the mean and the standard deviation of the sample in the grouped frequency distribution given in Table 3.8. The midpoint of a class interval usually is called the *class mark*. It is customary to use the letter *x* to stand for the class marks.

TABLE A.10

Tabulation for Long Method of Finding Mean and Standard Deviation for the Grouped Frequency Distribution in Table 3.8

Age at Marriage (Years) Class	No. of Men f	Class Mark (Years) x	(Years) xf	(Years)² x²f
18–20	5	19.5	97.5	1,901.25
21–23	38	22.5	855.0	19,237.50
24–26	73	25.5	1,861.5	47,468.25
27–29	56	28.5	1,596.0	45,486.00
30–32	26	31.5	819.0	25,798.50
33–35	19	34.5	655.5	22,614.75
36–38	11	37.5	412.5	15,468.75
39–41	3	40.5	121.5	4,920.75
42–44	3	43.5	130.5	5,676.75
45–47	3	46.5	139.5	6,486.75
48–50	1	49.5	49.5	2,450.25
51–53	2	52.5	105.0	5,512.50
54–56	1	55.5	55.5	3,080.25
57–59	1	58.5	58.5	3,422.25
Total	242		6957.0	209,524.50

$$m = \frac{6,957.0}{242} = 28.748 = 28.7 \text{ years, if we retain only one decimal place}$$

$$s = \frac{1}{242}\sqrt{242(209,524.50) - (6,957.0)^2} = \frac{1}{242}\sqrt{2,305,080.00} =$$

$$\frac{1,518.2}{242} = 6.27 \text{ years}$$

Notice that in finding the mean and the standard deviation of a grouped frequency distribution, the original class intervals of the variable were disregarded and the class marks were used instead. This is done because it can be proved by algebraic methods that no matter what the ages of the men in each class may have been, satisfactory values of the mean and the standard deviation can be obtained by assuming that all the men in each class were at exactly the age indicated by the class mark at the time of marriage.

Insofar as the mean is concerned, the errors involved in assuming that the men who were older than the class marks at the time of marriage were exactly at the class marks usually compensate for the errors involved in assuming that the men who were younger than the class marks at the time of marriage were exactly at the class marks.

If you prefer to think of it in another way, as we did in finding the median for Table 3.8, assume that the men in each age class are spaced evenly over the class interval. Then the average age of the men in each class would be at the midpoint of the class interval, that is, at the class mark.

The assumption that all the elementary units in a class have the value of the variable exactly at the class mark gives thoroughly satisfactory results when one is finding the mean, because positive and negative errors tend to compensate for each other in this case. However, in determination of the standard deviation, the deviations are squared and the results are all positive. Consequently, there is not the same kind of compensation of errors. The value obtained for the standard deviation after this assumption tends to be slightly too large. There is a formula, called Sheppard's correction, for correcting this systematic error in the standard deviation of a grouped frequency distribution; usually, however, the error is so small that it can be disregarded. If you have a reasonably large sample and if you do not make the class intervals unreasonably large, the error in the standard deviation may be considered to be negligible.

If you cover and disregard column one of Table A.10, the remainder of the table is similar in appearance to Table A.8. If you understand the procedure that was explained in section A.6.13, it will be obvious to you how to find the mean and the standard deviation for Table A.10, using the class marks as the values of the variable.

The general formulas for this long method of finding the mean and the standard deviation of a sample that has been tabulated in the form of a grouped frequency distribution are

$$m_x = \frac{\Sigma xf}{n} \tag{3.7}$$

$$s_x = \frac{1}{n}\sqrt{n\Sigma x^2 f - (\Sigma xf)^2} \tag{3.8}$$

An even longer method for finding the standard deviation would be to use the formula

$$s_x = \sqrt{\frac{\Sigma(x - m)^2 f}{n}}$$

A.6.16 Short method for finding *m* and *s* in grouped frequency distribution. If you understand how the short method in Table A.9 was developed from the long method in Table A.8, it will be obvious to you how the short method in Table A.11 is developed from the long method in Table A.10. The advantage of the short method is that the small numbers in the last two columns of Table A.11 are easier to deal with than the large numbers in the last two columns of Table A.10. Always choose one of the class marks as the constant *c* to be subtracted from every class mark. The constant step in the class marks is the class interval *i*. In Table A.11, *c* = 31.5, and *i* = 3.

TABLE A.11

Tabulation for Short Method of Finding Mean and Standard Deviation for the Grouped Frequency Distribution in Table 3.8

Age at Marriage (Years) Class	No. of Men f	Class Mark (Years) x	(Years) x−31.5=d	(Years) d/3=h	(Years) hf	(Years)² h²f
18–20	5	19.5	−12.0	−4	−20	80
21–23	38	22.5	−9.0	−3	−114	342
24–26	73	25.5	−6.0	−2	−146	292
27–29	56	28.5	−3.0	−1	−56	56
30–32	26	31.5	0.0	0	0	0
33–35	19	34.5	3.0	1	19	19
36–38	11	37.5	6.0	2	22	44
39–41	3	40.5	9.0	3	9	27
42–44	3	43.5	12.0	4	12	48
45–47	3	46.5	15.0	5	15	75
48–50	1	49.5	18.0	6	6	36
51–53	2	52.5	21.0	7	14	98
54–56	1	55.5	24.0	8	8	64
57–59	1	58.5	27.0	9	9	81
Total	242				−222	1262

$$m = 3\left(\frac{-222}{242}\right) + 31.5 = -2.752 + 31.5 = 28.748 \text{ or } 28.7 \text{ years}$$

$$s = \frac{3}{242}\sqrt{242(1262) - (-222)^2} = \frac{3}{242}\sqrt{256,120} = \frac{3(506.08)}{242} = 6.27 \text{ years}$$

These are the same values of *m* and *s* that we found by the longer method.

Some people dislike computations that involve negative numbers. If you always choose the smallest class mark as the constant *c* to be subtracted from every class mark, all the numbers in the resulting table similar to Table A.11 will be positive numbers or zeros. However, the numbers in the last two columns will be somewhat larger than they would be if you chose as *c* a class mark near the middle of the range of the class marks. Convince yourself of this fact by using 19.5 as the constant c to be subtracted from every class mark in Table A.11.

TABLE A.12

**Tabulation for Finding the Mean and the Standard Deviation
for the Ungrouped Frequency Distribution in Table 3.10**

Variable x	Frequency f	x − 90 = d	df	d²f
69	1	−21	−21	441
72	1	−18	−18	324
73	1	−17	−17	289
74	1	−16	−16	256
75	1	−15	−15	225
76	2	−14	−28	392
77	2	−13	−26	338
78	4	−12	−48	576
79	3	−11	−33	363
80	4	−10	−40	400
81	7	−9	−63	567
82	5	−8	−40	320
83	7	−7	−49	343
84	8	−6	−48	288
85	8	−5	−40	200
86	10	−4	−40	160
87	9	−3	−27	81
88	10	−2	−20	40
89	10	−1	−10	10
90	11	0	0	0
91	10	1	10	10
92	10	2	20	40
93	10	3	30	90
94	9	4	36	144
95	9	5	45	225
96	8	6	48	228
97	7	7	49	343
98	5	8	40	320
99	6	9	54	486
100	4	10	40	400
101	4	11	44	484
102	3	12	36	432
103	3	13	39	507
104	2	14	28	392
106	2	16	32	512
107	1	17	17	289
109	1	19	19	361
111	1	21	21	441
Total	200		9	11,317

If i is the class interval, c is one of the class marks, and $h = (x - c)/i$, the general formulas for the mean and the standard deviation of a sample that is tabulated in the form of a grouped frequency distribution are

$$m_x = i\left(\frac{\Sigma hf}{n}\right) + c$$

$$s_x = \frac{i}{n}\sqrt{n\Sigma h^2 f - (\Sigma hf)^2}$$

A.6.17 Error due to grouping usually is negligible. The following is an illustration of the fact that usually only negligible errors, if any, are caused in the mean and the standard deviation of a grouped frequency distribution by assuming that all the values of the variable in each class are equal to the class mark. Table 3.10 is an ungrouped frequency distribution of the data in Table 1.3 and Table 3.11 is a grouped frequency distribution of the same data.

To find the mean and the standard deviation of Table 3.10 by the short method, we construct Table A.12. Here $k = 1$ and $c = 90$.

$$m = 1\left(\frac{9}{200}\right) + 90 = 0.045 + 90 = 90.045 = 90.0$$

$$s = \frac{1}{200}\sqrt{200(11,317) - (9)^2} = \frac{1}{200}\sqrt{2,263,319} = \frac{1504.4}{200} = 7.52$$

To find the mean and the standard deviation of Table 3.11 by the short method, we construct Table A.13. Here $i = 3$ and $c = 90$.

TABLE A.13
Tabulation for Finding the Mean and the Standard Deviation
for the Grouped Frequency Distribution in Table 3.11

Class	Frequency f	Class Mark x	x − 90 = d	d/3 = h	hf	h²f
68–70	1	69	−21	−7	−7	49
71–73	2	72	−18	−6	−12	72
74–76	4	75	−15	−5	−20	100
77–79	9	78	−12	−4	−36	144
80–82	16	81	−9	−3	−48	144
83–85	23	84	−6	−2	−46	92
86–88	29	87	−3	−1	−29	29
89–91	31	90	0	0	0	0
92–94	29	93	3	1	29	29
95–97	24	96	6	2	48	96
98–100	15	99	9	3	45	135
101–103	10	102	12	4	40	160
104–106	4	105	15	5	20	100
107–109	2	108	18	6	12	72
110–112	1	111	21	7	7	49
Total	200				3	1271

$$m = 3\left(\frac{3}{200}\right) + 90 = 0.045 + 90 = 90.045 = 90.0$$

$$s = \frac{3}{200}\sqrt{200(1271) - (3)^2} = \frac{3}{200}\sqrt{254,191} = \frac{3}{200}(504.17) = 7.56$$

Grouping different possible values of the variable into classes tends to make the standard deviation slightly too large. In this example, the table with the grouped data gave us $s = 7.56$ and the table with the ungrouped data gave us $s = 7.52$, which is the true value for the serial distribution in Table 1.3. However, the difference between $s = 7.56$ and $s = 7.52$ is negligible compared with the possible effect of sampling error on the value of s as an estimate of σ.

A.6.18 Troublesome open-end intervals. Table A.14 is an illustration of a grouped frequency distribution in which the first class interval and the last class interval are said to be open at one end of the interval; there is no definitely specified lower limit for the first class interval and there is no definitely specified upper limit for the last class interval. For such a distribution, it is impossible to determine the range, the midpoint of the range, the mean, or the standard deviation, unless we are willing to make assumptions about the first class interval and the last class interval that may be difficult or even impossible to justify.

It is impossible to compute the mean or the standard deviation for the distribution in Table A.14 because there is no class mark for the first interval or for the last interval. Nevertheless, it is just as easy to compute the median for the distribution in Table A.14 as for any other grouped frequency distribution given in this book. And, of course, it is possible to indicate the modal class for the distribution in Table A.14.

TABLE A.14

The Number of Patients Seen by Doctors in a Certain Week

No. of Patients Seen by a Doctor During the Week	No. of Doctors
Less than 50	41
50–74	64
75–99	118
100–124	178
125–149	198
150–174	169
175–199	126
200–224	66
225–249	28
250 or more	12
Total	1000

Table 6.12 is another illustration of a frequency distribution in which the first class interval and the last class interval are open at one end.

Keep this troublesome matter in mind when you are planning a questionnaire or preparing to record measurements in a statistical investigation. For example, sometimes in a questionnaire the respondent is asked to place a check mark beside the classification that best represents him, say, in income. If either the first class interval or the last class interval in such a check list is open at one end, then there may be no way to compute the mean and the standard deviation of the sample after the field work of the survey is completed.

Sometimes, after the mean and the standard deviation have been computed from the original data, the data are thrown into a grouped frequency distribution with open intervals at the beginning and the end of the distribution for the final report. Usually, the computed values of the mean and the standard deviation will be shown in the report in such a case. Better still, all of the actually observed values of the variable in the open-end intervals will be given in a footnote of the table.

EXERCISES

1. What is the class mark for the second class in Table A.14? (*Ans.* 62.)

2. What is the class mark for the second class in Table 6.3? (*Ans.* 34.5.)

3. Compute the mean and the standard deviation for the distribution in Table 6.20. (*Ans.* $m = 15.295$, $s = 0.808$.) Remark: When relative frequencies are given in the form of percentages, it is easy to prove that we can compute s from the relative frequencies by using the formula $s = (i/100) \sqrt{100\Sigma h^2 f - (\Sigma h f)^2}$, because the sum of the relative frequencies is 100. This saves the labor of converting the relative frequencies into actually observed frequencies.

4. Compute the mean and the standard deviation for the distribution in Table 6.3. (*Ans.* $m = 73.2$, $s = 15.1$.)

5. Compute the mean and the standard deviation for the ungrouped frequency distribution that you constructed from Table 3.7 in exercise 2 on page 369. (*Ans.* $m = 55.4$, $s = 7.0$.)

6. Compute the mean and the standard deviation for the grouped frequency distribution that you constructed from Table 3.7 in exercise 3 on page 369. (*Ans.* $m = 55.2$, $s = 6.9$.) How much difference does the grouping make in the values of the mean and the standard deviation here and in exercise 5?

7. Here are the ages as of last previous birthday at the time of first marriage of the 242 men in the sample from which the grouped frequency distribution in Table 3.8 was constructed. Find the modal year, the median age at time of marriage, the mean age at time of marriage, and the standard deviation of the age at marriage for this grouped frequency distribution in which the class interval is $i = 1$ year. (*Ans.* modal year of age $= 25$, median $= 27.22$, $m = 28.76$, $s = 6.30$.) How much difference does the coarser grouping with the class interval $i = 3$ years, as in Table 3.8, produce in the values of the median, the mean, and the standard deviation?

Age at First Marriage	No. of Men	Age at First Marriage	No. of Men	Age at First Marriage	No. of Men
19	3	29	11	40	1
20	2	30	6	41	2
21	5	31	10	43	2
22	9	32	10	44	1
23	24	33	9	46	1
24	26	34	6	47	2
25	29	35	4	50	1
26	18	36	5	51	1
27	23	37	3	53	1
28	22	38	3	54	1
				59	1

8. Use the coefficient of variation to compare the variability in the neck size of the shirts in Table 6.20 with the variability in the sleeve length of the same shirts in Table 6.21. What is your conclusion about the relative variabilities of men in neck circumference and arm length?

9. Show that if we multiply (or divide) all the frequencies in a distribution by a constant which is neither zero nor negative, the values of the mean and the standard deviation are not changed. Consequently, if we divide all the frequencies in a distribution by the total frequency and multiply by 100 to express the results as percentages, thus producing relative frequencies, the values of the mean and the standard deviation that we obtain by using the relative frequencies must be the same as the values of the mean and the standard deviation that we would obtain by using the actual frequencies for the distribution.

10. Prove algebraically that $\Sigma(x - m)f = 0$ for every frequency distribution.

11. Prove algebraically that for every grouped frequency distribution

$$\frac{i}{n}\sqrt{n\Sigma h^2 f - (\Sigma hf)^2} = \frac{1}{n}\sqrt{n\Sigma x^2 f - (\Sigma xf)^2}$$

12. Prove algebraically that for every frequency distribution

$$\frac{1}{n}\sqrt{n\Sigma x^2 f - (\Sigma xf)^2} = \frac{1}{n}\sqrt{\Sigma(x - m)^2 f}$$

A.7 NOTES ON THE BINOMIAL DISTRIBUTION

A.7.1 Proof of the binomial theorem. In section 4.3, the binomial expansion was given in equation 4.1 without a formal demonstration of the proof of the binomial theorem. The method of proof known as *induction* is the simplest method of proof for the binomial theorem for any positive, integral value of the exponent n.

The first step in the proof by induction is to verify that the theorem is true for $n = 1$. Substituting 1 for n on both sides of equation 4.1 gives the result $Q + P = Q + P$, which obviously is true. The remaining step in the proof consists in assuming that equation 4.1 is true for the case with exponent n and then showing that for the exponent $n + 1$ the result must be

$$(Q + P)^{n+1} = Q^{n+1} + (n+1)Q^nP + \frac{(n+1)(n)}{1(2)}Q^{n-1}P^2 + \ldots$$

$$+ \frac{(n+1)(n)(n-1) \ldots (n-k+2)}{1(2)(3) \ldots (k)} Q^{n-k+1}P^k$$

$$+ \frac{(n+1)(n)(n-1) \ldots (n-k+1)}{1(2)(3) \ldots (k+1)} Q^{n-k}P^{k+1}$$

$$+ \ldots$$

$$+ \frac{(n+1)(n)}{1(2)}Q^2P^{n-1} + (n+1)QP^n + P^{n+1}$$

All that we need to do is multiply both sides of equation 4.1 by $(Q + P)$. Multiplying the left-hand side of equation 4.1 by $(Q + P)$ gives us $(Q + P)^{n+1}$. Now, multiply each term on the right-hand side of equation 4.1 by Q. Next, multiply each term on the right-hand side of equation 4.1 by P. Then add all these new terms on the right-hand side, collecting the pairs of terms that contain the same combinations of powers of Q and P. For example,

$$\frac{n(n-1)(n-2) \ldots (n-k+1)}{1(2)(3) \ldots (k)} Q^{n-k+1}P^k +$$

$$\frac{n(n-1)(n-2) \ldots (n-k+2)}{1(2)(3) \ldots (k-1)} Q^{n-k+1}P^k$$

$$= \frac{(n+1)(n)(n-1) \ldots (n-k+2)}{(1)(2)(3) \ldots (k)} Q^{n-k+1}P^k$$

The result of adding all the terms on the right-hand side is exactly the expression given above for $(Q + P)^{n+1}$ and the proof is complete.

A.7.2 Noncentral confidence limits for a proportion. If the value of the proportion p observed in a sample is small, satisfactory confidence limits for π usually can be obtained, especially if $n \geq 50$, by solving for π the equation $p - \pi = \pm z\sigma_p$. Substitute formula 4.7 for σ_p. Squaring both sides of this equation and then collecting the terms involving π and π^2, we obtain

$$(n + z^2)\pi^2 - (2np + z^2)\pi + np^2 = 0$$

As you probably know, every quadratic equation has two roots or solutions. The two values of π that satisfy the last equation are approximate confidence limits for π at the confidence level corresponding to the chosen value of z. In any elementary algebra textbook, you will find methods for solving quadratic equations. If we represent all quadratic equations by the form

$$ax^2 + bx + c = 0$$

then the two roots of any quadratic equation may be found by substituting the numerical values of a, b, and c into the formula

$$x = \frac{-b \pm \sqrt{b^2 - 4ac}}{2a}$$

All normal distributions are perfectly symmetrical and bell shaped. But the only perfectly symmetrical or balanced binomial distributions are those in which $P = Q = \frac{1}{2}$. As P and Q become farther and farther from $\frac{1}{2}$, binomial distributions become more and more unsymmetrical or lopsided. For example, in Figure 4.1, $P = \frac{1}{6}$ and $Q = \frac{5}{6}$ and the graph is unsymmetrical.

Consequently, you should not be surprised if you find that the upper confidence limit and the lower confidence limit for a proportion π are not symmetrical with respect to p, unless p is 0.50 or near to 0.50. For example, if 0.06 is the observed value of p in a sample containing 50 elementary units, we know for certain that the true value of π cannot be less than or equal to zero, which is not far below 0.06. But all that we know for certain on the other side of 0.06 is that the true value of π cannot be greater than or equal to 1, which is a long distance above 0.06. In other words, then, a good pair of confidence limits to be associated with the observed proportion $p = 0.06$ will not be expected to be equidistant above and below the value of p.

Here is the way in which this method is applied to a situation in which $p = 0.06$, $n = 50$, and we wish to find 95 per cent confidence limits for π. We must let $z = 1.96$. The equation becomes $53.84\pi^2 - 9.84\pi + 0.18 = 0$. Solving this equation, we find that

$$\pi = \frac{+9.84 \pm \sqrt{(-9.84)^2 - 4(53.84)(0.18)}}{2(53.84)} = 0.021 \text{ and } 0.162$$

Consequently, 95 per cent confidence limits for π when $p = 0.06$ and $n = 50$ are 0.021 and 0.162, or 0.02 and 0.16 if we retain only two decimal places. Notice that $p = 0.06$ is not at the midpoint between these two limits. If we had used formula 4.11 in this example, 95 per cent confidence limits erroneously would have appeared to be -0.01 and $+0.13$.

If N, the size of the universe, is not relatively much larger than n, the size of the sample, and if the sample is drawn without replacement, then formula 4.8 should be used for σ_p. The resulting quadratic equation is

$$[n(N-1) + (N-n)z^2] \pi^2 - [2n(N-1)p + (N-n)z^2] \pi + n(N-1)p^2 = 0$$

and the two roots of this equation are the required confidence limits for π.

A.7.3 Approximating binomial probabilities by Poisson probabili-

ties. Mathematicians have shown that if n is large, but either $n\pi$ or $n(1-\pi)$ is not greater than 5, the Poisson probability

$$P(x) = \frac{\mu^x e^{-\mu}}{x!}, \text{ where } x = 0, 1, 2, 3, \ldots$$

provides a good approximation to the true binomial probability of obtaining exactly x successes in a large number of independent trials or repetitions of an experiment for which the probability of success in a single trial is the small constant π.

In the Poisson distribution, μ is the mean number of successes expected in the n trials. Consequently, to approximate a binomial distribution, we must let $\mu = n\pi$ in accordance with formula 4.2. Also, e is the constant base of the natural logarithms; an approximation correct to six decimal places is $e = 2.718282$, but the exact value of e leads to a never-ending decimal number.

As an illustration of the use of the Poisson probability distribution to approximate binomial probabilities, let us think of a binomial situation consisting of 50 independent trials of an experiment in which the constant probability of success in a single trial is $\pi = 0.04$. Then we let $\mu = 50(0.04) = 2$. The Poisson probability of obtaining exactly x successes during the 50 trials of this experiment is

$$P(x) = \frac{2^x e^{-2}}{x!}, \text{ where } x = 0, 1, 2, 3, \ldots, 50$$

For example, the probability of obtaining exactly zero successes during the 50 trials is obtained by substituting zero for x. The result is

$$P(0) = \frac{2^0 e^{-2}}{0!} = \frac{1 e^{-2}}{1} = e^{-2} = \frac{1}{e^2} = \frac{1}{(2.718)^2} = \frac{1}{7.386} = 0.135$$

The probability of obtaining exactly one success during the 50 trials is obtained by substituting one for x. The result is

$$P(1) = \frac{2^1 e^{-2}}{1!} = 2e^{-2} = 2(0.135) = 0.270$$

The probability of obtaining exactly 2 successes during the 50 trials is obtained by substituting 2 for x. The result is

$$P(2) = \frac{2^2 e^{-2}}{2!} = \frac{4 e^{-2}}{2} = 2e^{-2} = 0.270$$

The probability of obtaining exactly 3 successes during the 50 trials is

$$P(3) = \frac{2^3 e^{-2}}{3!} = \frac{8 e^{-2}}{6} = \frac{4}{3} e^{-2} = \frac{4}{3}(0.135) = 0.180$$

The probability of obtaining either zero or one success, that is, the probability of obtaining not more than one success, during the 50 trials is $P(0) + P(1) = 0.135 + 0.270 = 0.405$. The probability of obtaining either zero or one or two successes, that is, the probability of obtaining not more than two successes, during the 50 trials is $P(0) + P(1) + P(2) = 0.135 + 0.270 + 0.270 = 0.675$.

The probability of obtaining at least one success during the 50 trials is $1 - P(0) = 1 - 0.135 = 0.865$. The probability of obtaining at least 2 successes during the 50 trials is $1 - [P(0) + P(1)] = 1 - 0.405 = 0.695$. The probability of obtaining at least 3 successes during the 50 trials is $1 - [P(0) + P(1) + P(2)] = 1 - 0.675 = 0.325$.

The Poisson distribution is a distribution of a discrete variable x. When we use the Poisson distribution to approximate binomial probabilities, the useful values of x, namely, the integral values of x from 0 to n, are the same as those we would use in the binomial distribution. Consequently, there is no correction for continuity when we use Poisson probabilities to approximate binomial probabilities. Figure A.1 is a graph of the Poisson distribution for which $\mu = 2$.

Notice that for a Poisson distribution, it is not necessary to know the numerical values of n and π separately, but only the product $n\pi$. That is,

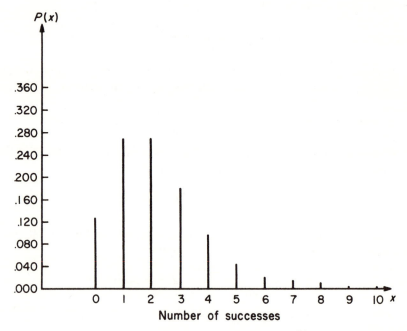

Fig. A.1.   The Poisson Distribution $P(x) = \dfrac{2^x e^{-2}}{x!}$

to use a Poisson probability distribution it is only necessary to know the numerical value of μ, the average number of successes to be expected in a very large but otherwise unspecified number of trials.

The Poisson distribution is an important probability distribution quite apart from the fact that it often can be used to approximate binomial probabilities. It is named for Siméon Denis Poisson (1781–1840), a French scholar who made many contributions to applied mathematics.

If you know how to use logarithms, it is not difficult to evaluate $e^{-\mu}$ for any specific value of μ. The following table gives many commonly used values of μ and $e^{-\mu}$.

Some Useful Values of $e^{-\mu}$

μ	$e^{-\mu}$	μ	$e^{-\mu}$	μ	$e^{-\mu}$
0.0	1.000	1.0	0.368	2.0	0.135
0.1	0.905	1.1	0.333	2.5	0.0821
0.2	0.819	1.2	0.301	3.0	0.0498
0.3	0.741	1.3	0.272	4.0	0.0183
0.4	0.670	1.4	0.247	5.0	0.00674
0.5	0.606	1.5	0.223	6.0	0.00248
0.6	0.549	1.6	0.202	7.0	0.000912
0.7	0.497	1.7	0.183	8.0	0.000335
0.8	0.449	1.8	0.165	9.0	0.000123
0.9	0.407	1.9	0.150	10.0	0.000045

EXERCISES

1. Assume that $\pi = 0.03$ is the binomial probability of success in a single trial of an experiment. If there are to be 10 trials of the experiment, use the Poisson distribution to find the probability of zero successes; one success; more than one success. (*Ans.* 0.741; 0.222; 0.037.) Plot a graph of this Poisson distribution.

2. A candidate in an election drops 10,000 leaflets from an airplane over a city that has 2000 blocks. Assume that each leaflet has equal probability of falling on each block. What is the probability that no leaflets will fall on a particular block? (*Ans.* 0.007.) Plot a graph of this Poisson distribution.

3. There are 1000 men doing the same job in a factory. Assume that each of them has the same probability of having an accident, namely, $\pi = 0.001$, in any day. What is the probability that there will be no accidents in the factory on a specified day? (*Ans.* 0.368.) Plot a graph of this Poisson distribution. What is the probability that there will be no accidents in the factory on three specific successive days of work? (*Ans.* 0.368^3.)

4. Assume that on the average only one person in a thousand has a particular rare type of blood. In a city of 10,000 people, what is the probability that no person has this blood type? (*Ans.* 0.00005.) How large a random sample would be needed to make the probability greater than one-half that the sample would contain at least one person with this blood type? (Hint: $\log_e 0.5 = -0.69315$; *Ans.* $n \geqq 694$.)

5. Assume that a typesetter makes an average of one mistake per 1000 words and that he is setting type for a book with 500 words to the page. Assume also that

the probability of a mistake is the same for each word. What is the probability that there will be more than two mistakes on a particular page? (*Ans.* 0.015.)

6. Assume that 500 blueberry muffins are to be made for a club luncheon. Determine the smallest number of blueberries that may be used if the baker desires that the probability is to be less than 0.01 that a muffin chosen at random from the 500 muffins will not contain any blueberries. (*Ans.* 2303.)

7. A seed company sells small seeds to gardeners in half-ounce packages. Each package contains a very large number of seeds but, of course, the number of seeds varies slightly from package to package. The seed company's quality inspection program shows there are on the average only 2.5 noxious weed seeds in each package. What is the probability that a half-ounce package selected at random and sold to a customer contains more than four noxious weed seeds? (*Ans.* 0.109.)

8. In manufacturing and selling large numbers of small items such as tacks, beads, labels, etc., it often is more economical to sell them by weight rather than by numerical count. If a manufacturer's quality inspection program indicates that on the average each one-pound package of such an item contains two defective items, what is the probability that a one-pound package chosen at random from the warehouse and sold to a customer will contain three or more defective items? (*Ans.* 0.325.)

9. An insurance company that has written a very large number of automobile insurance policies knows from past experience that it pays an average of 1.5 claims for damages per day. What is the probability that there will be no claims to pay on a particular day? (*Ans.* 0.223.) What is the probability that there will be more than three claims to pay on a particular day? (*Ans.* 0.066.) Plot a graph of this Poisson distribution.

10. A biologist found the following distribution of yeast cells in the 400 squares of a haemacytometer:

No. of yeast cells in a square:	0	1	2	3	4	5	6	7	8
No. of squares having the specified number of cells:	103	143	98	42	8	4	2	–	–

Assuming that this is a Poisson distribution, compute the theoretical relative frequencies and the theoretical frequencies for this sample.

11. In another sample, the biologist found the following distribution of yeast cells in the 400 squares of a haemacytometer:

No. of yeast cells in a square:	0	1	2	3	4	5	6	7	8	9
No. of squares with that number of cells:	75	103	121	54	30	13	2	1	0	1

Assuming that each of the samples in exercises 10 and 11 is a random sample from a Poisson universe and given that the variance of a Poisson distribution is equal to the mean of the distribution, test the hypothesis that the two samples were drawn from the same universe. Use the test for the significance of the difference in the means of two samples. (When μ is very large, the Poisson distribution can be approximated by a normal distribution.)

12. For every normal distribution, it is correct to state that approximately 68 per cent of the values of the variable fall in the interval from $\mu - \sigma$ to $\mu + \sigma$. Is it correct to state that for every Poisson distribution approximately 68 per cent of the values of the variable fall in the interval from $\mu - \sigma$ to $\mu + \sigma$? (*Ans.* No.) Why?

(Exercises 13–17 are listed here in order to record some of the important properties of Poisson distributions. The proofs should not be attempted by any student who has not studied calculus.)

13. Prove that

$$\sum_{x=0}^{\infty} \frac{\mu^x e^{-\mu}}{x!} = 1$$

14. Prove that

$$\sum_{x=0}^{\infty} \frac{x\mu^x e^{-\mu}}{x!} = \mu$$

15. Prove that

$$\sum_{x=0}^{\infty} \frac{x^2 \mu^x e^{-\mu}}{x!} = \mu + \mu^2$$

and from this result deduce the fact that the variance of x in the Poisson distribution is equal to the mean of x.

16. If the mean of a Poisson distribution is μ, show that a most probable outcome of the experiment is a value c of x such that $\mu - 1 \leq c \leq \mu$. Under what conditions will there be two most probable outcomes?

17. Prove that the Poisson distribution is a limiting case of the bionomial distribution.

A.8 DERIVATION OF THE FORMULAS FOR μ_m AND σ_m

Let the universe of values of the variable x consist of the N values x_1, x_2, x_3, . . . , x_N. The mean of this universe is μ_x and the standard deviation is σ_x. If we draw random samples, without replacement, each sample containing n values of x from this universe, there are $\binom{N}{n}$ different possible samples of size n. Compute $m_x = \Sigma x/n$ for each different possible sample of size n. We may consider these values of m_x as forming a new universe consisting of $\binom{N}{n}$ values of m_x. Let μ_m and σ_m be the mean and the standard deviation of this new universe. Then

$$\mu_m = \frac{\Sigma m_x}{\binom{N}{n}} = \frac{\frac{1}{n} \Sigma \binom{N-1}{n-1} x}{\binom{N}{n}} = \frac{1}{N} \Sigma x = \mu_x$$

because there are $\binom{N-1}{n-1}$ samples containing each value of x.

Subtract μ_x from each value of x, obtaining another new universe consisting of N values of v, namely, $v_1, v_2, v_3, \ldots, v_N$. Then $\mu_v = 0$ and $\sigma_v = \sigma_x$. Subtract μ_m from each value of m_x, obtaining a new universe consisting of $\binom{N}{n}$ values of w, namely, $w_1, w_2, w_3, \ldots, w_{\binom{N}{n}}$. Then $\mu_w = 0$ and $\sigma_w = \sigma_m$. Now,

$$\sigma_w^2 = \frac{\Sigma(w - \mu_w)^2}{\binom{N}{n}} = \frac{\Sigma w^2}{\binom{N}{n}} = \frac{\Sigma\left(\frac{\Sigma v}{n}\right)^2}{\binom{N}{n}} \quad \begin{array}{l} \text{summed over the } \binom{N}{n} \text{ sam-} \\ \text{ples, with } n \text{ values of } v \text{ in each} \\ \text{of the inside summations} \end{array}$$

$$= \frac{\Sigma(\Sigma v)^2}{\binom{N}{n}n^2} = \frac{\binom{N-1}{n-1}}{\binom{N}{n}n^2}\Sigma v^2 + \frac{2\binom{N-2}{n-2}}{\binom{N}{n}n^2}\Sigma v_i v_j$$

summed over the N values of v, with $i \neq j$. This is true because the square of each value of v appears in $\binom{N-1}{n-1}$ different samples and the product of each pair of values of v appears in $\binom{N-2}{n-2}$ different samples. For the universe consisting of N values of v, $(\Sigma v)^2 = \Sigma v^2 + 2\Sigma v_i v_j$, $i \neq j$, and, consequently, for $i \neq j$,

$$2\Sigma v_i v_j = (\Sigma v)^2 - \Sigma v^2 = -\Sigma v^2 \quad \text{because } \Sigma v = 0.$$

Therefore,

$$\sigma_w^2 = \frac{\binom{N-1}{n-1}}{\binom{N}{n}n^2}\Sigma v^2 - \frac{\binom{N-2}{n-2}}{\binom{N}{n}n^2}\Sigma v^2 = \frac{1}{n} \cdot \frac{\Sigma v^2}{N} \cdot \frac{N-n}{N-1} = \frac{\sigma_v^2}{n} \cdot \frac{N-n}{N-1} = \frac{\sigma_x^2}{n} \cdot \frac{N-n}{N-1}$$

But, as indicated above, $\sigma_m = \sigma_w$, and therefore

$$\sigma_m = \frac{\sigma_x}{\sqrt{n}}\sqrt{\frac{N-n}{N-1}}$$

EXERCISE

1. Show that

$$\frac{N\binom{N-1}{n-1} - N\binom{N-2}{n-2}}{\binom{N}{n}n^2} = \frac{N-n}{n(N-1)}$$

A.9 MEANS AND VARIANCES OF DISTRIBUTIONS OF DIFFERENCES AND SUMS

Consider a universe consisting of N elementary units for which there are N pairs of values of two variables x and y. Let μ_x and σ_x be the mean and the standard deviation of the values of x. Let μ_y and σ_y be the mean and the standard deviation of the values of y. Let ρ_{xy} be the linear correlation between the values of x and y. In other words, let $\rho_{xy} = \Sigma(x-\mu_x)(y-\mu_y)/N\sigma_x\sigma_y$. Multiply both sides of this equation by $\sigma_x\sigma_y$; the result is

$$\rho_{xy}\sigma_x\sigma_y = \frac{\Sigma(x-\mu_x)(y-\mu_y)}{N}$$

If each value of y is subtracted from the value of x with which it is paired, a new distribution consisting of N values of $x - y$ is formed. The mean and the standard deviation of this new distribution may be designated by μ_{x-y} and σ_{x-y}.
Then

$$\mu_{x-y} = \frac{\Sigma(x-y)}{N} = \frac{\Sigma x}{N} - \frac{\Sigma y}{N} = \mu_x - \mu_y \qquad (A.3)$$

and

$$
\begin{aligned}
(x-y) - \mu_{x-y} &= (x-y) - (\mu_x-\mu_y) = (x-\mu_x) - (y-\mu_y) \\
[(x-y) - \mu_{x-y}]^2 &= (x-\mu_x)^2 + (y-\mu_y)^2 - 2(x-\mu_x)(y-\mu_y) \\
\Sigma[(x-y) - \mu_{x-y}]^2 &= \Sigma(x-\mu_x)^2 + \Sigma(y-\mu_y)^2 - 2\Sigma(x-\mu_x)(y-\mu_y) \\
\frac{\Sigma[(x-y) - \mu_{x-y}]^2}{N} &= \frac{\Sigma(x-\mu_x)^2}{N} + \frac{\Sigma(y-\mu_y)^2}{N} - \frac{2\Sigma(x-\mu_x)(y-\mu_y)}{N}
\end{aligned}
$$

That is,

$$\sigma^2_{x-y} = \sigma^2_x + \sigma^2_y - 2\rho_{xy}\sigma_x\sigma_y \qquad (A.4)$$

Similarly, it can be proved that

$$\mu_{x+y} = \mu_x + \mu_y \qquad (A.5)$$

$$\sigma^2_{x+y} = \sigma^2_x + \sigma^2_y + 2\rho_{xy}\sigma_x\sigma_y \qquad (A.6)$$

If the two variables x and y are independent of each other, then the linear correlation between them is zero, that is, $\rho_{xy} = 0$. In that case,

$$\sigma^2_{x-y} = \sigma^2_{x+y} = \sigma^2_x + \sigma^2_y$$

If we replace x and y by p_1 and p_2, then $\mu_{p_1-p_2} = \mu_{p_1} - \mu_{p_2} = \pi_1 - \pi_2$
and

$$\sigma^2_{p_1-p_2} = \sigma^2_{p_1} + \sigma^2_{p_2} - 2\rho_{p_1p_2}\sigma_{p_1}\sigma_{p_2}$$

If n_1, n_2, p_1, p_2, q_1, q_2 are the values observed in two random samples, then it can be shown that an estimate $r_{p_1p_2}$ of the correlation coefficient $\rho_{p_1p_2}$.

may be computed by substituting the values of n_1, n_2, p_1, p_2, q_1, and q_2 into the formula

$$r_{p_1 p_2} = \frac{p_2 q_1 - p_1 q_2}{\sqrt{\dfrac{(n_1 p_1 + n_2 p_2)(n_1 q_1 + n_2 q_2)}{n_1 n_2}}}$$

Similarly, if we replace x and y by m_1 and m_2, then $\mu_{m_1 - m_2} = \mu_{m_1} - \mu_{m_2} = \mu_1 - \mu_2$ and

$$\sigma^2_{m_1 - m_2} = \sigma^2_{m_1} + \sigma^2_{m_2} - 2\rho_{m_1 m_2} \sigma_{m_1} \sigma_{m_2}$$

It can be shown that the value of the correlation coefficient $\rho_{m_1 m_2}$ is the same as the value of the correlation coefficient ρ_{xy} for the two variables x and y on which the two means are based. Consequently, an estimate r_{xy} of the value of $\rho_{m_1 m_2}$ can be obtained from the observed sample values of x and y.

A.10 THE METHOD OF AVERAGE EQUATIONS FOR FITTING A STRAIGHT LINE

Consider the five pairs of values of x and y given in Table 9.14, the corresponding five points on the graph in Figure 9.6, and the straight line representing the trend for the period from 1910 to 1950. If we substitute each of the five pairs of values of x and y given in Table 9.14 into the equation $y = c + dx$, the result is

$$
\begin{aligned}
76{,}508 &= c + d(1910) \\
143{,}664 &= c + d(1920) \\
288{,}737 &= c + d(1930) \\
344{,}977 &= c + d(1940) \\
\underline{463{,}495} &= c + d(1950) \\
1{,}317{,}381 &= 5c + d(9650)
\end{aligned}
$$

In the last line is shown the total of all the approximate equations in the set and the dot (approximation sign) is omitted over the equal sign. If we divide both sides of the equation by 5, the number of approximate equations in the set, we obtain our first average equation

$$263{,}476.2 = c + 1930d$$

There are two unknowns, c and d, in this equation. You probably know from elementary algebra that in order to find a unique solution, we need two independent equations when there are two unknowns.

In the set of five approximate equations arranged above, multiply both sides of the first equation by the first observed value of x, that is, by

1910. Multiply the second equation by the second observed value of x, that is, by 1920. Similarly, multiply both sides of the third equation by 1930, multiply both sides of the fourth equation by 1940, and multiply both sides of the fifth equation by 1950. The new equations are:

$$1910(76,508) = 1910c + d(1910)^2 \text{ or } 146,130,280 = 1910c + 3,648,100d$$
$$1920(143,664) = 1920c + d(1920)^2 \text{ or } 275,834,880 = 1920c + 3,686,400d$$
$$1930(288,737) = 1930c + d(1930)^2 \text{ or } 557,262,410 = 1930c + 3,724,900d$$
$$1940(344,977) = 1940c + d(1940)^2 \text{ or } 669,255,380 = 1940c + 3,763,600d$$
$$1950(463,495) = 1950c + d(1950)^2 \text{ or } 903,815,250 = 1950c + 3,802,500d$$
$$\overline{2,552,298,200 = 9650c + 18,625,500d}$$

The second average equation may be obtained by dividing the last equation by 5. However, for the step that follows, it will be more convenient to divide both sides of the last equation by the coefficient of c, namely, 9650. The two equations to be solved for c and d are then

$$264,486.86 = c + 1930.1036269d$$
$$263,476.20 = c + 1930.0000000d$$
$$\overline{1010.66 = \qquad 0.1036269d}$$

$$d = \frac{1010.66}{0.1036269} = 9752.87$$

Next, substitute this value of d into the first average equation and find the value of c.

$$c = 263,476.2 - 1930(9752.87) = -18,559,562.9$$

It is left to the reader to show that in general the two equations to be solved for c and d can be written in the form

$$\Sigma y = nc + d\Sigma x$$
$$\Sigma xy = c\Sigma x + d\Sigma x^2$$

and that the values of c and d that satisfy these two equations are given by formulas 9.2.

A.11 RANDOM NUMBERS

Example 1: Suppose that from a list of the names of ten students we wish to draw two students at random for a sample.

Step 1: The list numbers and the tabular numbers. Write the list numbers 1 to 10 beside the names of the ten students in the list. Then, beside the list numbers 1 to 10, write the tabular numbers 0 to 9.

Student's Name	List Number	Tabular Number
John Smith	1	0
Helen Jones	2	1
William Anderson	3	2
Henry Jackson	4	3
June Williams	5	4 ✓
Thomas Kent	6	5
Mary Carroll	7	6 ✓
James Johnson	8	7
Peter Allen	9	8
Sandra Alexander	10	9

Although the list numbers begin at 1 and end at 10, the tabular numbers must begin at 0 and end at 9. This is necessary because there are as many zeros as any other digit in Table A.15. If we had given the list numbers 0 to 9 to the students, then the list numbers and the tabular numbers would be identical. In the above table, tabular number 0 corresponds to student number 1, tabular number 1 corresponds to student number 2, and so on, until tabular number 9 corresponds to student number 10.

Step 2: The random starting point. Find a random starting point in Table A.15. One way to do this is to shut your eyes and aimlessly let your pencil point come to rest somewhere on Table A.15. Write down the digit nearest to the point of your pencil and also write down the next three digits to the right (or the left or above or below, if you prefer). For example, the digit nearest to the pencil point might be 2 and the next three digits to the right might be 7, 4, and 3. Write these four digits as follows:

$$27 \qquad 43 \qquad 27 + 1 = 28 \qquad 43 + 1 = 44$$

Choose as the starting point in Table A.15 the digit at the intersection of line 28 and column 44.

Step 3: The route. Choose a route to be followed in going through Table A.15. For this example, let us choose to read downward from the starting point. (You could choose to read upward, or horizontally to the right, or horizontally to the left, or diagonally in any one of several directions.) If we come to the bottom of the column before we have obtained enough random numbers, then we will read upward in the next column to the right.

Step 4: The first random number. The random starting point in this example is the digit 4. Consequently, the first random tabular number is 4. Therefore, the first random list number or student number is 5. The first student to be drawn into the sample is student number 5, that is, June Williams. A check mark beside the tabular number 4 may be used to indicate that this student has been selected for the sample.

Step 5: The second random number. The digit immediately below the random starting digit is 6. This is the second random tabular number, and a check mark has been placed beside the tabular number 6 in the list to indicate that the student whose tabular number is 6 and whose list number is 7 has been selected for the sample. Consequently, the second student selected for the sample is Mary Carroll.

Example 2: Suppose that we wish to draw the names of 42 students at random for a 10 per cent sample from a universe containing 420 students.

Step 1: The list numbers and the tabular numbers. Number the students' names from 1 to 420 and call these numbers the list numbers. Beside the list numbers, write the tabular numbers 000, 001, 002, . . ., 419. Because some of the list numbers here are three-digit numbers, we must use three digits in every tabular number in this example.

Step 2: The random starting point. Suppose that with your eyes closed you drop your pencil point onto the digit 8 that is on line 31 in column 11 of Table A.15. The three digits immediately to the right are 7, 9, and 1. Write these four digits as follows:

$$87 \qquad 91 \qquad 87 + 1 = 88 \qquad 91 + 1 = 92$$

This seems to suggest that the random starting point in Table A.15 might be taken at the intersection of line 88 and column 92. But there are only 50 lines and 70 columns in Table A.15. Subtract 50 from 88, obtaining 38. We shall use line 38 to determine the random starting point.

Perhaps you may be tempted now to subtract 70 from 92 to determine the column for the random starting point. Because there are fewer integers between 70 and 99 than there are between 0 and 69, we will not proceed by subtracting 70 from 92. Instead, we reject 91 because it is greater than $70 - 1 = 69$. Now, examine the pair of digits immediately below 91. They are 7 and 9 and we reject 79 also because it is greater than 69. The two digits immediately below 79 are 8 and 7 and we reject 87 because it is greater than 69. The two digits immediately below 87 are 5 and 7 and 57 is not greater than 69. Consequently, we may use $57 + 1 = 58$ as the column number for the starting point. The random starting point is, then, at the intersection of line 38 and column 58.

Step 3: The route. Let us choose here to read upward from the random starting point in Table A.15. If we reach the top of the table before we have obtained 42 random numbers for the sample, we will read downward in the next three columns immediately to the left, namely, columns 55, 56, and 57. If additional columns are required, we will read upward in the next three columns immediately to the left, namely, columns 52, 53, and 54. And so on.

Step 4: The first random number. The three-digit number on line 38 in columns 58, 59, and 60 is 389. Therefore, the first random tabular number is 389 and the first random list number is $389 + 1 = 390$. In other words,

TABLE A.15 Random Digits

Col. Line	1–5	6–10	11–15	16–20	21–25	26–30	31–35	36–40	41–45	46–50	51–55	56–60	61–65	66–70
1	10480	15011	01536	02011	81647	91646	69179	14194	62590	36207	20969	99570	91291	90700
2	22368	46573	25595	85393	30995	89198	27982	53402	93965	34095	52666	19174	39615	99505
3	24130	48360	22527	97265	76393	64809	15179	24830	49340	32081	30680	19655	63348	58629
4	42167	93093	06243	61680	07856	16376	39440	53537	71341	57004	00849	74917	97758	16379
5	37570	39975	81837	16656	06121	91782	60468	81305	49684	60672	14110	06927	01263	54613
6	77921	06907	11008	42751	27756	53498	18602	70659	90655	15053	21916	81825	44394	42880
7	99562	72905	56420	69994	98872	31016	71194	18738	44013	48840	63213	21069	10634	12952
8	96301	91977	05463	07972	18876	20922	94595	56869	69014	60045	18425	84903	42508	32307
9	89579	14342	63661	10281	17453	18103	57740	84378	25331	12566	58678	44947	05585	56941
10	85475	36857	53342	53988	53060	59533	38867	62300	08158	17983	16439	11458	18593	64952
11	28918	69578	88231	33276	70997	79936	56865	05859	90106	31595	01547	85590	91610	78188
12	63553	40961	48235	03427	49626	69445	18663	72695	52180	20847	12234	90511	33703	90322
13	09429	93969	52636	92737	88974	33488	36320	17617	30015	08272	84115	27156	30613	74952
14	10365	61129	87529	85689	48237	52267	67689	93394	01511	26358	85104	20285	29975	89868
15	07119	97336	71048	08178	77233	13916	47564	81056	97735	85977	29372	74461	28551	90707
16	51085	12765	51821	51259	77452	16308	60756	92144	49442	53900	70960	63990	75601	40719
17	02368	21382	52404	60268	89368	19885	55322	44819	01188	65255	64835	44919	05944	55157
18	01011	54092	33362	94904	31273	04146	18594	29852	71585	85030	51132	01915	92747	64951
19	52162	53916	46369	58586	23216	14513	83149	98736	23495	64350	94738	17752	35156	35749
20	07056	97628	33787	09998	42698	06691	76988	13602	51851	46104	88916	19509	25625	58104
21	48663	91245	85828	14346	09172	30168	90229	04734	59193	22178	30421	61666	99904	32812
22	54164	58492	22421	74103	47070	25306	76468	26384	58151	06646	21524	15227	96909	44592
23	32639	32363	05597	24200	13363	38005	94342	28728	35806	06912	17012	64161	18296	22851
24	29334	27001	87637	87308	58731	00256	45834	15398	46557	41135	10367	07684	36188	18510
25	02488	33062	28834	07351	19731	92420	60952	61280	50001	67658	32586	86679	50720	94953

26	81525	72295	04839	96423	24878	82651	66566	14778	76797	14780	13300	87074	79666	95725
27	29676	20591	68086	26432	46901	20849	89768	81536	86645	12659	92259	57102	80428	25280
28	00742	57392	39064	66432	84673	40027	32832	61362	98947	96067	64760	64584	96096	98253
29	05366	04213	25669	26422	44407	44048	37937	63904	45766	66134	75470	66520	34693	90449
30	91921	26418	64117	94305	26766	25940	39972	22209	71500	64568	91402	42416	07844	69618
31	00582	04711	87917	77341	42206	35126	74087	99547	81817	42607	43808	76655	62028	76630
32	00725	69884	62797	56170	86324	88072	76222	36086	84637	93161	76038	65855	77919	88006
33	69011	65795	95876	55293	18988	27354	26575	08625	40801	59920	29841	80150	12777	48501
34	25976	57948	29888	88604	67917	48708	18912	82271	65424	69774	33611	54262	85963	03547
35	09763	83473	73577	12908	30883	18317	28290	35797	05998	41688	34952	37888	38917	88050
36	91567	42595	27958	30134	04024	86385	29880	99730	55536	84855	29080	09250	79656	73211
37	17955	56349	90999	49127	20044	59931	06115	20542	18059	02008	73708	83517	36103	42791
38	46503	18584	18845	49618	02304	51038	20655	58727	28168	15475	56942	53389	20562	87338
39	92157	89634	94824	78171	84610	82834	09922	25417	44137	48413	25555	21246	35509	20468
40	14577	62765	35605	81263	39667	47358	56873	56307	61607	49518	89686	20103	77490	18062
41	98427	07523	33362	64270	01638	92477	66969	98420	04880	45585	46565	04102	46880	45709
42	34914	63976	88720	82765	34476	17032	87589	40836	32427	70002	70663	88863	77775	69348
43	70060	28277	39475	46473	23219	53416	94970	25832	69975	94884	19661	72828	00102	66794
44	53976	54914	06990	67245	68350	82948	11398	42878	80287	88267	47363	46634	06541	97809
45	76072	29515	40980	07391	58745	25774	22987	80059	39911	96189	41151	14222	60697	59583
46	90725	52210	83974	29992	65831	38857	50490	83765	55657	14361	31720	57375	56228	41546
47	64364	67412	33339	31926	14883	24413	59744	92351	97473	89286	35931	04110	23726	51900
48	08962	00358	31662	25388	61642	34072	81249	35648	56891	69352	48373	45578	78547	81788
49	95012	68379	93526	70765	10592	04542	76463	54328	02349	17247	28865	14777	62730	92277
50	15664	10493	20492	38391	91132	21999	59516	81652	27195	48223	46751	22923	32261	85653

the first student selected for the sample is the student whose tabular number is 389 and whose list number is 390.

Step 5:   The second random number. The three-digit number immediately above the digits that were used in step 4 is 517. We may reject this number because it is greater than $420 - 1 = 419$. The three-digit number immediately above 517 is 250. This number 250 is not greater than 419. Consequently, the second random tabular number is 250 and the second random list number is $250 + 1 = 251$. The second student selected for the sample is the student whose tabular number is 250 and whose list number is 251.

You may continue in this way until you have selected 42 students for the sample from the universe containing 420 students. In each instance, reject any three-digit tabular number greater than 419 and examine the next three-digit number along the prescribed route.

If the same list number happens to be drawn more than once, disregard it after the first time and draw a new random tabular number unless you wish to permit an elementary unit to be used more than once in the sample.

Remarks. You must use as many columns (or lines) as there are digits in the number $N - 1$, where N is the largest elementary unit number in the list.

There are other ways to select random numbers from a table of random digits. For example, many people interpret the tabular digit 0 as 10 when they are using one column of a table to select a random number in the range from 1 to 10. And they interpret the pair of tabular digits 00 as 100 when they are using two columns to select a random number in the range from 1 to 100. Similarly, they interpret the three tabular digits 000 as 1000 when they are using three columns to select a random number in the range from 1 to 1000, and so on. If you prefer this method, you may use it.

Sources of random digits. Table A.15 is a reproduction of one of the thirty pages of random digits in *Table of 105,000 Random Decimal Digits*, Statement No. 4914, File No. 261-A-1, Bureau of Transport Economics and Statistics, Interstate Commerce Commission, Washington, D. C. (May 1949). The table was prepared for the Commission by H. Burke Horton and R. Tynes Smith III.

For people who frequently need a large number of random digits, another source is *A Million Random Digits with 100,000 Normal Deviates*, published by The Free Press, Glencoe, Illinois, for the RAND Corporation.

A.12   THE MOST USEFUL FORMULAS

Number	*Formula*	*Used for determining*
3.1	$m_x = \dfrac{\Sigma x}{n}$	mean of a sample

Number	*Formula*	*Used for determining*

3.2 $s_x^2 = \dfrac{\Sigma(x - m)^2}{n}$ variance of a sample

3.3 $s_x = \sqrt{\dfrac{\Sigma(x - m)^2}{n}}$ standard deviation of a sample

3.4 $s_x = \dfrac{1}{n}\sqrt{n\Sigma x^2 - (\Sigma x)^2}$ standard deviation of a sample

3.5 $z = \dfrac{x - \mu_x}{\sigma_x}$ standard score

3.6 $\dfrac{\text{standard deviation}}{\text{mean}}(100)$ per cent   coefficient of variation

3.7 $m_x = \dfrac{\Sigma xf}{n}$ mean of sample frequency distribution

3.8 $s_x = \dfrac{1}{n}\sqrt{n\Sigma x^2 f - (\Sigma xf)^2}$ standard deviation of sample frequency distribution

4.1 $(Q + P)^n = \displaystyle\sum_{i=0}^{n} \binom{n}{i} Q^{n-i} P^i$ binomial probabilities

4.2 $\mu_x = nP$ mean number of successes in binomial distribution

4.3 $\sigma_x = \sqrt{nPQ}$ standard deviation of number of successes in binomial distribution

4.4 $\mu_p = P$ mean proportion of successes in binomial distribution

4.5 $\sigma_p = \sqrt{\dfrac{PQ}{n}}$ standard deviation of proportion of successes in binomial distribution

4.6 $\mu_p = \pi$ Mean of proportion p possessing attribute in samples

4.7 $\sigma_p = \sqrt{\dfrac{\pi(1 - \pi)}{n}}$ standard deviation of proportion p possessing attribute in samples of size n

4.8 $\sigma_p = \sqrt{\dfrac{\pi(1 - \pi)(N - n)}{n(N - 1)}}$ adjusted standard deviation of proportion p possessing attribute in sampling without replacement from finite universe

Number	Formula	Used for determining

4.9 $s_p = \sqrt{\dfrac{p(1-p)}{n}}$

estimated standard deviation of proportion p possessing attribute in samples of size n if π is unknown

4.10 $s_p = \sqrt{\dfrac{p(1-p)(N-n)}{n(N-1)}}$

estimated standard deviation of proportion p possessing attribute in sampling without replacement from finite universe

4.11 $p - zs_p < \pi < p + zs_p$

confidence limits for proportion π possessing attribute

4.12 $p - \dfrac{1}{2n} - zs_p < \pi < p + \dfrac{1}{2n} + zs_p$

confidence limits for proportion π possessing attribute, with correction for continuity

4.13 $z = \dfrac{p \pm \dfrac{1}{2n} - \pi}{\sigma_p}$

using $+ \dfrac{1}{2n}$ if $p < \pi$

using $- \dfrac{1}{2n}$ if $p > \pi$

test of hypothesis about proportion π

4.14 $n = \left(\dfrac{z}{k}\right)^2 \pi(1-\pi)$

sample size for estimating proportion π

4.15 $n_1 = n\left(\dfrac{N}{N+n-1}\right)$

sample size for estimating π in sampling without replacement from finite universe

5.1 $z = \dfrac{(p_2 - p_1) - (\pi_2 - \pi_1)}{\sqrt{\dfrac{\pi_1(1-\pi_1)}{n_1} + \dfrac{\pi_2(1-\pi_2)}{n_2}}}$

test of significance of difference in two proportions

5.2 $p = \dfrac{n_1 p_1 + n_2 p_2}{n_1 + n_2}$

$z = \dfrac{p_2 - p_1}{\sqrt{p(1-p)\left(\dfrac{1}{n_1} + \dfrac{1}{n_2}\right)}}$

test of hypothesis that $\pi_1 = \pi_2$

6.1 $\dfrac{3(\text{mean} - \text{median})}{\text{standard deviation}}$

skewness of a distribution of a variable

Number	Formula	Used for determining
6.2	$y = \dfrac{N}{\sigma\sqrt{2\pi}}\, e^{-(x-\mu)^2/2\sigma^2}$	normal distribution of variable x
6.3	$y = \dfrac{1}{\sqrt{2\pi}}\, e^{-z^2/2}$	standardized normal distribution
7.1	$\mu_m = \mu_x$	mean of distribution of means of samples
7.2	$\sigma_m = \dfrac{\sigma_x}{\sqrt{n}}$	standard deviation of distribution of means of samples of size n
7.3	$\sigma_m = \dfrac{\sigma_x}{\sqrt{n}}\sqrt{\dfrac{N-n}{N-1}}$	adjusted standard deviation of distribution of means in sampling without replacement from finite universe
7.4	$t = \dfrac{m-\mu_x}{s_x/\sqrt{n-1}}$	test of hypothesis about the mean μ_x
7.5	$m - \dfrac{ts_x}{\sqrt{n-1}} < \mu_x < m + \dfrac{ts_x}{\sqrt{n-1}}$	confidence limits for mean of universe
7.6	$n = \left(\dfrac{z\sigma_x}{k}\right)^2$	sample size for estimating μ_x
7.7	$n_1 = n\left(\dfrac{N}{N+n-1}\right)$	sample size for estimating μ_x in sampling without replacement from finite universe
7.8	$z = \dfrac{(m_2-m_1)-(\mu_2-\mu_1)}{\sqrt{\dfrac{\sigma_1^2}{n_1}+\dfrac{\sigma_2^2}{n_2}}}$	test of significance of difference in two means if σ_1 and σ_2 are known
7.9	$s = \sqrt{\dfrac{n_1 s_1^2 + n_2 s_2^2}{n_1 + n_2 - 2}}$ $t = \dfrac{(m_2-m_1)-(\mu_2-\mu_1)}{s\sqrt{\dfrac{1}{n_1}+\dfrac{1}{n_2}}}$	test of significance of difference in two means under the assumption that $\sigma_1 = \sigma_2$
8.1	$r' = 1 - \dfrac{6\Sigma d^2}{n^3 - n}$	rank correlation coefficient
8.2	$r = \dfrac{n\Sigma xy - \Sigma x \Sigma y}{\sqrt{[n\Sigma x^2 - (\Sigma x)^2][n\Sigma y^2 - (\Sigma y)^2]}}$	linear correlation coefficient

Number	*Formula*	*Used for determining*

8.3 $t = r\sqrt{\dfrac{n-2}{1-r^2}}, \; n' = n - 2$

test of significance for r

8.4 $\rho = \dfrac{N\Sigma xy - \Sigma x \Sigma y}{N^2 \sigma_x \sigma_y}$

linear correlation coefficient

9.1 $m'_{y \cdot x} = c + dx$

linear prediction equation

9.2 $d = \dfrac{n\Sigma xy - \Sigma x \Sigma y}{n\Sigma x^2 - (\Sigma x)^2}, \quad c = \dfrac{\Sigma y - d\Sigma x}{n}$

coefficients in linear prediction equation

9.3 $s'_{y \cdot x} = s_y \sqrt{1 - r^2}$

standard error of estimation

9.4 $m'_{y \cdot x} \pm t s'_{y \cdot x} \sqrt{\dfrac{1}{n} + \dfrac{(x - m_x)^2}{n s_x^2}}$

confidence limits for mean value of y with a given value of x

9.5 $m'_{y \cdot x} \pm t s'_{y \cdot x} \sqrt{1 + \dfrac{1}{n} + \dfrac{(x - m_x)^2}{n s_x^2}}$

confidence limits for individual value of y with a given value of x

9.6 $d = r\dfrac{s_y}{s_x}, \quad c = m_y - r\dfrac{s_y}{s_x} m_x$

coefficients in linear prediction equation

A.1 $\text{P.R.} = (n - R + 0.5)\left(\dfrac{100}{n}\right)$

percentile rank

A.2 $P_p = L + \left(\dfrac{pn}{100} - \Sigma f_{BL}\right)\dfrac{i}{f_c}$

pth percentile in a frequency distribution

A.3 $\mu_{x-y} = \mu_x - \mu_y$

mean of distribution of difference between two variables

A.4 $\sigma_{x-y}^2 = \sigma_x^2 + \sigma_y^2 - 2\rho_{xy}\sigma_x\sigma_y$

variance of distribution of difference between two variables

A.5 $\mu_{x+y} = \mu_x + \mu_y$

mean of distribution of sum of two variables

A.6 $\sigma_{x+y}^2 = \sigma_x^2 + \sigma_y^2 + 2\rho_{xy}\sigma_x\sigma_y$

variance of distribution of sum of two variables

INDEX